Small-Craft Harbors: Design, Construction, and Operation

by

James W. Dunham and Arnold A. Finn

Fredonia Books
Amsterdam, The Netherlands

Small-Craft Harbors:
Desing, Construction, and Operation

by
James W. Dunham
Arnold A. Finn

ISBN: 1-58963-965-0

Reprinted from the 1974 edition

Fredonia Books
Amsterdam, The Netherlands
http://www.fredoniabooks.com

PREFACE

This report is published to assist design and construction engineers and operators of small-craft harbors. The work was carried out under the coastal construction research program of the U.S. Army Coastal Engineering Research Center.

This report is the first in a series to be published to form a Coastal Engineering Manual.

This report was prepared by Moffatt and Nichol, Engineers, under CERC Contract No. DACW72-72-C-0011. Cooperation and assistance were provided by Corps of Engineers Divisions and Districts and various boating agencies of the 50 States. Preparation of the report was under the direction of James W. Dunham with the assistance of Arnold A. Finn. Consultation and technical advice were rendered by William J. Herron, Jr., and John M. Nichol. The report was reviewed by Lawrence W. McDowell, retired director of the Long Beach Marina, whose excellent suggestions were incorporated in the final draft. Federal, State and local government agencies, consulting engineers, marina-product manufacturers, and marina managers throughout the United States and its territories were contacted to obtain data on marina construction and boating customs that would make the report as universally applicable as possible. Although differences of opinion were encountered on much of the subject matter and in many geographical areas, it is believed that the information presented represents the current concensus of all authoritative sources on the subject matter covered.

Special appreciation is extended to all of the marina managers and operating personnel who answered the inquiries of the task group or who completed the rather lengthy and detailed marina-survey questionnaire. The survey participants listed below were particularly helpful in their contributions to the data supporting the case studies.

Bahia Mar . . . . Mr. J. Yeager
Galveston Yacht Basin . . . . Mr. R. E. Smith
Garrison Resort and Trophy Room . . . . Mr. E. P. Lee
Hawaii-Kai . . . . Mr. L. V. Andrade
Lighthouse Bay Marina . . . . Mr. L. A. Stadel
Marina Del Rey . . . . Mr. J. W. Quinn
Ohio Projects . . . . Mr. R. G. Stanhope, Stanley Consultants, Inc.
Salmon Harbor . . . . Mr. H. Ludwig
Spruce Run Reservoir . . . . Messrs. D. L. Somma and A. D. Stasi, Edwards and Kelcey, Inc.
Still Waters Marina . . . . Mr. J. R. Wilbanks
Wahweap Marina . . . . Mr. C. A. Greene

The special contributions of Mr. A. M. Filak of Marina Associates, Mr. T. C. Hardwick of Floating Dock Systems, and Mr. S. Sedlmaier of Frederic R. Harris, Inc. are gratefully acknowledged.

Mr. Robert A. Jachowski, Chief, Design Branch, was the CERC contract monitor for the report, under the general supervision of Mr. George M. Watts, Chief, Engineering Development Division.

Comments on this publication are invited.

Approved for publication in accordance with Public Law 166, 79*th* Congress, approved 31 July 1945, as supplemented by Public Law 172, 88*th* Congress, approved 7 November 1963.

JAMES L. TRAYERS
Colonel, Corps of Engineers
Commander and Director

CONTENTS

CONTENTS—Continued

TABLES

CONTENTS–Continued

FIGURES

FIGURES—Continued

FIGURES—Continued

FIGURES—Continued

FIGURES—Continued

SMALL-CRAFT HARBORS: DESIGN, CONSTRUCTION, AND OPERATION

by
James W. Dunham and Arnold A. Finn

I. INTRODUCTION

The objective of this report is to enable anyone with a basic engineering background, aided by other general engineering handbooks, to plan and design a small-craft harbor, or to do so with the help of one or more specialists. The report outlines methods of investigating the problems involved and the various engineering, economic, and environmental criteria to be applied. It also points out the technical areas in which the assistance of specialists is needed and the kinds of information they are capable of supplying.

The size, shape, meteorological environment, and hydraulic characteristics of the body of navigable water to which access by boat is to be provided usually dictate the requirements for many of the harbor features, including the entrance, the interior facilities, and any special protective works needed to deflect currents or dissipate wave energy. Harbors along the open seacoast require protection far in excess of that required for harbors naturally protected in bays, rivers, and small lakes. Floating boat slips in areas of extremely large tidal range and in reservoirs subject to large seasonal water level fluctuations must be anchored differently than those in lakes and along the seacoast where water level fluctuations are only a few feet. Therefore, a major purpose of this report is to provide the designer with the means for selecting the right site, developing the most functional layout plan, providing adequate protection, and installing the proper facilities for the water area served.

Local weather is an important factor to consider. Winds frequently reach hurricane proportions in some regions, but seldom exceed gale force in others; snow or ice complicate planning for some regions, whereas the hot sun is the only problem in others. Thus, another objective of this report is to show the designer how to cope with the specific weather problems of the region, e.g., the rigorous climate of Alaska or the mild climate of the subtropics.

Environmental factors, including ecology, and related effects on economic analysis are beginning to have a much greater impact upon engineering design. Hence, this study includes a discussion of financing marinas and ensuring their economic sufficiency, of limitations imposed by laws and ordinances, and of environmental considerations (U.S. Army, Corps of Engineers, 1972).

Some of the terms used in this report may be unfamiliar to the reader or may have different meanings here than elsewhere. A glossary is presented in Appendix A to show the intended definition of each word or term that might otherwise be misconstrued.

If engineering assistance with the design of a small-craft facility is desired, see the current publications by The National Association of Engine and Boat Manufacturers (NAEBM) and

The Outboard Boating Club of America (OBC). These publications are periodically updated and contain lists of consultants. The full title and publisher's address for each are given under "Literature Cited." These references are not intended to be an endorsement of the firms listed, nor is it intended to exclude from consideration any firms not included. Before retaining the services of any consultant the owner should first determine the qualifications of that consultant for the type of engineering work required.

II. TYPES OF HARBORS

1. Commercial. Consideration of commercial facilities in this report is limited mainly to harbors for commercial fishing fleets, barges, and small-craft transportation terminals, including berths for excursion craft of various kinds. Figure 1 shows a typical small-craft commercial harbor used primarily by a west coast fishing fleet and a few charter boats. Small-craft facilities are often within or adjacent to harbors built primarily for deep-draft cargo or passenger vessels. In such cases, large ships and small craft will move through the same waters. Planning criteria must be adopted to reduce the collision hazard to a minimum without curtailing the activities of either class more than is essential for navigational safety. Otherwise the planning of such small-craft facilities should follow the guidelines contained herein.

Figure 1. Depoe Bay, Oregon. A typical small-craft harbor used primarily for berthing commercial fishing and charter boats.

2. Recreational. Small-craft harbors are designed for various recreational craft, including sailboats, rowboats, pedal-craft, air-cushion, and vehicle. Other exotic craft are not specifically covered, although the basin and entrance planning techniques described will, in most instances, be found satisfactory for all classes of using craft.

3. Harbor of Refuge. Most of the boats and small vessels that fall within the general classification of small craft have relatively short cruising ranges. To take these craft on trips that extend beyond half of their safe cruising range in any large body of water may be extremely hazardous unless harbors are available along the route at safe cruising intervals. Such harbors may be needed only for replenishment of fuel or provisions, for shelter during unexpectedly bad weather and when equipment fails, and for sick or injured boaters. Depending on the class of boat and characteristics of the region, the safe cruising distance is usually between 20 and 40 miles, or 2 hours' cruising time. When a remote harbor is provided specifically to accommodate transient craft rather than as a home port for the local craft of the immediate area, it is designated as a *harbor of refuge*. Such a harbor need not have all the refinements of a home port, but must have an entrance that is navigable in adverse weather, access to emergency aid, and appropriate facilities to accommodate the transient boater. In remote areas, harbors of refuge meeting just the needs of the transient boaters often are subsidized. In these instances, the harbor of refuge may possibly be made self-sustaining by berthing a small number of home-based craft in addition to meeting the periodic needs of transient craft; it may not survive economically on either type of craft alone.

III. SITE ANALYSIS

1. General. In selecting a harbor site, two basic needs must be fulfilled regardless of the body of water on which it is to be located. The site must provide safe navigation access to cruising waters, and have adequate land access, including approach roads, for boat owners to conveniently reach their craft. Other important factors in site selection are: (a) enough protected water area or low land that can be excavated to navigable depth, and areas for future expansion; (b) adequate perimeter land or low land that can be filled for vehicle parking, harbor service structures, roads, and ancillary facilities, including land needed for future expansion; and (c) utility service to the site, such as electric power, potable water, telephone, gas, and sewerage.

In addition to these physical factors, several environmental, economical, and sociological factors may have a governing influence on site selection. These factors are discussed in detail later in this report, but are mentioned here because of their importance in the early phases of harbor planning. They include legislative restraints, zoning ordinances, permit requirements, land ownership problems (also those involved with submerged lands), and water quality and ecological preservation factors. Before any site is finally selected, all of these factors must be considered and any conflicts resolved (Sec. VII).

2. River Mouths. The area on either side of a river just upstream from its mouth has certain advantages as a site for a small-craft harbor (Fig. 2). The river current is often minimal, and the site is usually protected from ocean or lake waves. Also, the river has usually scoured a deep channel that may be adequate for navigational access. However, the site may be endangered by occasional river floods, and sediments moved by river currents or tidal action may shoal the harbor basins and channels. Much depends on the characteristics of the river: whether it is relatively slow-flowing and stable; has a steep gradient and is subject to flash floods and channel meandering; and if the sediment load may be transported into the water areas of the harbor. These factors require in-depth study by qualified coastal engineers. At times, the river flows through a delta estuary or marshland before reaching the ocean or lake; then the harbor can be situated in a basin and isolated from the main flow of the river. This type of river-mouth site is safer from direct flood damage, but the channel leading to the harbor from the main flow of the river may be subject to shoaling by river floods.

3. Dredged Lowlands. Low, marshy areas are often found adjacent to ocean and lake shores where no river exists. Ecological factors permitting, these areas are good sites for small-craft harbors (Fig. 3). Marshes can usually be dredged to navigable depth at reasonable costs. If the dredged material is suitable, it may be used to fill perimeter areas to design levels for roads, parking, and other harbor-support uses. Although the low cost of using this marginal land may be attractive, this advantage may be offset by the high cost of maintaining an entrance to navigable waters in the lake or ocean if the rate of longshore movement of littoral material is high. Protection against shoaling due to littoral drift is possible by using various methods. If the longshore transport is predominantly in one direction, provision must be made for sand bypassing (discussed in Sec. V). If the shore of the lake or ocean is frequented by high waves, protective structures may be required to reduce surge and wave action in the entrance and interior basins, as discussed in Section IV. There are advantages to siting an ocean harbor opposite the head of a submarine canyon. The deep water of the canyon refracts waves toward its flanks, making the canyon head a relatively calm area of wave divergence. Thus, the harbor is ensured not only of a deep, maintenance-free channel, provided littoral transport is excluded, but a minimum of wave energy penetration into the berthing basins. The need for adequate exchange of water in a dredged interior basin to preserve its quality may be another important factor.

4. Bays. Some natural embayments provide excellent sites for small-craft harbors (Fig. 4). However, the mouth of the bay should be small or protected by islands to exclude most waves from entering the bay from open water. Preferably, the bay should be small enough so that waves of undesirable heights cannot be generated by winds within the bay's confines. In a large bay, some type of protective structure may be required to exclude larger, locally generated waves from the berthing areas, or for secondary protection from large storms.

Figure 2. Hastings Small-Boat Harbor, Minnesota

Figure 3. Merrill's Marina, Sunset Beach, California

Figure 4. Hyannis Inner Harbor, Massachusetts

Several types of natural bays present a deceptive first impression to harbor siting. The first type is a natural bay protected by an offshore spit on an open coast. Many of these bays along the Atlantic and gulf coasts appear quite fragile, but provide excellent sites for small-craft harbors. Sounds, on the other hand, appear to provide excellent harbor sites because they offer protection from high waves approaching from open waters. However, their great lengths usually are conducive to high, locally generated wind waves and possible resonance with waves (seiching), which can result in extreme ranges of water level fluctuations. Many sounds are too deep and anchorage of slips and docks becomes a severe problem. Where a sound is located between headlands and is narrow like a fjord, perimeter lands often rise abruptly from the water, and access routes and provision of adequate land area for supporting a harbor complex become critical factors.

5. Roadsteads. A good site for a small-craft harbor may often be found close to the open ocean in an area that is shielded from the prevailing ocean waves by natural terrain features. An example is a bight in the shoreline behind projecting headlands (Fig. 5). Site selection here must include consideration of occasional wave episodes from directions other than those of the normal wave regime and from which the site is exposed. Studies of wave refraction may be needed to ensure that open-sea swell from certain offshore directions does

Figure 5. Port San Luis, California. A typical harbor in a roadstead. This site is protected from the prevailing westerly waves but is exposed to occasional high wind waves from the south.

not penetrate the area, although the site may appear to be well protected from those directions. Before selecting a harbor in a roadstead location, the vulnerable approach directions must be carefully analyzed by thorough examination of historical wind and wave records and by wave refraction study techniques. A special case of roadstead siting is a harbor on the lee side of an offshore island, where the island acts as a breakwater against the prevailing waves.

6. Open Shorelines. Some factors controlling harbor location dictate that a site be selected along an open shoreline. In this case, protection against wave action must be achieved by breakwater-type structures (Fig. 6). Two steps that will provide guidance on site selection are: (a) to make a statistical study of the historical wave climate coupled with wave refraction analysis to pinpoint areas of frequent wave convergence; and (b) to make subsurface probes to detect bedrock levels that can be used to ascertain the possibility of keeping the harbor close to shore by dredging into the shallow nearshore bottom to obtain navigable depth. Both steps require the assistance of specialists. The resulting information will be useful in the detail design of the harbor complex after the site has been selected.

7. Riverside Sites. Rivers often provide excellent water courses for small-craft cruising. Harbor sites may be found along or just behind a river bank, the choice usually depending on good road access to the site and places where the current near the bank is minimal. Floating docks may be anchored to dolphins or boomed out from the shore on the open river, where they are not in violation of navigation-encroachment regulations and flood control or other restrictions (Fig. 7). Some protection from floating debris and ice may be required, depending on the physical environment and characteristics of the river. Also, provision must be made for adjustment of the floating system to fluctuations in the water level of the river.

In some instances, channels and basins may be excavated into and behind the river bank to provide safe small-craft harbors. The navigation entrance to the basin should be at a point along the river bank where the current is minimal and preferably connect with the downstream end of the basin. The site must be located in an area where it can cope with extreme high and low river levels both with respect to slip anchorage and access from the land. Many off-river basins have entrance shoaling, as suspended sediment carried by the river is deposited on the bottom in the quiet entrance waters. Only experience will determine the severity of this problem along a river and may dictate whether an off-river basin or a complex of riverbank slips is best.

IV. ENVIRONMENTAL SITING CONSIDERATIONS

1. Local Weather Factors.

a. Precipitation. Normal rainfall or light snowfall present no serious problem in small-craft harbor design, provided an adequate surface-drainage plan is adopted. Such a plan must provide for a facility capable of draining the waters from a maximum probable rainfall without eroding the perimeter land; and also for the diversion of any possible

Figure 6. Shilshole Bay Marina, Seattle, Washington. **A typical offshore harbor protected from the prevailing waves by a breakwater.**

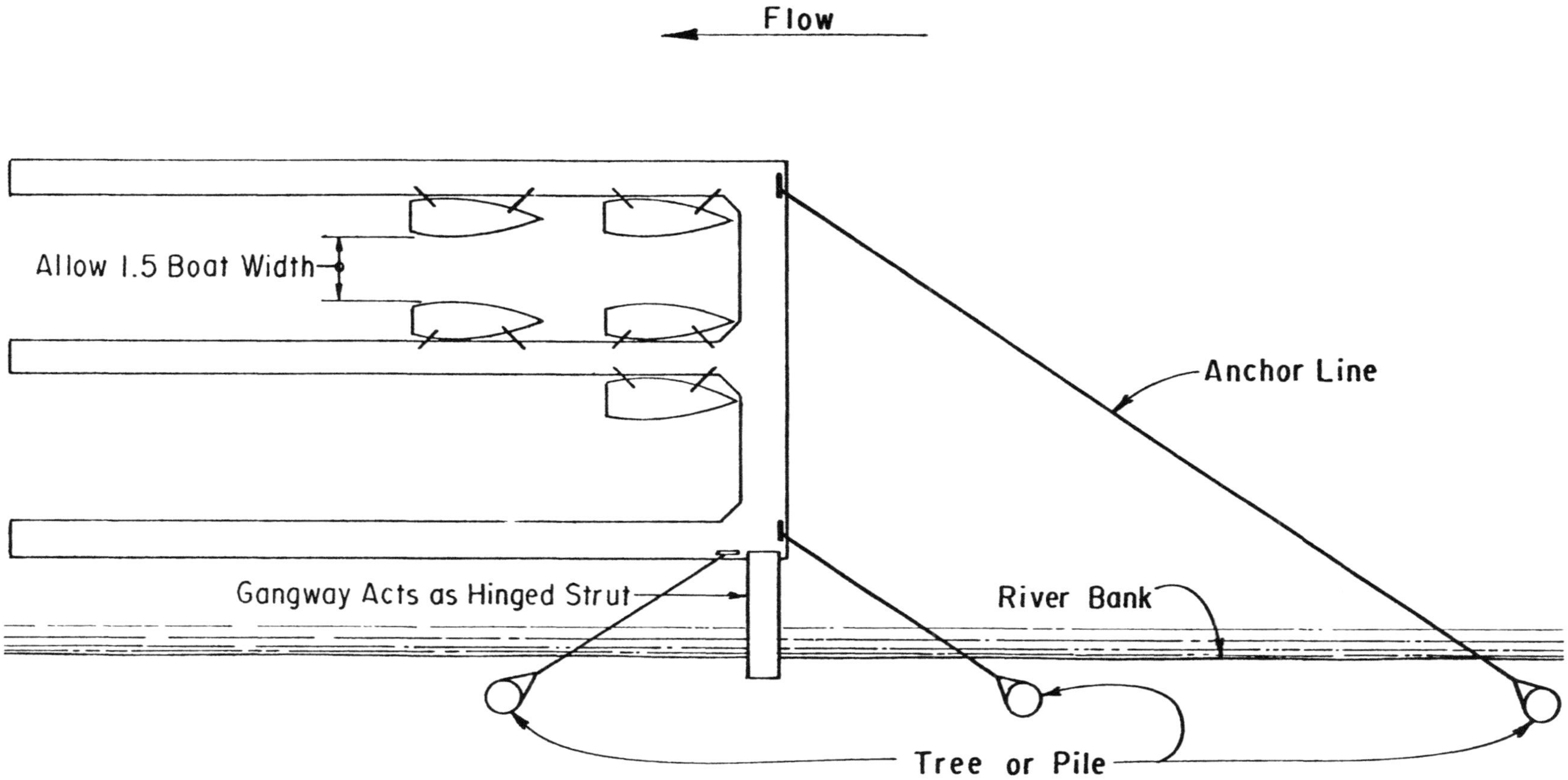

Figure 7. Trailing slips in a river. A schematic arrangement of floating docks anchored to the adjacent bank.

inflows from land around or safely through the harbor complex. Although most small craft are designed to shed water, covered slips are sometimes provided to keep the craft dry above the waterline, and to shed snow, prevent hailstone damage, and shield the craft from excessive exposure to sunlight (Fig. 8). In regions where snowfall is heavy, landside structures and slips must be designed to shed snow or to carry a heavy snow load.

Figure 8. Covered slips in Bellingham Harbor, Washington.

b. Wind. In most regions, the design windloading of structures is specified by local building and safety codes. If not specified, historical meteorological records must be examined to determine the most severe wind conditions that might occur. Floating slips must be designed to withstand the horizontal thrust of the berthed craft during the design wind condition; also, the slips must be anchored securely so that the entire complex with its berthed craft does not drift under wind stress.

Where winds may be strong and even reach hurricane strength, the water area structures must be designed to withstand the induced forces, and the land structures anchored securely to their foundations (American Society of Civil Engineers, 1961). In addition, substantial increase or decrease of the water level due to wind stress must be considered in the layout and design of the harbor. Where strong winds blow landward over a shallow offshore shelf or across a shallow lake, wind stress effects may be greatly amplified, raising or depressing the water level at the harbor site by several feet in a few hours. A fair assessment of the water level fluctuation can be obtained in some areas by examining historical records. However, the design maximum and minimum water levels must usually be determined by computing the wind setup or depression in accordance with instructions in the Shore Protection Manual (U.S. Army, Corps of Engineers, Coastal Engineering Research Center, 1973).

In regions where tornados occur, little can be done beyond following the precautions recommended for hurricane-prone areas. The destructive force of a tornado exceeds the practical design safety factor used for most structures. Some damage may be prevented by designing buildings for quick exchange of atmospheric pressure by the use of large sacrificial doors and windows to avoid complete collapse.

The movement of sand by the prevailing winds of a coastal area creates special problems that must be considered. Windblown sand may shoal harbors and cover land areas, requiring frequent dune and drift-sand removal from roads, parking lots and other areas. Evidence of dune formation at a harbor site should be examined, and an estimate made of its seriousness. At some sites the windblown sand problem has been solved by paving, by landscape-planting the dry sandy areas windward of the harbor complex, or by stabilizing with sand fences and dune grass (U.S. Army, Corps of Engineers, Coastal Engineering Research Center, 1973).

c. Ice. Sheet ice can cause damage to a small-craft harbor. The best and most popular precaution against ice damage is to remove boats from the water in winter. They can either be removed to dry storage or hoisted out of their slips and left suspended above the water surface.

Damage to fixed and floating slips occurs in two ways. As sheet ice forms, it expands and can crush floats and cut into piles. A secondary effect of this action is that if the water level rises after freezing has begun, the ice sheet hugging the piles exerts an upward force tending to jack them up and reduce penetration into the soil. Repeated freezing and thawing may eventually lift the piles completely out of the ground. Most ice damage is caused by the impact of drifting floes on structures as the ice melts in spring.

In areas where freezing does not produce a thick ice sheet, ice formation can be prevented near piles, floating slips, and boats by forced-convection currents. This system is discussed in Section V. Steel or metal-cladded timber piles can be driven deep enough in some soils to develop great withdrawal resistance, so that the ice will slide along the pile as it rises. Floating slips can also be designed with tapering or rounded bottoms so that the pinching effect of the ice squeezes them upward (Fig. 9).

Natural basin perimeters and revetted slopes can be severely eroded by expanding ice sheets. One method of preventing this erosion is to line the perimeter slope with smooth concrete. Vertical perimeter walls may be pushed back into soil behind them in winter and then spring back when the ice thaws. The extent of weakening that results will depend on the type of construction.

Figure 9. Floating docks being pinched upward by forming ice in Lake Michigan (Courtesy of Harbor Host Corporation).

Frost formation at night followed by thawing in the daytime may cause damage in some regions. Frost or freezing of water trapped in crevices between structural members may pry them apart or loosen connections. Eliminating the crevices through careful design and construction can avoid this type of ice damage. Figure 10 shows how frost and ice action can damage a concrete parking area.

Damage from ice-floe impact is common in rivers, but it may also occur in harbors as a result of floes drifting with the wind or with circulatory currents in lakes or arctic waters. Deflecting booms made of logs or heavy timbers can often be devised to protect the berthing area from drifting ice. Fixed-pile structures that break the ice sheets into smaller pieces are discussed later in Section V.

d. Fog. Although fog causes little damage in a small-craft harbor, the reduced visibility is a serious navigational problem. Most water areas have occasional foggy conditions and most recreational boaters have limited skill in navigating their craft under poor visibility conditions. For this reason, entrance channels and main fairways in a harbor should be as straight as possible. There is no method of dispelling fog at a reasonable cost, and although radar and infrared devices have helped the navigator, such sophistication is seldom available to the small-craft operator. Small-craft harbor entrances and fairways should therefore be designed so that they can be navigated in dense fog by following marker buoys and other channel-marking devices, with as few turns as possible.

2. Wave Factors.

a. Sea and Swell. These factors have been mentioned as they apply to site selection. Once the site has been fixed, the harbor must be planned so as to reduce wave action from the entrance and interior basins to acceptable heights. This is done through a combination of entrance-channel orientation, protective breakwaters and jetties, and interior wave-dissipating devices. Where a harbor opens into the ocean or a large lake, the entrance should be oriented for a boat to enter without turning broadside to the incoming waves and thus risk broaching or being "surf-boarded" into a jetty during high wave conditions. Both historical wave data and the statistical hindcast data required for site selection can be useful in orienting the entrance and designing the protective structures. Wave-dissipation structures can often be used in reducing to acceptable heights the waves that do find their way into the entrance. The normal criteria for acceptable maximum wave heights are about 2 to 4 feet in the entrance channel and 1 to 1.5 feet in the berthing areas, depending on the characteristics of the using craft.

Generally, if waves can be attenuated to a height of about 1 foot in the berthing areas, their horizontal oscillations will not be troublesome, and any longer-period resonant effects will go unnoticed. In initial planning, the best orientation of the entrance and location of protective structures can be obtained by refraction and diffraction diagram analysis (U.S. Army, Corps of Engineers, Coastal Engineering Research Center, 1973), (Fig. 11).

Figure 10. Ice- and frost-damaged parking lot, Boston Harbor Marina, Massachusetts

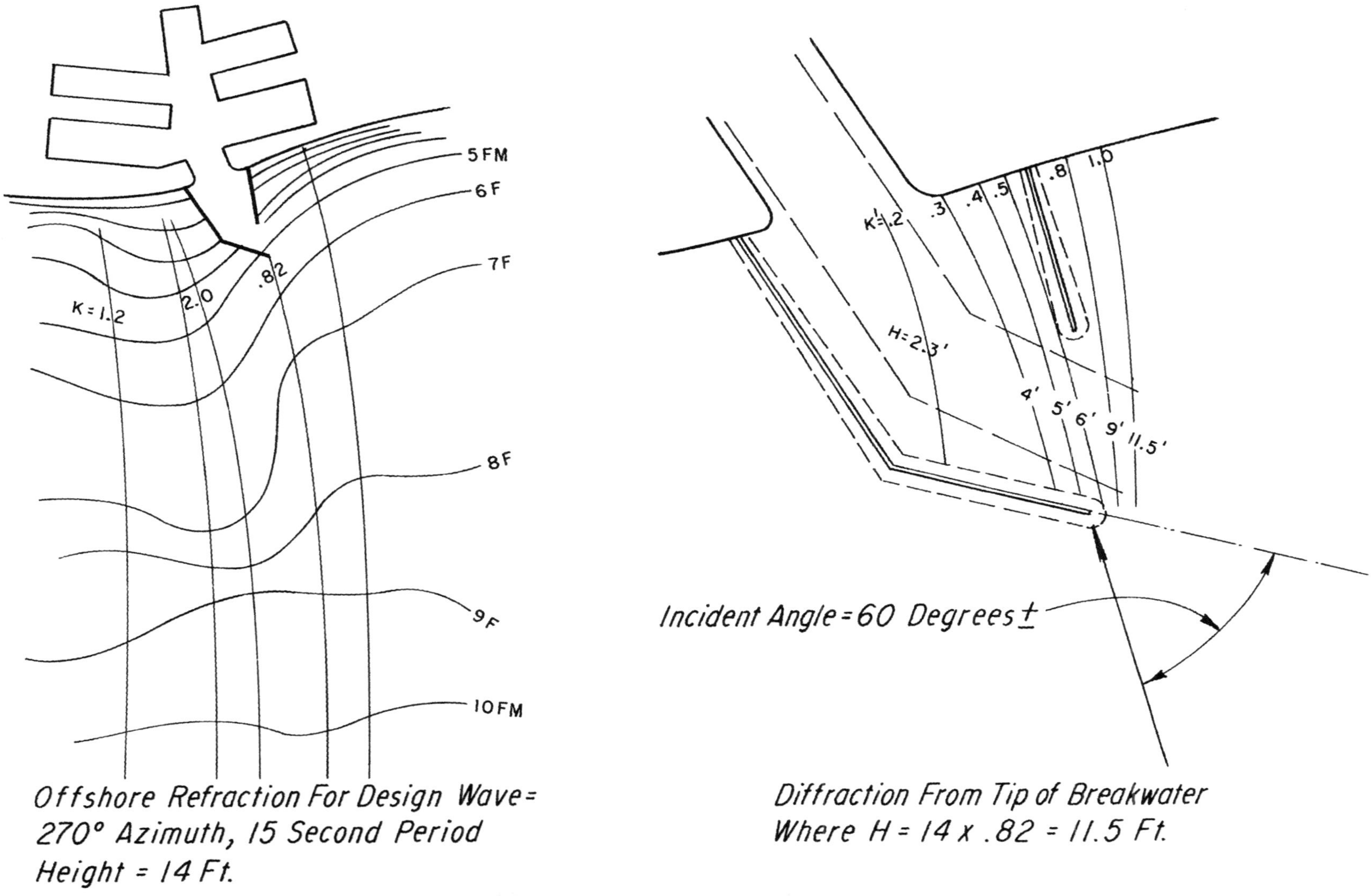

Figure 11. Sample refraction-diffraction diagram. Coefficients indicate how waves will be attenuated by breakwater.

A few diagrams tracing the divergence factors from deep water into the entrance for the most critical wave periods revealed by statistical hindcast analysis will indicate the relative effectiveness of any entrance plan for attenuating waves. Once the approximate layout plan is determined by this method, it may need to be checked by model analysis and the protective structures adjusted to achieve maximum attenuation of the waves.

Some harbor sites have natural entrances that cause a reduction of the height of waves entering from the main water body. This can occur as a result of island shielding in an embayment or an entrance located within a series of prominences in a roadstead or sound. In this case, the need for protection is reduced in proportion to the degree of wave attenuation at the site. The historical records may show that only minimal protective works, if any, are required.

If the site is close to a large water body and an artificial entrance must be provided, a model study may be needed to ensure that the required degree of wave attenuation in the inner basins is achieved (Fig. 12). If wave height just outside the proposed artificial entrance exceeds 10 feet, construction of the entrance and its protective structures will be costly and a harbor capacity of several hundred boats must be provided to ensure enough revenues to defray the initial cost outlay. Cost of such a harbor may exceed a million dollars and the additional $50,000 to $100,000 cost of a hydraulic model study is worth the ensurance that the inner basins will be free from troublesome wave action.

Some harbors are so large that troublesome, short-period waves may be generated within the harbor confines by strong winds blowing down long fairways or the wakes of numerous power craft may combine to produce undesirable turbulence in the berthing areas. Where small craft and large commercial vessels share the same harbor, the wakes of the large vessels may penetrate the small-craft berthing areas. In such cases, the probability of troublesome interior wave generation must be determined in the planning stage and corrective action taken through the proper arrangement of moles or the provision of wave baffles as described in Section V.

b. Surge. Waves of long period and great length with low height are termed *surge.* The most troublesome aspect of surge in a harbor is its horizontal water motion or oscillation, which causes stress in mooring lines and anchorage systems and can make the maneuvering of boats into slips difficult. Ordinary wind-generated waves that penetrate the entrance of a harbor may acquire the characteristics of surge in the inner basins. If the period of the surge oscillations in the basins correspond to the period of the sea or swell outside the entrance, then the exterior sea or swell is the cause of the surge within the harbor.

If the harbor surge period differs from the wave period outside, the oscillations may be due to harbor geometry with respect to the wave input at the entrance. Most waves that penetrate the entrance and reach the interior basins with a significant amount of residual energy will usually have a length that is much greater than the water depths of these basins.

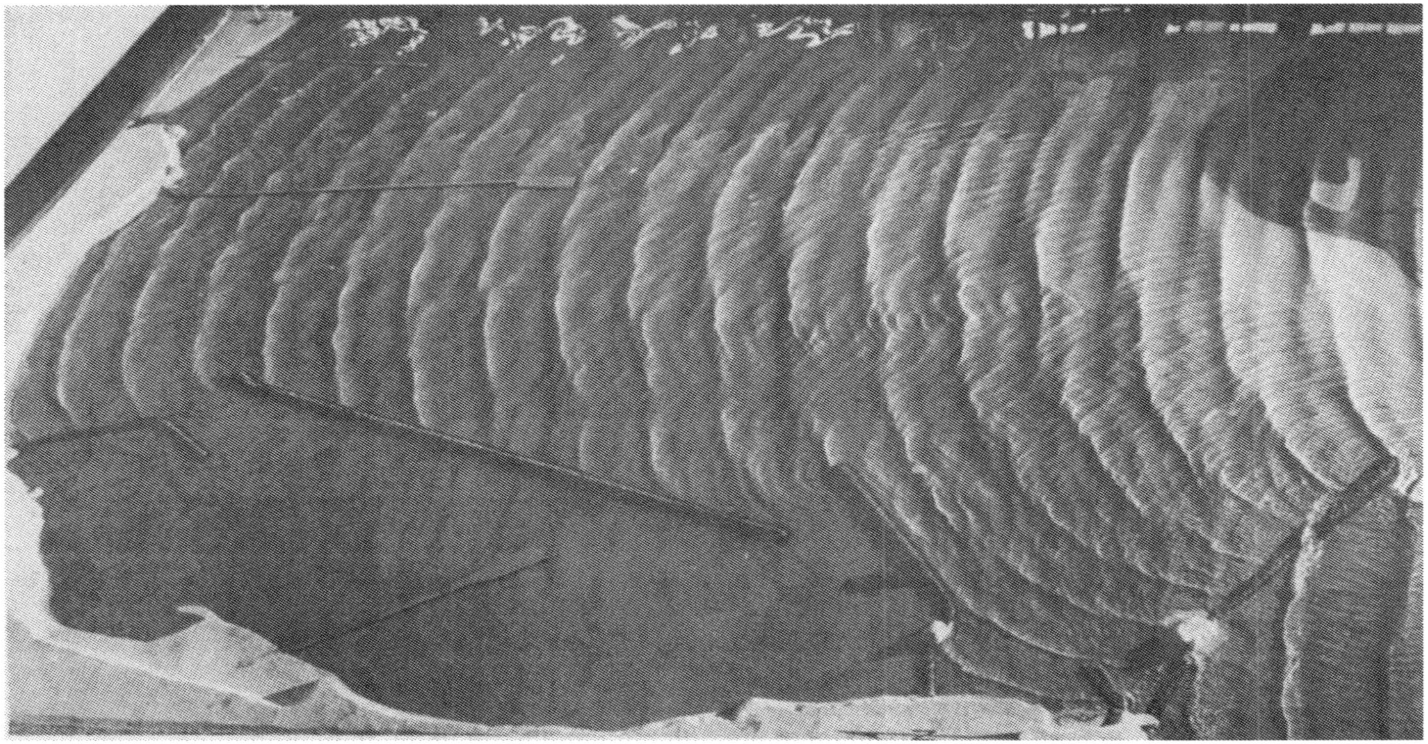

Figure 12. Model study for proposed harbor at Port San Luis, California. Note how proposed breakwaters attenuate waves from the south. Pier in lower left represents prototype near center of Figure 5.

All the waves will then travel according to shallow water wave theory at a speed that may be calculated by the formula:

$$C = \sqrt{gd}$$

where

C = the wave propagation speed
g = the acceleration due to gravity

and

d = the basin depth .

If a rectangular basin with vertical walls is closed at both ends, a wave starting from one end will travel to the opposite end and be reflected back to its starting point in a fundamental period of time that may be calculated by the formula:

$$T = \frac{2b}{\sqrt{gd}}$$

where

T = the fundamental resonance period
b = the length of the basin

and

d = the basin depth.

For example, if a basin is 650 feet long and 15 feet deep, the fundamental period of the wave will be about 1 minute. If small waves were propagated through a narrow entrance at one end of the basin at 1-minute intervals, each would add energy to the wave being reflected back and forth until the amplitude of that wave would become considerably larger. This would be resonant surging to the fundamental period of the basin. Since most wind waves and swell that penetrate harbor entrances have periods of a fraction of a minute, such resonance is unlikely to occur. However, the harmonic wave periods of the basin are 1/2, 1/3, 1/4, 1/5–1/n of the fundamental, and if wind waves or swell arrive at the entrance with periods equaling any of those harmonics, they may induce harmonic resonant surging in the basin. The wave periods of concern would therefore be computed by the formula:

$$T = m \frac{2b}{\sqrt{gd}}$$

where

$$m = 2, 3, 4, \cdots n \ .$$

The value of *m* would be the number of nodes in the resultant harmonic surge-wave system between the two reflecting walls. Because resonant surging is more readily induced at the fundamental or lower harmonic periods of a basin, small, deep basins (with short fundamental periods) triggered by outside waves are more susceptible to surge problems than are large, shallow basins. Also, the wave input at one end may induce a lateral resonant surging in the direction of the short axis of the basin.

If the ends and sides of the basin are poor wave reflectors, resonant surging will be difficult to induce. Also, if the shape of the basin is irregular or its depths vary irregularly, resonant surging is less likely to occur. Unfortunately, vertical basin walls are usually more desirable than poorly reflective basin perimeters, and rectangular basins are more efficient than irregular basins for berthing arrangements. As a result, some kind of compromise will often be the best solution.

In large, shallow lakes surging effects may be triggered by wind stress and often produce fluctuations of the water surface level similar to the lunar tides of the ocean. Surge in a river-connected harbor may be the result of navigation-lock operations, and hydrogenerator operations of a major cooling-water withdrawal system. The body of water served by the harbor site should be examined for past records of water level fluctuations and these fluctuations should be considered in the harbor design. Regardless of the cause, surging oscillations are difficult to reduce once the harbor has been built. Therefore, every effort should be made during harbor design to ensure that surging does not reach troublesome proportions in the berthing areas.

The best way to study harbor surge is by hydraulic-model analysis. This will determine the proper entrance orientation, protective structure positioning, and inner basin configurations required to minimize internal water motion. Progress is being made on new techniques in mathematical analog analysis from computed programming (LeMehaute and Hwang, 1970) to study harbor designs. Although the analysis is not as complete as hydraulic modeling, it does provide a technique to evaluate the basic features of a design. Another method that gives promise of detecting flaws in preliminary design is the acoustic-analogy technique (Morrow, 1966). Acoustic analogy has been used in designing resonator basins at the harbor entrance, which are tuned to specific troublesome wave periods (James, 1970). Such basins resonate to waves with periods within a given bandwidth of the spectrum, regardless of incident direction, and reflect them back out to sea.

Most recreational boats in a small-craft harbor are insensitive to surge with periods in excess of those normally encountered in sea and swell. Large cruising and commercial craft may experience fender and mooring line difficulties under long-period surging, and any harbor with these vessels should be designed to eliminate its cause.

c. Tsunamis. These long-period waves of seismic origin are relatively infrequent and normally occur only in the major oceans. The major fault zones that rim the Pacific Ocean have produced most of the tsunamis of record, some of which have caused great damage in the coastal harbors of the U.S. Pacific coast and Hawaii. These waves usually have periods of 10 to 20 minutes and travel at speeds of several hundred miles per hour. Because of a low steepness factor they cannot be visually detected in deep water; when these waves arrive at a distant shore they may cause fluctuations of several feet in the water surface elevation.

Tsunami waves behave like a very rapid tide impressed on the lunar tide at whatever stage it happens to be. The rapid flushing of channels and basins that results may cause unusual amplifications and channel velocities that scour banks and make navigation of any size vessel extremely hazardous. The tsunami wave may rise above the adjacent land surface of a harbor and flood marginal lands.

Little can be done to prevent a tsunami wave penetration in a small-craft harbor, but precautions can be taken to minimize the damage. A site should be selected where the perimeter land surface is above the level to which tsunamis might raise the water surface. If this is not practical, the floors of buildings in the perimeter area should be kept above this level. Another precaution is to keep the tops of the anchor piles of floating slips well above this level so that the floats will not rise above them.

Anchorages should be designed to resist the lateral loads that may be imposed by tsunami-generated currents. The trough of a tsunami wave might lower some of the berthed craft to the bottom. To prevent damage from such an occurrence, shearing off the barnacles from guide piles below the normal low tide level by a cutting device could prevent a pile guide from sticking on their rough surface in a tsunami drop and capsizing the berthed craft on the tsunami crest. In regions where tsunami flooding has been serious, a warning signal and emergency evacuation system should be devised.

d. Tides. Lunar tides vary from place to place along the shores of all oceans. Tide prediction tables, published annually by the National Oceanic and Atmospheric Administration, are available for most locations where small-craft harbors may be built. Ocean tides may extend many miles upstream from the mouths of large rivers and are semipredictable for most potential harbor sites. If a coastal site is chosen that is remote from any of the stations listed in the tables, it is possible to interpolate predictions for the site from values given in the tables for the two nearest stations. The extremes of the predicted spring tides provide criteria for designing the harbor to accommodate any water level fluctuation that may occur.

Interior surge, boat-wake waves, tsunamis, and hurricane surge should be considered in conjunction with the tidal fluctuations. In determining the project bottom depths of interior channels and basins, consideration must also be given to the drafts of the largest vessels that will be using the facility and to effects of squat, roll, pitch, and heave on these vessels in maneuvering areas. For example, the berthing basins of one marina were dredged to what was thought to be sufficient depth below the predicted extreme low tide level to provide adequate clearance for the keels of the deepest-draft boats to be accommodated; yet, after a short period of operation, there were complaints of groundings. Investigations showed that the maneuvering habits of the larger power craft in some areas resulted in a combined squat and resonant pitch that "kicked up" bottom materials into submerged "windrows." In other instances, designers have neglected to take into account internal wave action, and during the first extreme low water, the normal wave penetration into the berthing area caused some boats to strike bottom.

Astronomical tides and tidelike effects of surge in stillwater bodies, due to any of the causes previously enumerated, often play an important role in water quality control. The current-producing exchange of water between the harbor basins by tidal action may be essential to the marine ecology and the prevention of a stagnant condition. Consideration should be given in harbor planning to the achievement of the best possible circulation of water through the basins and channels by effective use of the tidal prism. Water level fluctuations on a slower cycle often occur in inland lakes and rivers for various reasons, and although they usually occur too slowly to produce beneficial water-exchange effects, they must be accounted for in design. Typical examples are: (a) seasonal drawdown of water conservation reservoirs, (b) annual changes in the water levels of the Great Lakes, and (c) periodic water level changes in many natural lakes and rivers due to increases and decreases in precipitation over their tributary areas.

3. Water Area Shoaling Factors.

a. Littoral Drift. A principal cause of shoaling at entrances to harbors along the shores of oceans and large lakes is littoral drift (Fig. 13). Because the longshore movement of sand is due mainly to wave action, any structures that change the normal regimen of waves breaking along a coast may influence the littoral movement. If an unprotected channel is dredged through a beach into an inner basin, the wave impinging on either side at the mouth will be refracted in such a way as to cause abnormalities in the wave pattern approaching the lips of the channel. If the approach of the prevailing waves is normal to the shore, the initial effect will be a movement of the littoral material from the lips inward along each flank of the channel, thus eroding the lips and shoaling the inner channel. As the process continues, the channel banks accrete toward the center of the channel, fed by material from the beach on either side of the entrance. Unless tidal currents are strong enough to maintain an opening against the forces tending to shoal the entrance, the channel will soon be closed. Where the prevailing wave approach is oblique to the shoreline, sediment being transported along the shore by littoral currents will be interrupted at the channel opening near the updrift lip and that lip will soon begin to accrete. As the wave-induced longshore current again begins to "feel" the shore downdrift of the channel mouth, it attempts to reacquire its sediment load. As a result, the downdrift lip of the channel will erode at about the same rate the updrift lip accretes, and the channel mouth will migrate in the downdrift direction.

In each of these cases, the forces of nature are attempting to reestablish the littoral balance that was present before the channel was excavated. The above account is an oversimplified version of an extremely complex process and excludes consideration of the effects of sandbar formation, eddy currents, and tidal channel meandering.

The customary solution to entrance shoaling is the construction of jetties along each flank of the entrance channel from the lips of the mouth seaward beyond the breaker zone. The structural features of the jetties must be such that the materials will not be washed through or over the structure into the channel. A typical section of a sandtight rubble-mound jetty is shown in Figure 14. If the littoral transport from one direction predominates and the entrance is stabilized by jetties, accretion will occur along the updrift shore and erosion along the downdrift shore. Mechanical bypassing of this littoral material must be instituted to minimize the influence of the jetty structures on longshore transport. Since littoral transport rates in excess of one million cubic yards annually have been recorded along some reaches of the continental coastline, bypassing may be costly but necessary for harbor maintenance.

Longshore-type sediment movement can also occur inside a harbor where any segment of the perimeter is left as a beach for wave absorption or recreational purposes. If such a beach is near the entrance, residual wave action entering the harbor may cause longshore movement of the beach material. The best means of avoiding this problem is to determine, before harbor construction, the approach direction of the waves that will pass through the entrance to the vicinity of the beach, and then to aline the beach normal to this direction. If a revetted slope or bulkhead wall is to be located on either or both ends of the beach sector, the beach must be recessed behind the general alinement of the revetment or wall so that the beach toe does not extend beyond the toe of the revetment or wall. Because the residual wave action inside the harbor is usually diverging, the beach alinement should be made so that the incoming wave will be normal to the beach at all points.

Boat-wake generated waves can also cause sediment movement along beaches within a harbor. This may be a minor problem that can be solved at low cost by using mechanical equipment to periodically deepen the channel and restore the beach. If the beach existed before the harbor was developed, and it was not intended for recreational use or esthetic

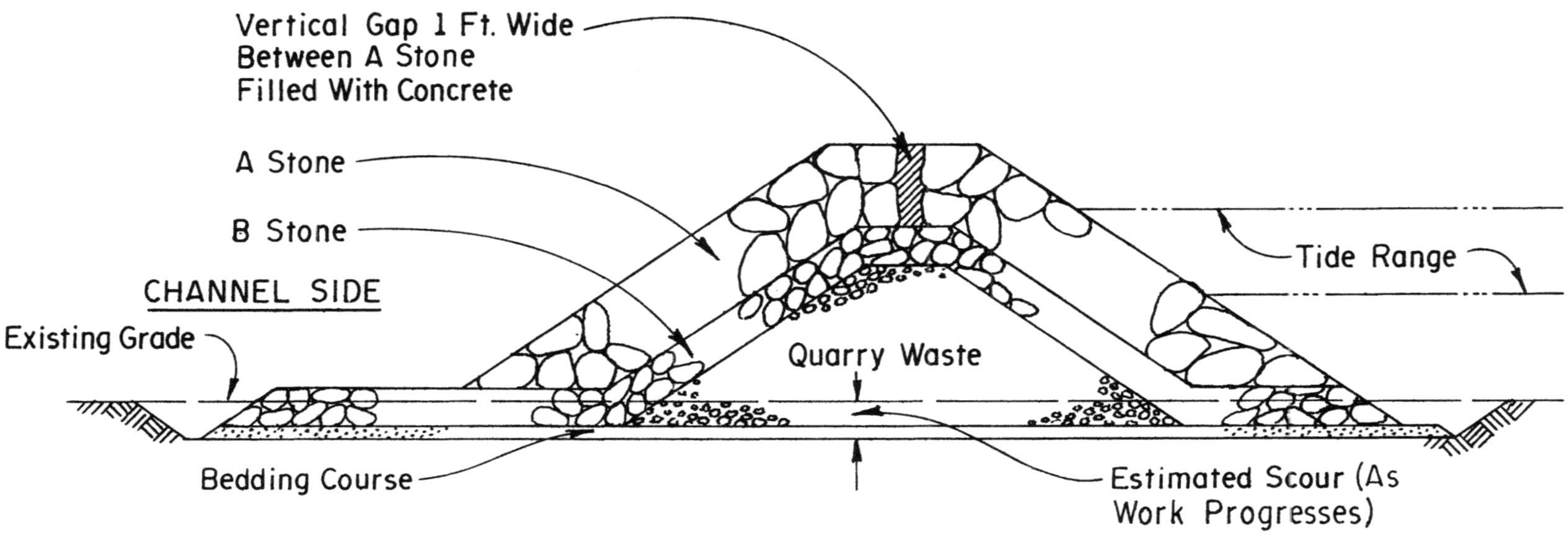

Figure 14. Nearshore section of rubble-mound jetty. Grout dike in crown makes jetty relatively sandtight.

enhancement, the best means of preventing the shoaling of nearby harbor waters by material eroded from this beach is to revet the beach face. To perserve the beach, an inharbor wave baffle or fence may be required between the navigation channel and the beach to control the sediment migration. Another control is to effectively implement a slow-speed zone in the reach of channel opposite the beach. If the beach is only for recreational purposes, it should be located at the inner end of a basin where boat-wake waves are seldom generated, provided that the water quality in that location is adequate. In this location, the beach would also help to reduce surge oscillations in the basin.

b. River Discharge. Harbors in off-river basins are subject to shoaling because of sediment deposition in the quiet water area and by eddy currents that may be created by the entrance configuration and the flowing water in the river. Although shoaling can not be prevented, it is often reduced by proper entrance design. Along the Colorado River, for example, the banks are friable and sandy, without much cohesive property. The river silt-load increases rapidly with slight increases in flow velocity; the suspended material is coarser than normal, and it settles out rapidly in the quiet waters of the entrance channel to any off-river basin. The best solution is to provide a flat area on the downstream lip of the entrance from which a dragline can excavate deposits from the bottom of the entrance channel and cast them into the river downstream from the entrance (Fig. 15). The entrance must be kept narrow to permit such an operation, and a training dike off the upstream lip is helpful in reducing the deposits.

c. Nearby Water Area Structures. Structures in the water area outside the harbor entrance may also cause harbor shoaling, especially along shorelines where littoral transport is a problem. An example is a groin or jetty located in an area that is normally downdrift from a harbor entrance. This type of structure tends to impound or stockpile sediment under normal conditions. However, on those occasions when waves approach from the opposite direction and cause a reversal of direction of littoral transport, some of this sand may be transported back into the entrance and cause shoaling. Relatedly, periodic maintenance operations of another harbor located updrift from the problem harbor may result in large quanities of sand being transported downcoast toward it. Beach-fill operations along the nearby coast may increase the transport rate of material into the harbor entrance.

On rivers, any structure upstream or across from the harbor may alter the current flow and cause excessive shoaling at the harbor site. Current-deflecting dikes, bridge piers, and pile-supported structures are examples of manmade works that may affect the shoaling characteristics of the river at the site. Bank realinements and erosion-control works undertaken after a harbor has been built may have a profound effect on the harbor. Thus, it is important first to consult with controlling authorities about any plans for future construction or corrective works near the site, or upstream from it, before harbor construction on any river. Such information should be considered in the siting and design of the harbor.

d. Redistribution of Bottom Materials. In some water areas, the shifting of bottom materials by natural processes poses a severe threat to any harbor built along the margins of

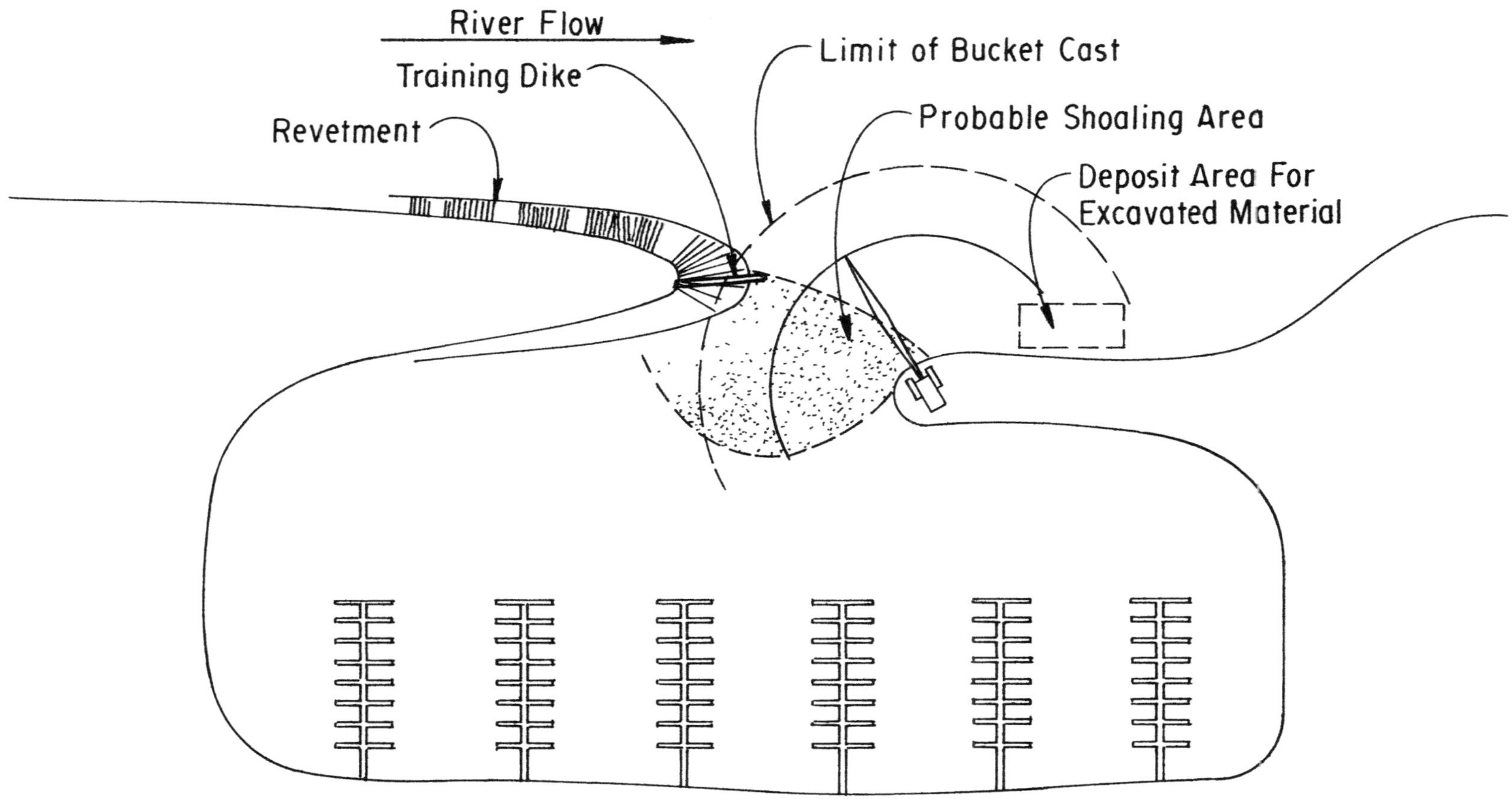

Figure 15. Maintenance of entrance to off-river basin with land-based equipment.

those waters. A good example is San Francisco Bay, mostly rimmed with a wide shelf of soft bay silts and mud, the surface of which lies but a few feet below mean sea level. Any harbor constructed along the margin of this bay must have an access channel to deep water through this shelf. Wind waves generated within the bay frequently stir this muddy bottom material into suspension and move it long distances; most of the material is redeposited in navigation channel and harbor entrances. As a result, the channel characteristics are continually changing and the channels must be marked with buoys and redredged frequently to maintain navigation.

Rivers meander from the natural process of flood plain and delta aggradation. River flow continuously erodes material from concave bends where the currents concentrate; the material is then deposited on convex bars farther downstream where the current is slower. Examples of this meandering are prevalent throughout the Mississippi Valley and other major river systems. The harbor designer must study the proposed site carefully to determine if this natural channel migration and its related movements of bottom sediments pose a severe threat, or if their effect can be overcome by maintenance efforts within the financial capabilities of harbor operations. If the problem is serious, he should seek the advice of a river-hydraulics specialist familiar with the particular area involved.

4. Geological Factors.

a. Basin Excavation. On rare occasions, the harbor site provides basins that are already the correct size and shape and have adequate depth throughout. However, some excavation is usually required, and a knowledge of the characteristics of the substrata to be removed must be obtained to determine the best method of removal. This is usually done by taking borings and core samples. The core samples are analyzed to determine if the materials can be moved by normal dredging or excavating equipment, or if ripping or blasting is required.

In areas where the material is granular and unconsolidated, wash-borings may be satisfactory for determining the nature of the materials and can be done at less cost than core-borings. However, wash-borings tend to segregate the fine materials from the coarse materials and give an unreliable indication of substrata characteristics. Borings should be done only by personnel familiar with the local geology and with wash-boring techniques; they must be able to distinguish the different strata by their relative resistance to probing with the jetting rod, by the color of the effluent brought to the tops of the casting, and by the feel and appearance of the sediments entrained in the effluent.

Occasionally, the subsurface characteristics of the area are so well known or uniform that it is only necessary to know the depth of bedrock below the overlying strata of unconsolidated materials. Sonic probings can make this determination at less cost than any boring procedure. Sonic probings can complement the knowledge gained by a few strategically placed core-borings; geologic interpretations are possible over a large area, and thus reduces the overall cost of the subsurface exploration. In some instances, a satisfactory delineation of the mud-sand or the sand-rock interface may be obtained simply by probing with a long steel rod.

b. Foundations. Data on soil mechanics for the various substrata of the site are needed to determine the holding resistance of guide piles, the probable lengths of bearing piles and sheet piles required for bulkhead walls and building foundations, the bearing capacity of spread footings, and the active earth pressures to be resisted by bulkheads and retaining walls. Such data can only be obtained by testing core-borings or other undisturbed samples of the substrata in a soils laboratory.

c. Seismic Activity. Some areas are more prone to earthquakes than others, and all structural works must be designed to resist these seismic forces. The United States has been subdivided into seismic-risk zones according to historical seismic records, and local building codes use this zoning in their design standards. Thus, harbor and related facility construction in any metropolitan area will automatically be governed by the applicable codes. Where local codes have not been prescribed, the Uniform Building Code (1970) gives seismic-design data for buildings; the State of California *Bridge Planning and Design Manual* (1969) gives seismic-design data for bulkheads and piling in water areas. If possible, areas of active faults should be avoided in siting the harbor.

d. Material Sources. Stone for breakwaters, jetties, and revetments is usually obtained from the nearest developed rough-stone quarry. If a developed quarry is not near the site, it may be advisable to explore the comparative cost of transporting stone from distant quarries versus conducting a geologic exploration for a potential new quarry. Quarrystone should be sound, durable, hard, and free from laminations, and of such character that it will not disintegrate from the action of air, water, or conditions to be met in handling and placing. Suitable fill material for the marginal area of the harbor is generally available from the harbor excavations. Sometimes this is insufficient to meet the need for landfill or it is of poor quality. Fill material then may have to be obtained from other nearby "borrow" areas. Fill is usually not as difficult to obtain as stone, but, like stone, it must be of adequate structural quality. Most clays and fine silts are too plastic for good foundations; the best is a sandy material with the right amount of binder for good cohesion. A variety of qualities may be found in between. Sand and silty sand can be compacted to maximum density by flooding. Other soils must be compacted mechanically to a carefully regulated optimum moisture content. If aggregates for concrete are not available commercially, gravel pits and sand deposits may have to be located for this purpose.

5. Project Impact on Environment.

a. Water Quality. This subject has been discussed briefly in relation to site selection, tidal, and riverflow factors. Water is particularly important for health and environmental quality, especially in warmer climates where biological processes are accelerated. Successful control of water quality is usually dependent on periodic exchanges of harbor water with the main water body that the harbor serves. In flowing rivers, the problem is minimized because the river currents will induce circulatory flow even in off-river basins. Tidal or pseudotidal fluctuations are important factors in adequate water exchange in harbors. For single-entrance harbors, an average daily exchange of water equivalent to about one-third of the harbor's mean tide volume is usually sufficient to prevent water stagnation. Although such exchanges penetrate only part of the harbor, sufficient diffusion occurs to maintain adequate water quality.

Where water level fluctuations are small, special arrangements may be necessary to ensure adequate water exchange. Sometimes this can be done by providing two entrances to the harbor so that wind-generated currents or tidal currents, feeble though they may be, move continuously through the harbor. It is important, however, to make sure that the outside source of exchange water is not already polluted. Complete exchange of water about every 10 days is usually adequate for proper control. If this is not done naturally, a mechanical-flow generator may be required. Such devices are manufactured especially for this purpose (Fig. 16) and may be installed to circulate the water within a single basin or to move it from one basin to another through a large conduit.

Water exchange does not always ensure good quality, especially in the back basins of a multibasin harbor. A significant factor in water quality control in any harbor is elimination of the direct sources of pollution. These sources may be: (a) discharges of industrial wastes and sewage into the harbor, (b) local surface runoff, (c) discharge of water from storm drains from tributary areas, (d) flushing of boats' sanitary facilities within the harbor, (e) inadequate control of bilges, or (f) dumping of garbage and trash in the harbor waters. Sanitary-sewer and industrial waste discharges into harbor waters must be eliminated in

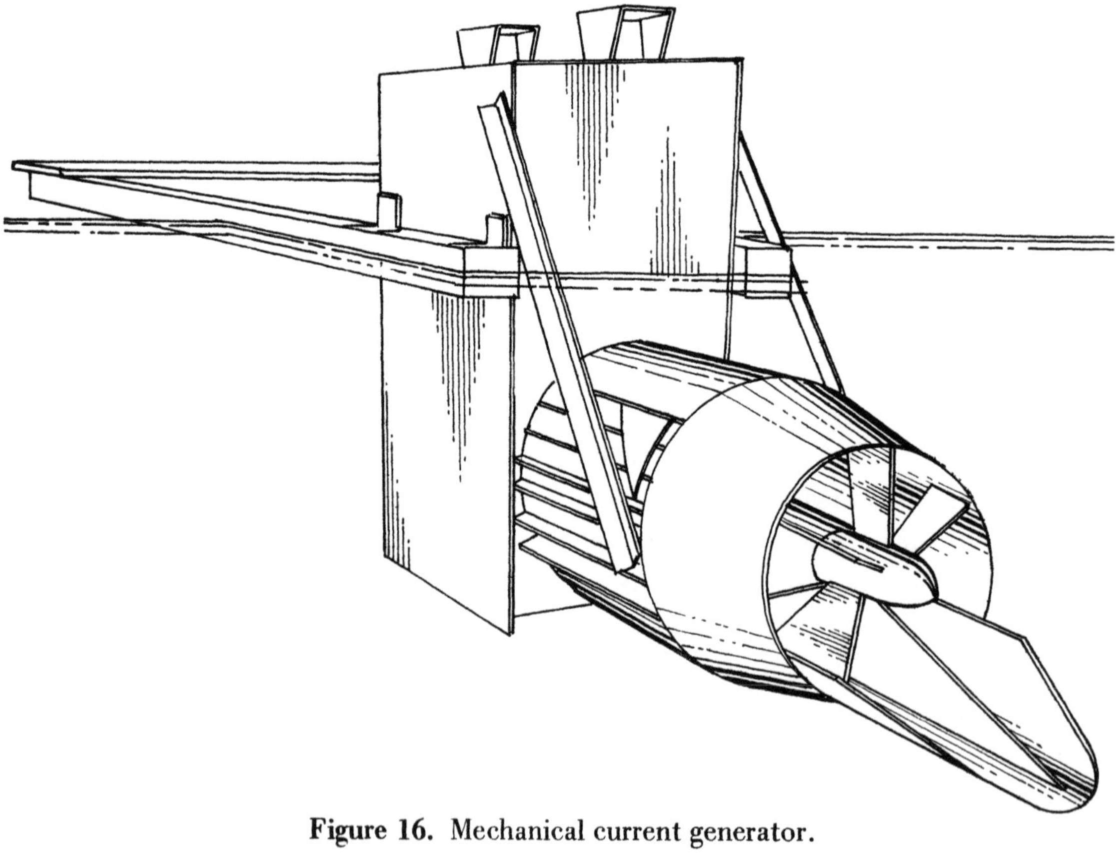

Figure 16. Mechanical current generator.

harbor planning. The flushing of sanitary facilities and dumping of pollutants must be controlled by ordinance and by provision of pumpout stations and garbage and trash collection services at convenient locations. Surface runoff waters accumulate contaminants, and their discharge directly into the harbor should be prevented as much as possible. Although storm drains should be diverted away from the harbor, some local surface water and drains may have to discharge into it. In this case, extra care should be taken to keep harbor streets, parking lots and other marginal surfaces reasonably clean, and to prevent fertilized landscapes from overflowing when watered.

The biostimulated deterioration of water quality is largely a function of biochemical oxygen demand (BOD). The input of nitrates, phosphates, and silicates in sewage and surface runoff stimulates aquatic plant life, which then dies and decomposes, using up large quantities of dissolved oxygen in the process. This in turn leads to the demise of water-dwelling animal life, which needs oxygen to survive. The result is turbid and often foul-smelling water. This is an oversimplified account of a rather complex chain of events; its many ramifications are best known to the marine biologist. However, an occasional check on dissolved oxygen levels in various parts of a harbor will indicate whether corrective action is needed. In most States, control agencies establish the level of water quality that must be maintained in harbor areas and reservoirs and provide the necessary testing.

The problems and the plant and animal life involved differ in freshwater and saltwater. Seawater is considerably more aseptic than freshwater, and fairly low rates of exchange may be capable of controlling large amounts of contamination. When man interferes with natural processes near the ocean where freshwater rivers become tidal, care should be exercised not to upset the salinity balance. The U.S. Federal Water Quality Control Administration recommends that, "For the protection of marine and estuarine organism, no changes in channels, in the basin geometry of the area, or in freshwater inflow should be made that would cause permanent changes in isohaline patterns of more than ±10 percent of the natural variation."

b. Preservation of the Ecology. In recent years, considerable attention has been given to the effects of man's enterprise in coastal areas on aquatic life, from the smallest one-celled creatures to fish, birds, and water-oriented animals. In the construction of artificial harbors, the habitat of many living creatures may be disturbed. As a result, fish and wildlife administrations of most States now exercise considerable control over such activities and often require special permits and comprehensive environmental impact statements for the construction of all new facilities and for the expansion of existing ones. The National Environmental Policy Act of 1969 (Public Law 91-190) requires an environmental impact statement on all actions or projects partly or wholly funded by Federal allocations and on major structures requiring Federal permits. This statement must describe: (a) the environmental impact of the proposed action, (b) any adverse environmental effects that cannot be avoided should the proposal be implemented, (c) alternatives to the proposed action, (d) the relationship between local short-term uses of man's environment and the maintenance and enhancement of long-term productivity, and (e) any irreversible and irretrievable commitments of resources that would be involved in the proposed action should it be implemented.

The intent of these statements is to present to the reviewing authorities information with which to evaluate the probable effects of each project on the environment to the end that the nation may: (a) fulfill the responsibilities of each generation as trustee of the environment for succeeding generations, (b) ensure for all Americans safe, healthful, productive, and esthetically and culturally pleasing surroundings, (c) attain the widest range of beneficial uses of the environment without degradation, risk to health or safety, or other undesirable and unintended consequences, (d) preserve important historic, cultural, and natural aspects of our national heritage, and maintain, wherever possible, an environment which supports diversity and variety of individual choice, (e) achieve a balance between population and resource use that will permit high standards of living and a wide sharing of life's amenities, and (f) enhance the quality of renewable resources and approach the maximum attainable recycling of depletable resources. Not only must the effects on living creatures be considered, but in wooded areas, limitations may be placed on clearing-out trees and shrubs, and on rocky coasts, the alteration of scenic landscapes.

The excavation of an entrance channel through a beach or the placement of rock and earthfill on the offshore bottom may affect the habitat of bottom-living creatures and perhaps some of the creatures themselves, such as shellfish and worms. During construction, some adverse effect may prevail; however, in many cases the constructed or completed works greatly enhance the habitat and existence of bottom-living creatures. The runoff from a dredge-fill placed on land areas normally carries large volumes of suspended fines that ultimately settle out on the bottom in the quiet waters of the effluent-runoff basin. It is becoming a common and required practice to implement techniques and means of controlling the runoff effluent to meet water quality standards. Construction practices and procedures must be followed during the design and planning stages so that compliance with environmental standards will be met.

In coastal estuaries, the tidal flats are usually teeming with lower-life species upon which many aquatic birds and fish depend for their existence. Tidal flats are also spawning grounds for many species of fish. Considerable pressure by environmental protection groups is now being exerted to prevent the further encroachment of harbor construction into these areas; legislative action has been instituted in many areas to convert them into permanent preserves. The ultimate solution in such cases may be to locate the harbor offshore. Where protective structures such as rubble breakwaters and jetties must be built, an important marine biological factor concerning their construction may be found in the excellent habitat they provide for fish life. The small fish are able to swim into the interstices in the stonework, which provides protection against predators. Also, in seawater, marine growth on the stones soon supports a wide variety of shellfish and marine plantlife.

c. Disposal of Dredged Material. Most materials excavated in the dry can be used directly for backfill, for raising ground levels, and for various other beneficial purposes; other materials excavated by dredging below the water surface are difficult to dispose of beneficially. Without careful planning, the disposal of dredged materials may be detrimental to the environment, yet dredged sand and coral may be highly useful. Because such material settles out of suspension quickly, it regains the same characteristics it had before being dredged. It can be used to replenish or build new beaches by controlling the discharge from the pipeline. By manipulating the runnoff waters properly, it can also be used to raise ground levels over large areas.

Silts, clays, and loams are often transformed by the mixing process of a hydraulic dredge into substances that present numerous problems in disposal. Extremely fine particles with certain chemical properties that occur in many soils often form colloids that remain in suspension indefinitely in the waters into which they are discharged. Fine silts settle out slowly and may be diffused over a wide area of the water body into which they are released before finally dropping to the bottom. Clays form in balls during their passage through the disposal pipe and create undesirable amorphous pockets wherever they are discharged. Nonsandy material contained by dikes or training walls in a landfill after being discharged tends to segregate with the coarsest materials settling out near the end of the discharge line and with the fines being carried toward the spillway end of the enclosure. The fine silts and colloids remaining in suspension at the spillway are then discharged into the receiving waters where they present further problems.

There are no simple solutions to the disposal problem, but some of the more successful efforts have been as follows:

(1) In firm, clayey material, use a clamshell dredge in lieu of a hydraulic dredge, and transport the material to a well-drained drying area. Spread it in a 2- or 3-foot layer, and when reasonably dry, rehandle it to some beneficial disposal area such as a grade-raising project, the core of a dam or dike, or a sanitary landfill.

(2) Discharge the material from a hydraulic dredge into a diked area and spill the effluent into another basin for holding until clarified. This system may have to be expanded to include two or more clarifying basins if the settling process is exceptionally slow. As the diked area fills, the good material may push a deposit of sludge toward the spillway. This sludge may have to be removed and disposed of at times, but the disposal of these minor amounts of poor material is not difficult. After the diked area is filled with reasonably good material to about 110 percent of the design-fill thickness, it can be dried in any one of several ways. Ditching at intervals to draw off the water that seeps out will help. Running over the area with heavy equipment will speed the process. Rehandling the material into high stockpiles will also help it to drain faster. If the base material over which the material is spread is sandy, well points may be used to extract the water that leaches down into it from the new fill. The dried-out result should shrink to about the desired grade. If not, vary the amount of initial overfill until it does.

(3) If a layer of very soft mud overlies firmer substrata on the bottom, remove this soft material first with a "dustpan" suction head or some other device that screens the local dredging area from the surrounding water, discharging the slurry into a series of clarifying ponds. After as much mud as possible has been removed in this manner, proceed with the usual cutterhead operation as described earlier.

(4) If alternate good and bad strata are encountered, control the depth of each successive cut and dispose of each layer of material in a manner best suited to that material. This procedure is usually too expensive for most projects, but it may be the only solution if the environmental requirements are sufficiently demanding.

(5) Dike off the area to be dredged, unwater it, and excavate with dryland equipment. Allow the excavated basin to refill slowly so as not to wash the dike material into it, and then remove the dike progressively with a dragline working along the crown.

(6) Where a major concern is that basin waters clouded by dredging operations will diffuse into the main water body, screen off the area where the dredge is working with a vertical skirt of flexible impermeable plastic material. The skirt is suspended from a plastic pipe floating on the surface, and the lower end is held on the bottom by rocks or sand bags placed by divers. When dredging begins, enough water will be drawn in over the top of the plastic pipe to replace the water used by the dredge. After dredging has been completed and the basin waters have clarified, the screen may be removed.

(7) Where dredge fill is to be pumped onto a beach or into the water adjacent to a beach, the disposal site may be screened off (described earlier) to prevent the escape of suspended fines. However, once the screen is removed, wave turbulence may again stir the fine material into suspension. A better solution is first to bulldoze the beach sand into a dike that is far enough offshore to contain all the dredged material between it and the shoreline. A separate basin may be needed to receive and store effluent from the disposal site until it has clarified sufficiently. If desired, the top of the resulting fill may be left a foot or two below the adjacent ground surface and covered with beach sand or other indigenous material.

(8) If the material is dredged from a basin adjacent to the ocean or a deep sound, it may be possible to obtain permission to carry the dredge slurry via floating pipeline offshore to a disposal area deep enough to be undisturbed by surface waves. If the material contains a high percentage of fines or colloids, however, the danger of widespread diffusion of clouded waters may prevent issuance of such a permit.

(9) A new Italian product called the *Hydocyclone*, when installed at the discharge end of dredge pipe, is capable of separating the heavier grains from the lighter silts, clays, and colloids by vortex action. Although not tested extensively in this country, it appears to be capable of extracting the good material from the undesirable fraction for use in fills. If not too costly to operate, this device may offer a solution to those disposal problems that require selective placement of certain fractions of the dredged material.

d. Esthetics. A small-craft harbor can readily be made esthetically attractive. With good architectural treatment of the buildings and imaginative landscaping of the marginal land, any harbor can be made pleasing to the eye. There are a few factors that may reduce a harbor's attractiveness if not properly taken into consideration. All marginal lands on which people walk or vehicles travel should be paved or treated to eliminate mud and dust. Perimeters should either be bulkheaded, slope revetted, or beached. Work areas where boats are repaired or maintenance equipment is kept should be located in a remote part of the harbor, preferably shielded from view by fencing. Utilities should be kept underground in the marginal lands and below deck level in the berthing area.

Good maintenance is also a requirement for esthetic appearance. Banks of basins and channels that are left natural must be maintained against erosion and unsightly weed growth. Provision must be made for the removal of floating debris from pockets on the lee side of basins and channels or wherever debris accumulates. Painted surfaces should be repainted as required, worn or damaged facilities repaired promptly, landscaping properly cared for, and streets, parking areas, and walks kept clean. With these few precautions in planning and maintenance, any harbor will be a credit to the community and a focal point of civic pride.

6. Sociological Factors.

a. Relation to Adjacent Development. Any site selected for a small-craft harbor within or close to a city should be compatible with other development in its general area. Some sites may be satisfactory otherwise, but not desirable from a sociological standpoint. Examples of undesirable adjacent development might be waste disposal areas. Zoning restrictions may rule out harbor construction at some highly desirable site, and conversely, may prescribe the harbor location in accordance with a municipal development plan. Examples of desirable adjacent development are parks, large estates, shopping centers, condominiums, historical monuments, forest and wildlife preserves, and amusement centers.

b. Related Recreation. Other types of recreation are compatible with boating activities, and facilities for them will be complementary to the harbor itself whether they are sited within or adjacent to the harbor complex. Examples of such facilities are gymnasiums, saunas, swimming pools, scuba diving schools, tennis courts, golf courses, rowing courses, and hiking and bicycling trails (Fig. 17). The provision of such facilities either by a municipal recreation department or by private enterprise should be encouraged to help round out the harbor complex as a recreational center.

c. Transportation Facilities. Although most boaters will commute to and from the harbor in their personal cars, many employees of the harbor and its ancillary facilities may require public transportation. Transient boaters from other harbors also may desire to use the local public transportation system. If it is possible, arrangements should be made to include one stop at the harbor on the scheduled route or routes of a local bus or rail system. Where the harbor complex is isolated from such a system, providing an internal shuttle service either by franchise or by operation under harbor management should be considered.

Figure 17. Adjacent park complements harbor development.

V. FUNCTIONAL PLANNING AND DESIGN

1. Primary Objectives. Once the site has been selected and environmental factors evaluated, the harbor must be planned with certain primary objectives as follows:

The entrance should be safe to enter by all using craft under all conditions of winds and waves that may occur in the water body served, except for conditions under which the craft cannot navigate. This objective may not always be achieved, as other factors may place limits on the orientation, size of opening, controlling depth, and perimeter features of the entrance that preclude the development of the best entrance plan for all-weather safety. Nevertheless, this basic objective must be kept in mind throughout the planning stage.

The interior channels, fairways, and berthing areas must be adequate for the sizes, types, and numbers of craft to be berthed. Long, tortuous channels should be avoided. Channels must be wide enough to accommodate the anticipated peak-hour traffic without undue hazard to navigation. Fairways must be wide enough to permit the maneuvering of boats in and out of slips and to and from the entrance channel. Berthing areas must be large enough for the planned number of berths without encroaching on established clearance standards and fairways.

The water area use allocation must be planned to accommodate all classes of boats, such as privately owned rental, charter, and commercial fishing, and where possible, to avoid interferences with the use of perimeter lands, fairways, and turning basins.

The initial cost of the harbor, including protective structures, channels, basins, docks, and administrative and service facilities, must be consistent with: (a) achievement of reasonably long-lived construction, (b) avoidance of components with unusually high maintenance requirements, and (c) provision of an esthetically pleasing and functional installation.

Conversely, the harbor design, types of materials used, and workmanship involved in the initial construction should be of a quality no higher than is necessary to reduce maintenance to the minimum level possible without overspending to achieve slight advantages in maintenance reduction.

The overall harbor area must be adequate for all purposes, with a proper balance between land and water area. Enough perimeter land must be available for streets and parking, for ancillary facilities that produce the revenues needed to supplement slip rental fees, and for all harbor support facilities. Where future expansion is anticipated, adequate undeveloped land or water area must be available and reserved for this purpose.

A master plan should be adopted as a guide for all future developments.

2. Design Criteria for Protective Features and Entrances.

a. Breakwaters. Where site selection and environmental considerations indicate a breakwater to be necessary, the type of construction and design will depend on availability and cost of materials, and on the amplitude of the waves to be resisted. The most common type of breakwater used in the open ocean and most large lakes is the rubble-mound breakwater, sometimes armored with concrete units of various shapes (Fig. 18). This breakwater is intended to prevent or reduce the transmission of wave energy into the harbor by absorbing some of this energy and by reflecting as much of the remaining energy as possible back toward the main water body. Design criteria for this type of breakwater can be found in the Shore Protection Manual (U.S. Army, Corps of Engineers, Coastal Engineering Research Center, 1973), and Quinn (1972). If a rubble breakwater is too porous, it will allow transmission of a high percentage of the longer period wave energy through it, and excessive wave disturbance will occur within the interior channels and berthing areas. Under this condition the orbital motion within the advancing wave is destroyed at the breakwater, but the potential energy is transmitted by hydrostatic pressure differentials through the voids in the stonework and a new wave of the same period is generated in the lee of the structure. For this reason, voids in rubble mounds should be reduced by incorporating a low porosity or impermeable core built to as high an elevation as possible in the structure.

A breakwater is seldom built to a height that will not be overtopped by the design wave. A certain amount of overtopping can usually be tolerated, but only to the extent that regenerated waves causes only minor acceptable disturbance in the protected area. For example, if an open water area such as an entrance channel or main fairway lies immediately behind the breakwater, overtopping waves can usually be tolerated up to the point that the

Figure 18. Quadripod armor units being placed on breakwater

interior waves they regenerate are about 3 feet in height, provided that these interior waves do not in turn cause excessive waves in the berthing areas. Moreover, as high overtopping waves will occur infrequently, the slight damage or inconvenience they cause will usually be minor in comparison with the extra cost of building the breakwater to nonovertopping dimensions. On the other hand, the overtopping waves will subject the breakwater to disruptive forces that would not occur under normal conditions, and the structure must, of course, be designed to withstand such forces.

If the breakwater is in shallow water, large waves may break before reaching it. Then it is only necessary to design the structure to resist the highest wave that can reach it without breaking. As a first approximation, waves break when the depth of the water is about 1.3 times the wave height. For design purposes, the wave should be determined by the method described in the Shore Protection Manual (U.S. Army, Corps of Engineers, Coastal Engineering Research Center, 1973) which indicates that the height of a breaking wave is a function of wave period, slope of the seabed, and the water depth at the structure. The ratio of the water depth to breaking wave height can vary from 0.7 to 1.5 depending on the above factors.

If the breakwater is in deeper water, records of the measured deepwater waves during the highest wave episode ever recorded may be used to determine the design wave height. If no records are available, wave hindcasts for the site based on recorded wind speeds or analysis of barometric pressure patterns in the wave-generating area can be used. In this event, a general rule for rubble-mound design is to use as the design wave height the significant height of the one-tenth-percent-occurrence wave episode, i.e., will not be exceeded in wave height (for any direction within a 90° sector centered on the perpendicular to the breakwater's axis) more than one-thousandth of the time, or about 9 hours each year. Some displacement of armor units may occur during the exceedance-design wave episodes, but the cost of their replacement is usually small in comparison to the extra initial expense of designing against the highest wave that might occur.

The integrity of a rubble breakwater is largely dependent on the stability of the stones and armor units of which it is built. Large stones, if placed directly on a soft bottom, may sink and lose their usefulness. This can be avoided by first covering the entire base with filter cloth and then covering the cloth with a bedding layer of spalls or quarry waste to a thickness of about 1½ times the average dimension of the average stone in the bedding layer. Filter cloth made of woven monofilament plastic yarns have proved to be the best for breakwaters, jetties, and shore-protective structures (Barrett, 1966; and Calhoun, 1972). Tests have shown that these filter cloths are superior to perforated plastic membranes, cloth made of woven multifilament plastic yarns, and glass fiber filters. Several methods of placing the cloth under water have been devised, each suitable for a specific bottom condition or type of water agitation. The filter manufacturers can furnish information as to methods of placement.

The toe of a rubble-mound structure in water shallower than about twice the design-wave height may be subjected to severe scouring currents caused by wave turbulence. It is important that the bedding layer be carried well beyond the toe stones to prevent these currents from scouring sand or other soft bottom material from under them. If the toe stones are dislodged, they may allow the armor units above them to slide down the slope. For this same reason, a berm of secondary armor stone about two stones wide is usually placed at the toe of the armor units.

When a rubble-mound breakwater is subjected to massive wave overtopping, the armor units in the crown and back slope of the structure are as much in danger of being dislodged as those in the front slope. Therefore, the crown should be three and preferably four armor units wide and the units on the back slope down to about wave height distance below the surface should be as large and as well placed as those on the front slope. In designing a rubble breakwater for overtopping, this fact must be pointed out in the project reports so that there will be no misunderstandings or recriminations if minor damage occurs in later stages of the project. It should be noted that damage to rubble mounds is not usually sudden but progressive, and even a badly damaged structure retains much of its original wave breaking capability and can be easily repaired. Van de Kreeke (1969), and the Shore Protection Manual (U.S. Army, Corps of Engineers, Coastal Engineering Research Center, 1973) present data obtained from model studies that can be used to estimate damage to rubble-mound structures under design-exceedence wave attack.

Timber, steel, and concrete sheet-pile or masonary-block breakwaters, or composite structures made with these materials, differ from rubble-mound structures in that they may fail or be severely damaged by a single wave of more than design proportions. It is essential that rubble-mound breakwaters be designed to withstand either the breaking wave or the highest 1 percent nonbreaking wave, although the wave may overtop the structure by several feet. A wall must be designed for hydrodynamic shock forces if it presents a continuous flat surface to the sea with no pressure-release features, and is in water to a depth where waves break directly on its face. Hydrodynamic shock forces may be highly localized and of extremely short durations, but pressures within the shock zone may exceed by several fold the normal hydrostatic pressures exerted by nonbreaking waves.

Design for timber, steel, and concrete breakwaters take many forms. The most common in low wave areas is the sheet-pile structure, but its use is limited to soft bottoms into which piles can be easily driven. Sheet-pile structures must be strengthened by wales or a substantial cap, and unless the design wave is small, it must be braced to prevent overturning. A variation of the sheet-pile breakwater is the diaphragm wall supported by king piles (Fig. 19). In hard bottom, the king piles may have to be drilled and grouted in place.

Figure 19. Wave and surge barrier, Monterey Harbor, California. An outer breakwater intercepts most of the wave action, but timber diaphragm with braced king-pile supports keeps large diffracted waves from reaching berthing area.

Other breakwater types include: (a) sand-filled sheet-pile or prefabricated concrete cells set side by side and securely fastened together, (b) concrete superstructure built on a submerged rubble base where large pieces of stone are not available for an all rubble structure, and (c) the perforated *swiss-cheese* breakwater, which partially dissipated wave energy in a chamber behind a perforated wall (Fig. 20). Design details for these structures may be found in the Shore Protection Manual (U.S. Army, Corps of Engineers, Coastal Engineering Research Center, 1973), and Chaney (1961).

Regardless of the type of construction used, special attention must be given to the foundation and bottom materials. Any wave resisting, vertical-faced structure built on a bottom of unconsolidated materials can cause the waves to generate toe-scouring currents and undermine the structure or make it unstable. Caissons or other large structural units that rest on the bottom rather than imbedded, are particularly vulnerable; they have been known to topple seaward into their own scoured-toe trench. Sheet-pile walls have sometimes lost so much embedment from toe scour as to threaten their integrity. Such scour can be prevented by dumping stone of appropriate size (large enough not to be carried away by the toe currents) along the toe of the newly placed structure (Fig. 21). The individual stone size and overall fillet size will vary with the height of the design wave.

Figure 20. Baie Comeau Breakwater, Quebec

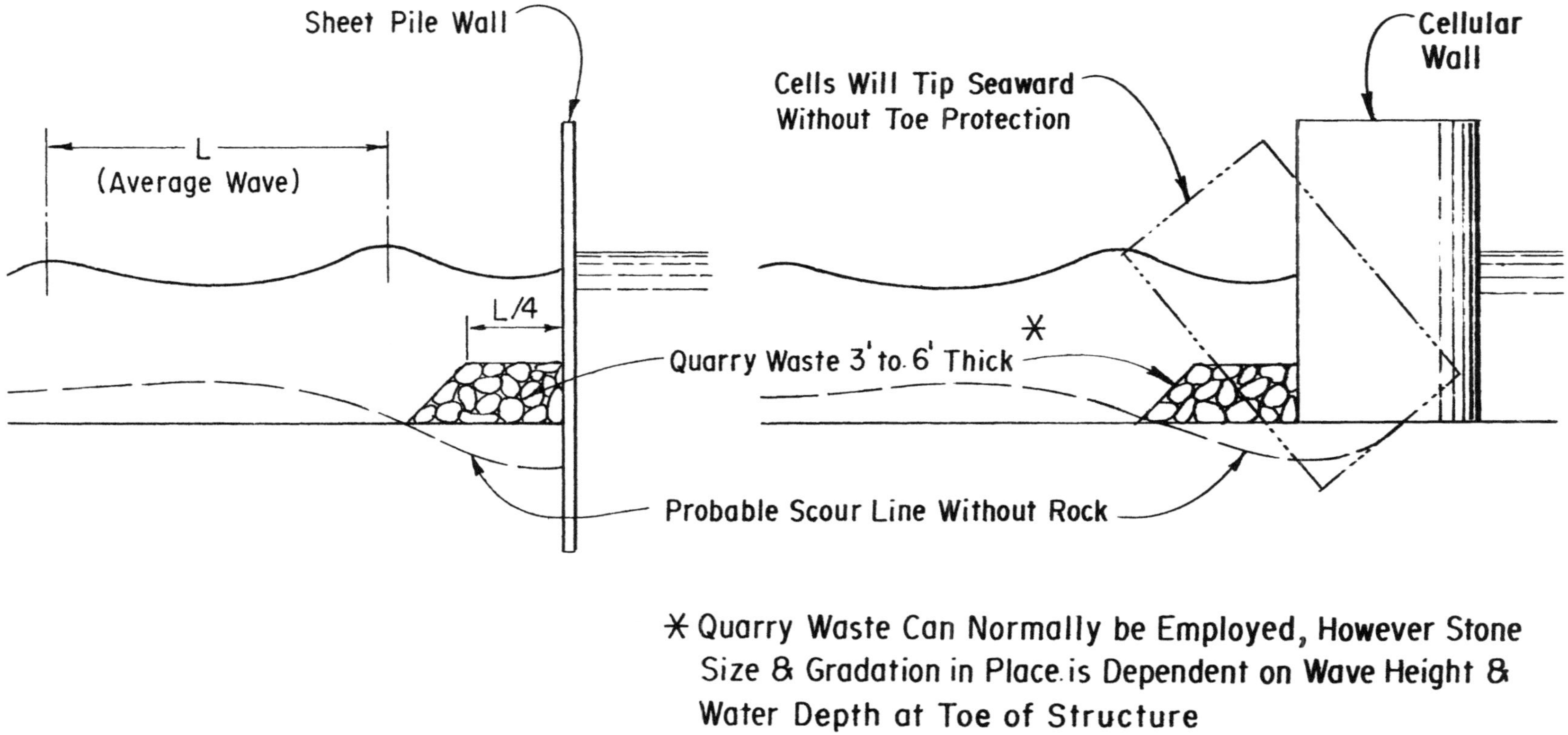

Figure 21. Use of toe stone to prevent wave scour at the base of vertical walls.

The precise positioning of a breakwater usually requires careful study. In the absence of other controlling factors, the alinement should be roughly normal to the primary direction of wave approach to intercept the maximum amount of wave energy with the shortest possible longshore length of structure. The breakwater should be as close to the shore as possible, because, as the water depth increases, the structure cost usually increases. However, the breakwater should not encroach on water area needed for an entrance and fairway in its lee during normal and peak traffic conditions; it should be only as long as is required to effect suitably quiet water for a safe entrance. An exception to these general criteria is the occasional need to make an offshore breakwater long enough to trap littoral drift and to hold it for periodic bypassing operations. A low-height breakwater may have to be located farther off shore to allow for the turbulence that overtopping waves will generate in its lee. A submerged reef may provide a good base for a lower-height structure required at optimum distance from shore. Many coral reefs with wide lagoons in their lee fall in this category.

Where a harbor is built entirely offshore rather than in a basin behind the shoreline, its entire outer perimeter except for the entrance must generally be a breakwater. The breakwater location is then determined by the harbor configuration as well as by wave characteristics and depth requirements. An example of such a harbor is shown in Figure 22. In this event, transmitted and overtopping waves along the seaward leg of the perimeter may require a second line of defense, such as a smaller inside breakwater or a revetted mole. These secondary structures will usually be designed to protect the berthing areas in their lee from surge and low-height waves that penetrate the outer structure. This configuration leaves a parallel-to-shore entrance channel between the inner secondary structure and the outer breakwater in which intermediate-height waves are acceptable.

An important factor in breakwater design is construction planning. The designer must be aware of construction problems and design within the capabilities of the equipment that is available. He must be aware of potential sources of materials and lead times required to get them to the site, and he must study possible routes and means of transportation. For example, an offshore breakwater must be constructed with a floating plant, or a temporary trestle must be built to the offshore site for the use of land-based equipment. During early stages of construction, there will be no protection against waves; therefore, the work should be done during the most favorable season of the year. Provisions must be made to get personnel and equipment to safety in a sudden storm or wave attack.

In rubble-mound construction, the bedding layer and internal layers of smaller stone must be placed before the armor layer is applied. The designer must know how this is done, and analyze the difficulty of holding a slope of small stones exposed to sea and swell until it can be protected with armor stone or concrete armor units. He must know boom lengths of cranes and not design sections that can only be built with special equipment that will create excessive costs. A transfer point will be required for moving land-source stone onto floating

Figure 22. Dana Point Marina, California. Note the double wave-defense system. The breakwater is overtopped by high waves but the moles across the main channel protect the inner basins

equipment, and it may be important to provide the place and perhaps some of the means for doing this. A shore-connected breakwater can be built entirely with land-based equipment, but progress will be slowed by the single-point operation and congestion of haul traffic over the breakwater crown. A two-stage construction program may be advisable, with the crown being kept low (and wider) in the first stage and the capstone being applied in the final stage (Fig. 23).

A shore-connected breakwater is usually an attractive structure to fishermen. During the design stage, it must be determined if the top of the breakwater will be made available for fishing and the conditions for this use. Additional provisions may be required to reduce the hazards to those who avail themselves of such a concession. A level footpath along the top of a rubble mound or a handrail along the cap may be required. Because of unexpectedly high waves, people have been swept off these structures, especially along ocean shores. Users must be warned of the danger by posted signs. This same consideration also applies to jetties.

The wave diffraction pattern caused by a breakwater has a significant impact on the adjacent shoreline, and must be considered in project planning. A schematic analysis of this diffraction effect on a typical sand beach is shown in Figure 24. This analysis should be made whenever a breakwater is constructed off a shore area comprised of friable materials, and the indicated effects should be accounted for in project design. The Shore Protection Manual (U.S. Army, Corps of Engineers, Coastal Engineering Research Center, 1973) contains instructions and data for analyzing wave diffraction under most conditions encountered in harbor design.

b. Entrance Channel and Structures. The design of harbors and entrance channels must be related to the site selection and positioning of the protective structures. The jetties and breakwater must permit construction, maintenance, and passage of tidal flows. All these aspects must be considered in harbor design.

The channel alinement should be as close to the natural channel alinement as possible. Any bends that are necessary should be gradual. The number and size of vessels must be known to determine the design width and two-way traffic that must be accommodated. The minimum width for small boat traffic should be about 50 feet or 5 times the beam of the widest boat expected to be berthed in the harbor. If sailboats use the harbor, extra width should be provided for tacking. When small craft are combined with a commercial fleet, separate channels should be considered. Channel depth is usually measured from low water datum and depends on many factors, including size and types of vessels, traveling speed, and wave magnitude. Traveling speeds govern the degree of squat which can be determined. An overdepth of 1 foot in soft material and 2 feet in rock should be allowed for dredging irregularities. The channel depth should be the sum of: (a) the draft, (b) squat, (c) one-half the wave height, and (d) overdepth. A minimum of 6 feet is suggested for the channel depth.

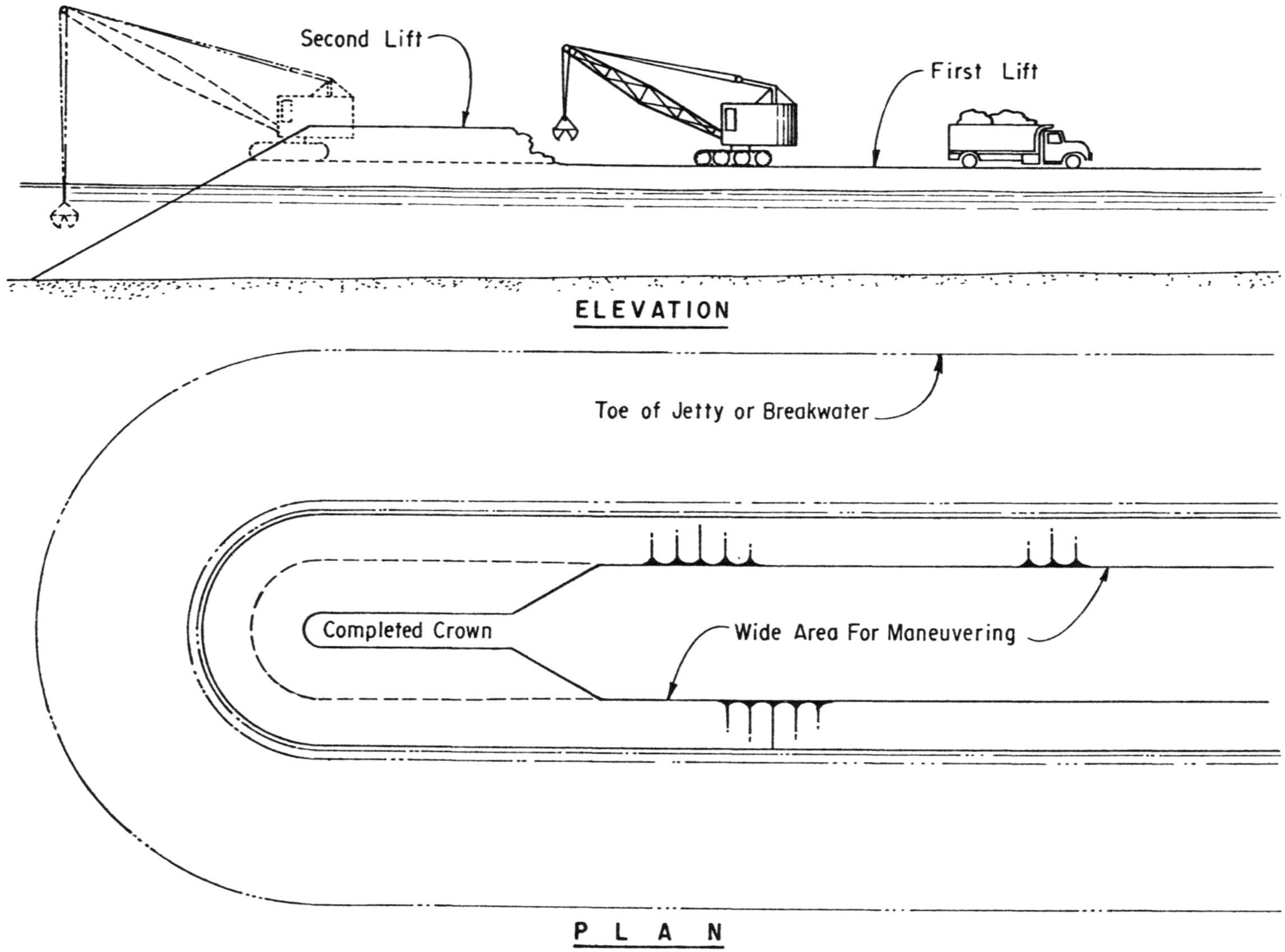

Figure 23. Two-stage construction of rubble-mound structure.

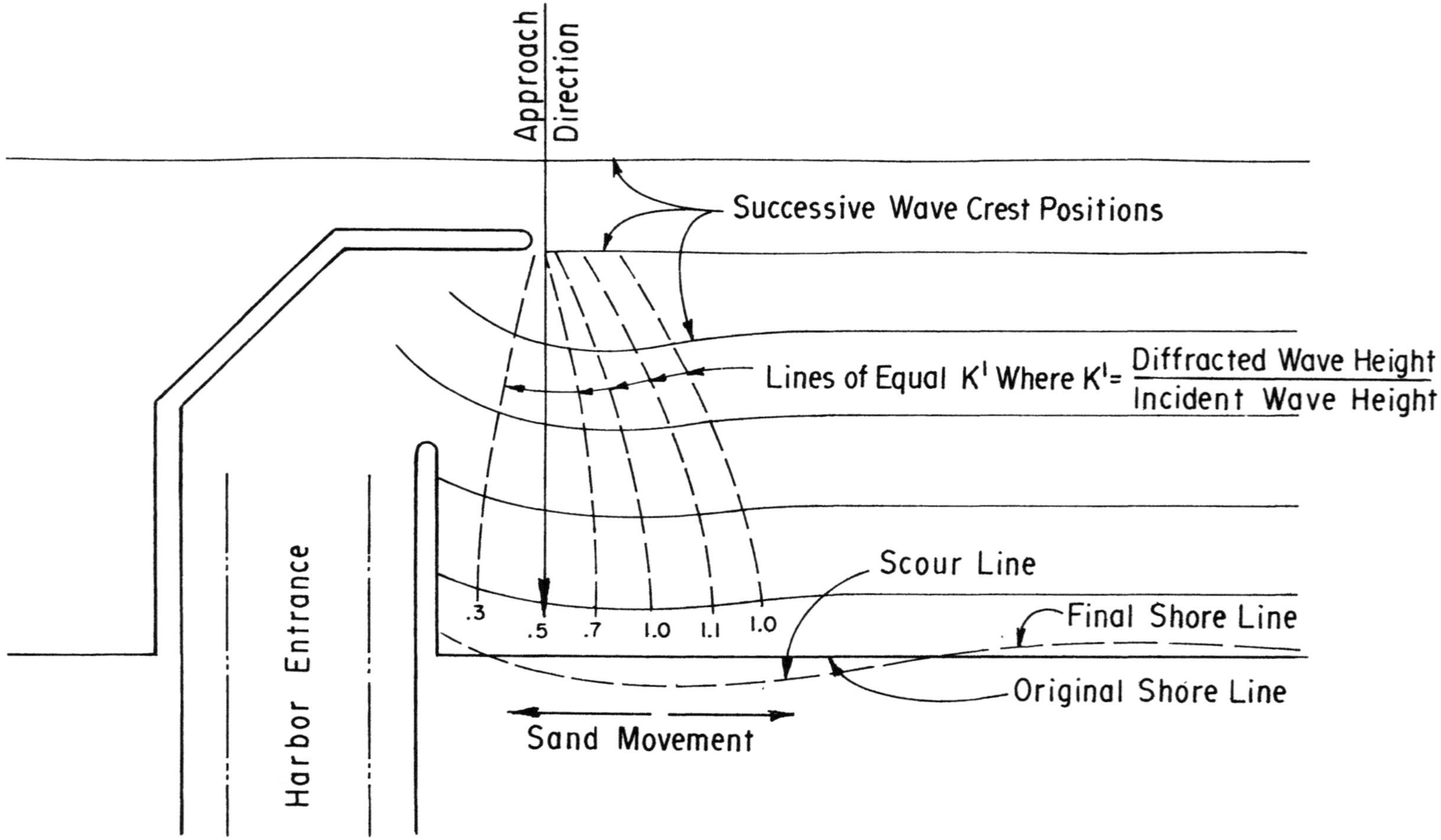

Figure 24. Littoral transport and beach erosion due to wave diffraction.

In large marinas with entrance problems, it may be necessary to prohibit the tacking of sailboats and require them to negotiate the entrance channel on auxiliary power under peak-hour conditions.

Although no criteria have been established for determining the width of an ocean entrance channel where boat traffic is a controlling factor, a good practice is to provide a navigable width of 300 feet for the first 1,000 boats, plus an additional 100 feet for every additional 1,000 boats berthed in the harbor including the daily launching capacity of operational ramps and hoists. In 1973, Marina Del Rey (see Sec. VIII) berthed about 6,000 boats; its entrance has a navigable width of about 700 feet (versus 800 feet by the above rule), and has recently begun to experience an entrance-congestion problem on peak days. The harbormaster has solved this problem successfully, by creating lanes to separate powercraft and sailcraft entering and leaving the harbor. The system is not regulated by ordinance, but requires the voluntary cooperation of the boaters. Newport Bay, California (Fig. 25) is a natural harbor with a jettied entrance channel of about the same length as that of Marina Del Rey, but skewed to the general alinement of the coast so that the daily sea breeze is quartering. The navigable channel width ranges from about 500 to 700 feet, and the bay contains about 8,000 powercraft and sailcraft. Long reaches of cruising waters within the bay satisfy many small sailboat owners so that they seldom use the entrance. Assuming that 1,000 boats are in this class, the required entrance channel width would be 900 feet. Although the channel is overcrowded on peak days, its orientation with respect to the prevailing winds reduces the requirement for sailboat tacking. Also, along a reach just inside the 1,800-foot jetties, small sailboats can range well outside the marked channel boundaries into marginal waters. Of general interest, Newport Bay traffic counts made during typical summer weekends show that an average of about 27 percent of the harbor's oceangoing fleet uses the entrance on Saturdays and 32 percent on Sundays.

Every entrance has its own special characteristics that may modify the tentative entrance width determined by the above general rule. A short reach of constricted channel with more area for maneuvering at either end can be considerably narrower than would be desirable for a long channel of uniform width. Where boating characteristics of the harbor tenants spread the daily entrance-use pattern uniformly over several hours rather than to concentrate in a few peak hours, a narrower entrance may be acceptable. At times, the need to exclude as much wave energy as possible from the harbor may override the congestion consideration; then, an exceptionally narrow entrance must be provided and its use restricted in some manner during peak hours. However, if the harbor berths a high percentage of sailboats and requires a long entrance channel that for geographic reasons must be alined parallel to the prevailing wind direction, the channel may have to be even wider than the general rule would indicate.

One method of entrance-congestion analysis (Ely, 1972) considers the passage of boats through entrances by assuming each boat to have a rectangular "blockading area" larger

Figure 25. Entrance to Newport Bay, California. Prevailing winds are across the channel, hence sailboats need not tack

than its actual dimensions. The probable number of interferences of the tacking sailboats with the axially moving craft are calculated mathematically for various speeds and numbers of craft per hour in channels of various widths. Although a small percentage of such interferences would result in actual collisions, each would require some kind of safety maneuver; the more frequently they occurred the greater would be the likelihood of an accident. The harbor planner must use a considerable amount of judgment in determining the entrance channel width, orientation, and configuration.

The primary function of a jetty is to protect the entrance channel from waves in the main water body and from littoral drift entering the channel from the flanking beaches. Because the entrance must usually be alined nearly normal to shore, a jetty is usually required on each flank. Occasionally, a natural protection (e.g., a rocky headland) will obviate the need for a jetty on one side or the other. Although the jetties may be alined nearly parallel to the prevailing direction of the waves, occasional waves from other directions must be anticipated. For this reason, jetties must also be designed to serve as breakwaters. The direction limits can be determined by a review of historical records or by wave hindcast and refraction analysis. In small lakes, jetties can often be designed as freestanding sheet-pile structures, but wherever large waves are anticipated, they must be quite massive. Rubble-mound or broad-based concrete construction is usually required to protect entrances from the open ocean or large lakes.

Since wave energy decreases as the water becomes shallower, considerable savings in construction cost can be affected by making the jetty section progressively smaller from the outer end toward shore (Fig. 26). The outer end must be designed to withstand the full force of a wave from any direction. Rubble mounds may be decreased progressively toward shore, both in height and size of armor stone in accordance with design criteria for breakwaters. Sheet-pile structures require progressively less penetration as they approach shore. Refraction and diffraction analysis (or model study) will be required to determine design wave heights both on the channel side and the exposed flank of each jetty.

Jetties are usually constructed from the shore end outward for equipment-access purposes, and before channel dredging for dredge-protection and drift-exclusion reasons. If the work is done during wave action, the turbulence at the working end tends to scour a hole in unconsolidated bottom material whether the jetty is of wide-base or sheet-pile construction. Therefore, in a rubble structure, a quarry-waste bedding layer should be placed on the bottom before construction to prevent scouring. If the bottom is soft, this bedding layer should be placed on filter cloth. The outer end of a sheet-pile jetty will usually require stabilization against scouring for the same reasons. If construction proceeds from the outer end toward shore, a similar scouring effect will take place at the unfinished inner end, particularly as it nears the shoreline. It is important that scouring be anticipated in the design and accounted for in the cost estimate and the construction procedure.

Spacing of the jetties must accommodate both the entrance channel and a protective berm of appropriate width on either side of the jetty (Fig. 27). The berm deters

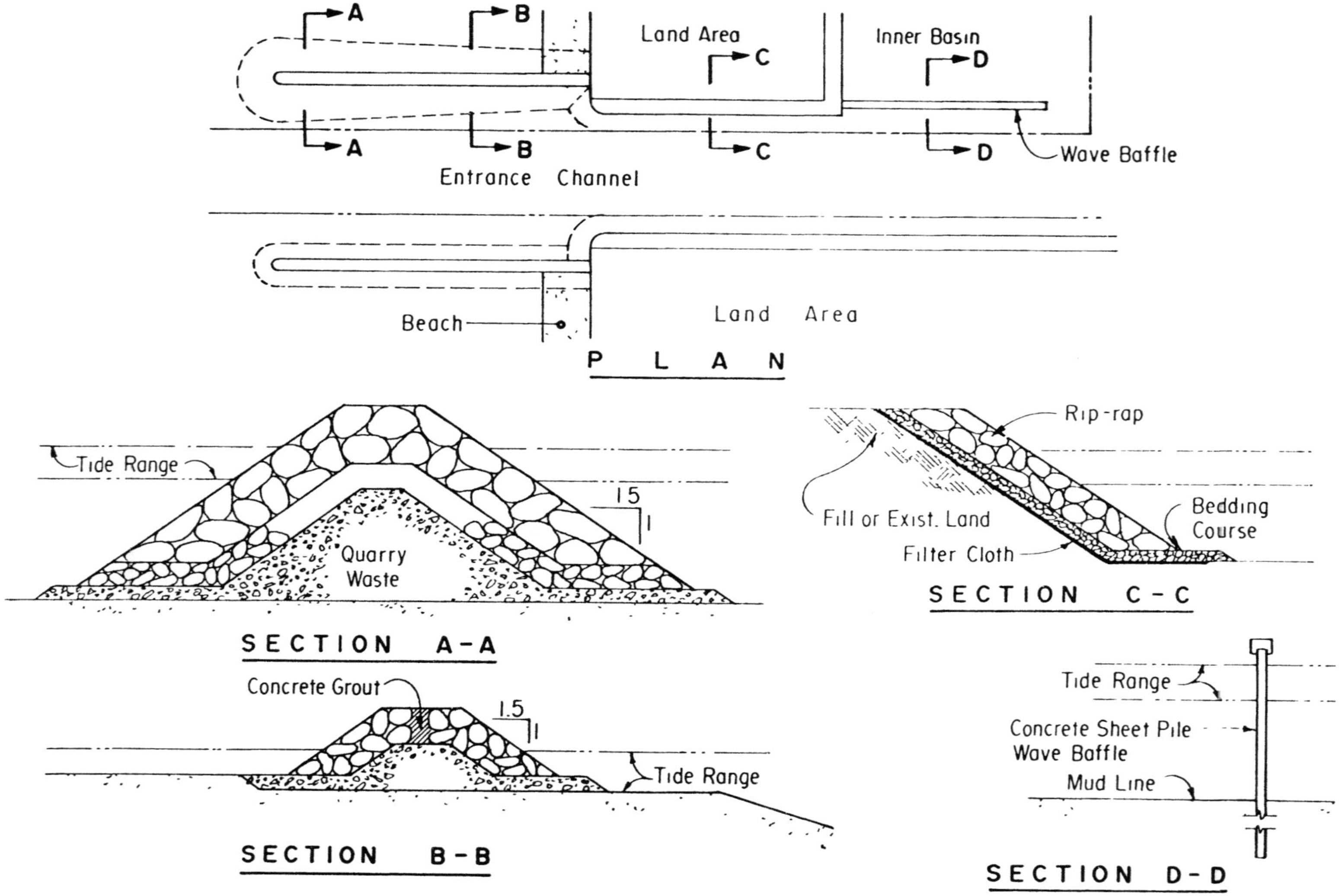

Figure 26. Changing cross sections of channel-flank protection.

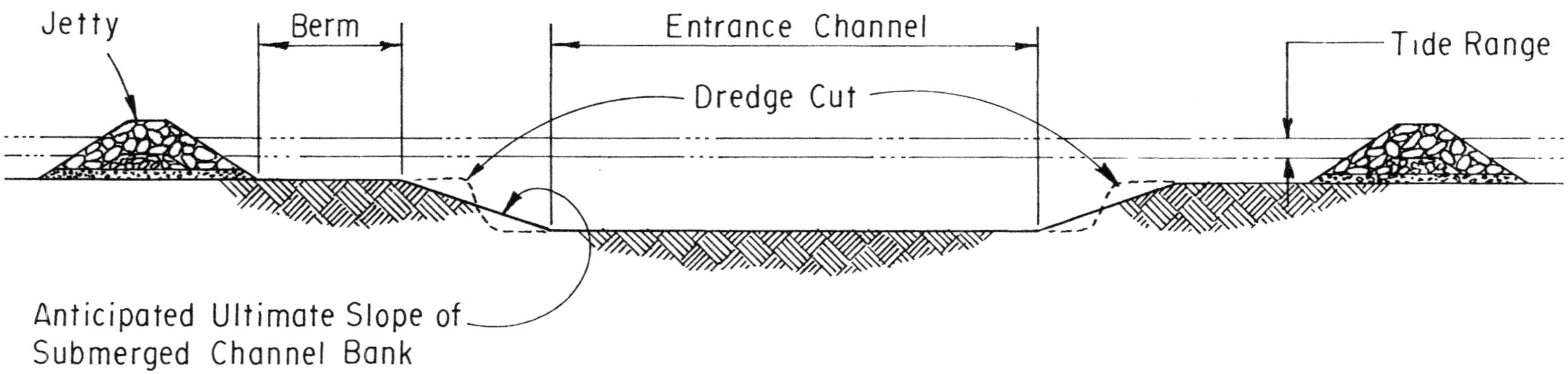

Figure 27. Typical dredge cut through harbor entrance.

undermining of the jetties. The side slopes of the channel cut must be stable for the type of material required, and for the characteristics of wave action and currents that are anticipated within the jetty confines. When a cutterhead dredge is used, it is customary to cut the channel to full project depth slightly outside the design bottom-of-slope limits. This leaves the steeper slope that the material will initially stand on and allows wave action to flatten it to an approximation of the design slope later. The width of the berm will vary with construction methods and characteristics of the bottom material. A good general rule is to make the berm width either 25 feet, or the cut-depth multiplied by the horizontal-to-vertical ratio of the design side slope of the channel, whichever is greater; however, considerable judgment should be used in actual design. The width may be allowed to vary from a quite narrow width at the outer end, where the depth of excavation is minimal, to a generously wide width at the shoreline where the cut is deep.

Where a jetty nears the shoreline, a transition must be made in the entrance section from that of a jetty-flanked channel to a protected-bank channel (Fig. 26). This protection bank usually consists of a revetted slope, but it may be a sheet-pile wall or poured-in-place concrete wall as shown in Figure 28.

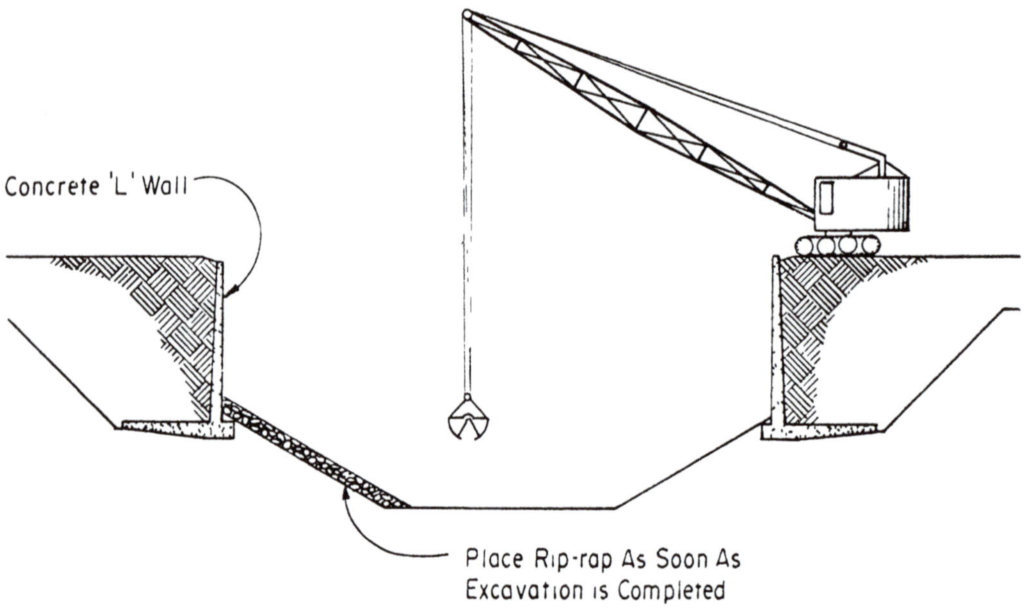

Figure 28. Narrow channel construction by land-based equipment.

The channel may be trimmed to design side slopes where narrow entrances are flanked by retaining walls or riprapped slopes and behind which land equipment can be maneuvered to excavate the channel (Fig. 28). Whatever type of bank protection is used, it should extend at least 5 feet below the extreme low water level to avoid wave erosion. A berm may be left between the toe of the wall of revetment and the top of the design side slope of the navigation channel, using the above general rule if desired. Land or water area may be saved, however, by carrying the revetted slope or exposed wall face to the channel bottom, thus obviating the need for a berm. In this event, care must be exercised during maintenance dredging to avoid damaging the toe of the revetment or wall.

The depth of the inner entrance channel must provide adequate clearance below the hulls of the largest user craft, taking into account possible water level fluctuations, wave trough depressions and vessel squat. Near the outer end of the channel, it is important to eliminate breaking waves. This is a factor only when the channel outlet has no offshore breakwater protection and waves can enter directly from long offshore fetches. In this case, the depth may have to be increased to prevent the design wave from breaking as it enters the channel.

Where river or tidal ebb-flow currents are present, the period of each incoming wave is shortened, but its energy remains the same, and will break at greater bottom depth. This effect must be analyzed and accounted for in the entrance-depth calculations. Moreover, such currents, coupled with the incoming wave turbulence, may cause asymetric erosion of the channel side slopes with resultant meandering of the channel itself (Fig. 29). This is a special problem that may require the help of a stream-flow analyst.

c. *Wave and Surge Dissipation.* Site selection and environmental considerations have indicated that it is not always possible to prevent undesirable swell and surge from entering a harbor, particularly along the Pacific coast. However, there are several methods of reducing swell and surge to acceptable heights before they reach the interior basins. The provision of harbor-resonator basins just inside the entrance has been discussed but they are still experimental and, if used, might occupy land sorely needed for other purposes. A very effective energy dissipator is a nonuniform array of large stones placed on a flat slope facing the outer entrance at the first turn in an entrance channel (Fig. 30). Waves are broken by the stones and their energy dissipated in turbulence and heat rather than being reflected off a wall or revetted slope into the inner basins.

A variation of this type of energy dissipator is the wave-absorption beach, recessed in the elbow of the first bend in the channel. Another variation is the wave reflector, designed to reflect the waves back toward the first leg of the entrance channel rather than to break them up. The wave reflector is a series of short reflecting walls or revetted slopes set in echelon around the first bend. Each is aimed back toward the entrance and separated from its neighbor by a short training wall alined in the wave direction. The reflected waves are out of phase with each other and undergo diffraction as they pass the ends of the training walls, they then diverge in a scattered pattern of harmless wavelets as they move erratically back through the channel. These reflectors have not as yet (1973) been installed in a prototype channel, but their effectiveness in model analysis is encouraging.

Some dissipation of wave energy can be achieved by the combination of a trapezoidal channel-bed section and upper slopes roughened by large stones in the revetment. In theory, the waves diverge away from the channel axis by refraction and are partly destroyed by the

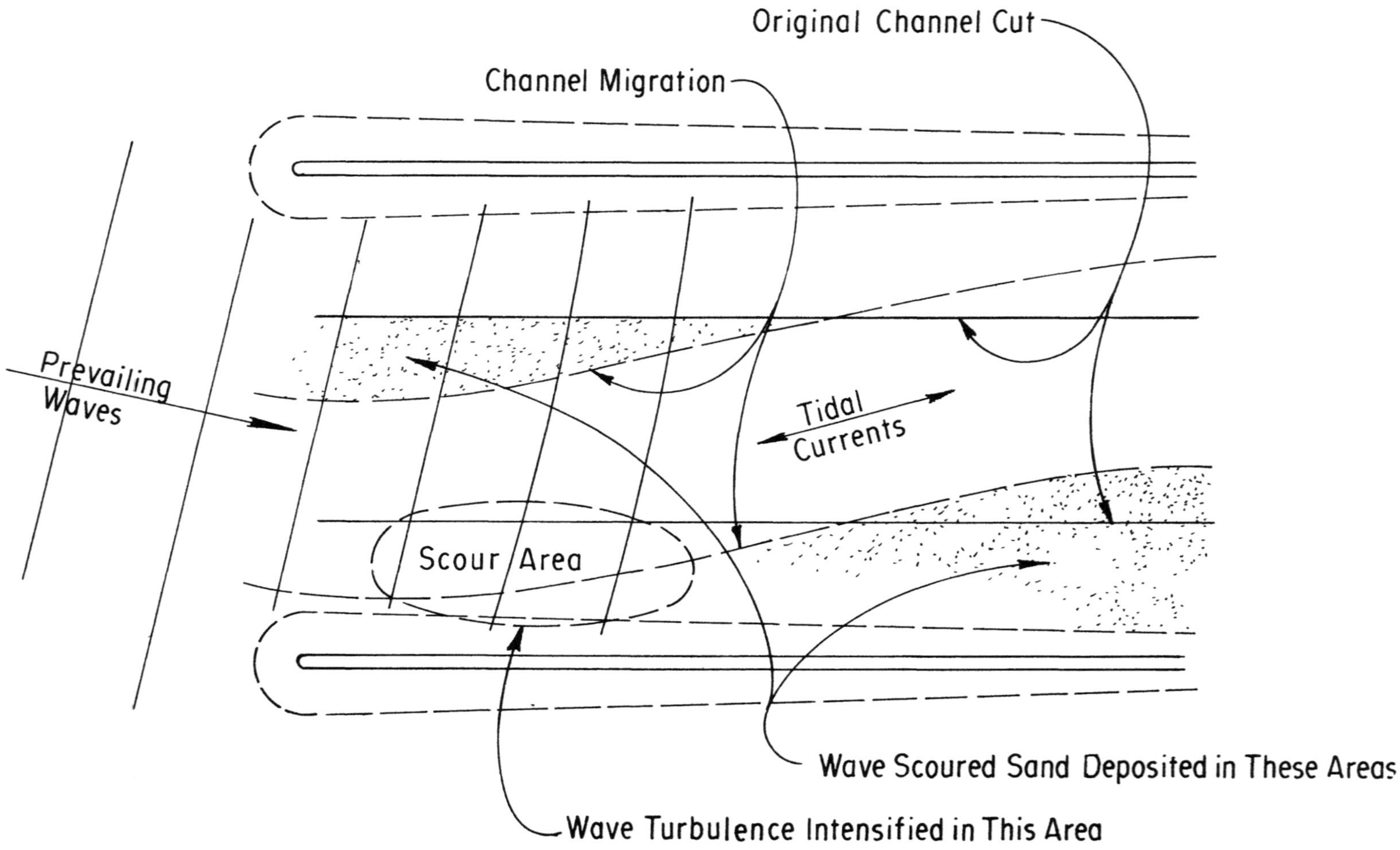

Figure 29. Channel bottom erosion and shoaling due to waves and tidal currents.

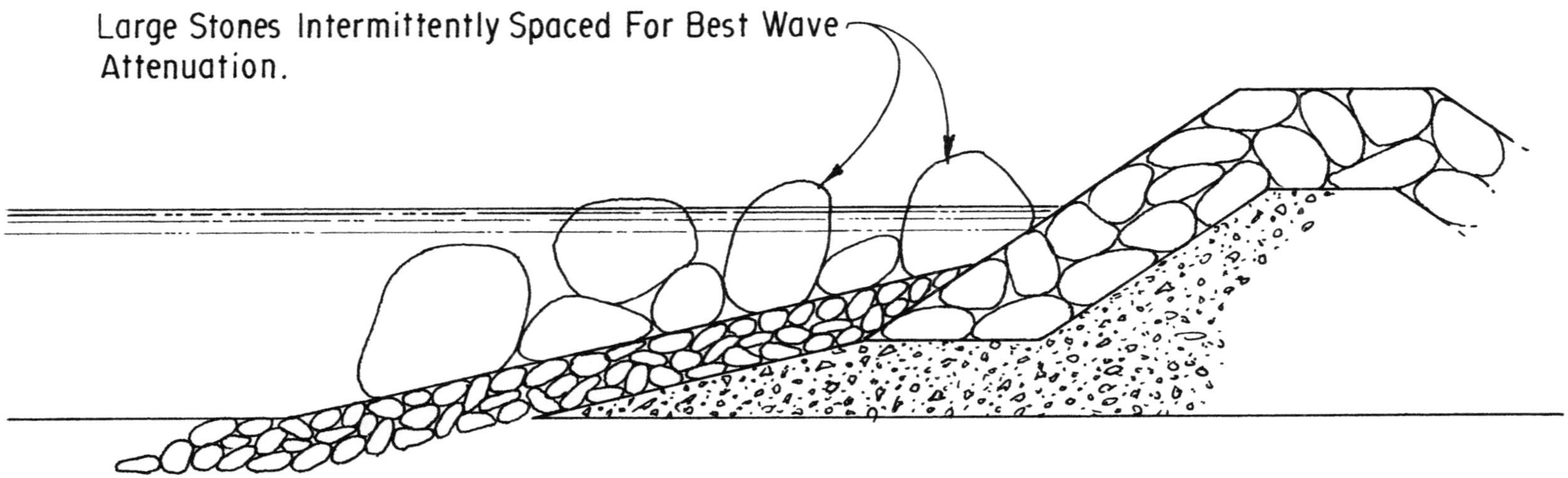

Figure 30. Entrance to Oceanside Harbor, California, showing wave breaker design.

roughness of the sides. In actual practice, the dissipation of long-period or unusually large waves in this manner is small, and the cost of roughening the sides sufficiently to be effective is not justified by the results. If long-period waves (16 seconds or more) or unusual height (15 feet or more) are common to the water body served, a detached breakwater set athwart the entrance will probably be necessary to keep surge in the harbor within acceptable limits, regardless of any interior dissipating devices that might be tried. Only a hydraulic model study should be used to make this determination.

d. Bank Protection. Landward of its jetty-protected reach, the channel requires no protection from the main water body, but does require protection from waves and currents in the channel itself. Erosion of the channel banks is the main concern, and several types of revetted slope have been used successfully to prevent erosion. The least expensive, where quarrystone is readily available, is a rubble layer (Fig. 31) placed on a filter layer. Each bank must first be built or trimmed to a slope that will be stable under the conditions of placement of the revetment. The slope will depend primarily upon the nature of the bank materials. If it is entirely a cut-slope in undisturbed natural soil, it may be stable with an inclination of 1 on 1.5. If any part of the bank is filled, it must be compacted and trimmed not steeper than about 1 on 2. Some soils will require a flatter slope, and this determination should be made on the basis of laboratory soil tests.

Figure 31. Well protected entrance to boat basin, Harwick, Massachusetts. Rubble revetment protects exposed bank.

The filter material must be porous enough to allow water to pass through and not build up a dangerous hydrostatic head behind the rrevetment. but tight enough to prevent wave action from pumping the underlying material through the voids. A graded gravel or crushed stone blanket can be used as a filter but the gradation of the blanket material is dependent on the size and gradation of the bank material. The filter layer should be spread in a uniform layer 8 inches to 1-foot thick on the trimmed slope. Woven plastic filter cloth, placed in accordance with the manufacturer's directions, is equally satisfactory and often less expensive (Fig. 32). It has the added safety feature of ensured continuity, as it cannot be worked through voids in the protective armor layer by eddy currents or wave turbulence. The cloth should be of woven monofilament polyvinylidene or polypropylene.

Below the water surface the filter also serves as a bedding layer for the armor stone. It should be placed with extra care to ensure reasonable uniformity of gravel thickness or full-design overlapping of cloth strips. The revetment stone should be placed on the underwater part immediately after the filter has been layed to prevent being disturbed by wave or currents.

The revetment stone may be designed in accordance with formulas for breakwater armor layers, using the anticipated maximum wave height in the basin or channel whose perimeter is being protected. Since patrol boats may traverse a basin or channel at high speeds in emergencies, and generate large wake waves, a lower limit of about 4 feet should be placed on the design-wave height, and the thickness of a rubble revetment should seldom be less than 3 feet. The result will be an armor that can withstand almost any tidal or river current. The top of the revetment must be above the limit of uprush of the design wave at extreme high tide, and the bottom must be the design-wave height below extreme low water. Where strong currents are anticipated, the revetment should continue to the bottom. Prefabricated, overlapping concrete blocks may be used for the armor layer where stone is too expensive (Fig. 33). The blocks should be sized according to the manufacturer's recommendations for the design wave, and the toe should be well secured to prevent the entire revetment from sliding down the slope. If the armor material does not extend to the top of the bank, the remainder of the slope must be paved or otherwise stabilized to prevent rainwash and possible displacement by pedestrians or animals.

An alternative to the rubble or concrete-block revetment is the grout-filled cloth mattress. Some proprietary fabrics manufactured especially for this purpose, may simply be unrolled empty into place and then pumped full of grout (Fig. 34). The fabrics have the advantage of requiring no bedding layer ensuring continuity of the protection and facilitating placement. Also, they can be placed on a slope of 45° if the underlying material will safely stand at that angle considering possible static-head fluctuations. Filter-point mattresses are available for those parts of the revetment that must be "weep-holed" to release hydrostatic pressures.

Figure 32. Rubble-slope revetment being placed directly on filter cloth (Courtesy of Carthage Mills, Incorporated)

Figure 33. Slope revetment of prefabricated overlapping block, Florida

Figure 34. Grout-filled cloth mattress bank protection with weep holes (Courtesy of Construction Techniques, Inc.).

In limited spaces, a composite type of bank protection may be used, with a sheet-pile or a poured-in-place concrete wall at the top and a revetted slope from the toe of the wall to the channel bottom. The least costly combination is usually a poured-in-place L-wall with footing at about water table level (Fig. 28). The revetted slope should just cover the toe projection of the footing and extend downward on as steep a slope as the material will safely stand, considering the surcharge loading of the wall and the material it retains. The wall should be provided either with weep holes and a French drain just above the footing slab, or with continuous drainage from the backface of the wall, under the footing slab, to the revetment filter layer. If sheet piling is used, the joints will usually be permeable enough to relieve hydrostatic pressures without weep holes. The design of the wall part of a composite section generally follows the design of an ordinary bulkhead or retaining wall; several versions are covered in Chaney (1961) and various general engineering texts.

e. Riverfront Protection. To berth boats directly in a river, several means are available for providing some protection to the berthing area. A current deflector placed upstream will divert the main force of the river current away from the area and prevent floating debris and ice floes from impinging at full-current speed on the berths and boats. The deflector is usually built like a pile-dike fence inclined downstream, with heavy beams for horizontal members in the area of surface level fluctuations to fend off floating debris. Enough current is allowed to pass through the fence and keep the silt in suspension, but does not make maneuvering into and out of slips difficult; a solid-diaphragm fence would result in undesirable eddy currents and perhaps deposition of silt in the berthing area. A floating log camel along the outside of the fence may be needed to fend off exceptionally heavy floating debris. Bank protection may also be advisable, both for protection against wave and current scour and to prevent damage by floating debris and ice.

One method of protection along a waterfront is to excavate a shallow basin into the riverbank (Fig. 35). The main current will still follow the old riverbed, but some residual current may flow through the basin. A series of piles along the old bankline will help keep debris out of the basin. They can normally be the guide piles at the outer end of floating piers, thus serving a dual purpose. If debris flow is heavy, an overlapping mole may be left at the upstream end of side basin, enclosing as much of it as necessary. The banks of the mole may have to be revetted to prevent erosion. The ultimate step in such a modification is to leave the mole long enough to enclose all of the basin but the entrance, so that it becomes an off-river basin.

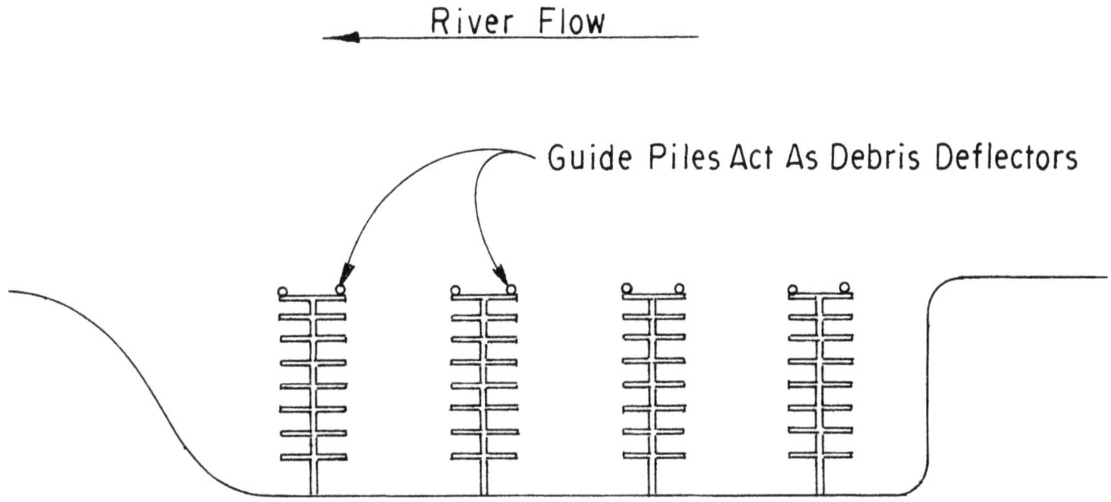

Figure 35. Shallow boat basin marginal to a river.

Another version of the off-river basin is the utilization of one arm of a river bifurcation around an island as a harbor. Usually, the smaller arm is selected, allowing the main current to follow the larger channel. Debris deflectors may be required where the river bifucates to prevent large pieces of debris from entering the boat basin, or the upstream end of the smaller arm may be diked off as shown in Figure 36. Some widening of the river arm selected for harbor use may be necessary to obtain enough water area for the desired number of slips. If the river carries a heavy silt load, such enlargement should be kept to a minimum to avoid siltation problems.

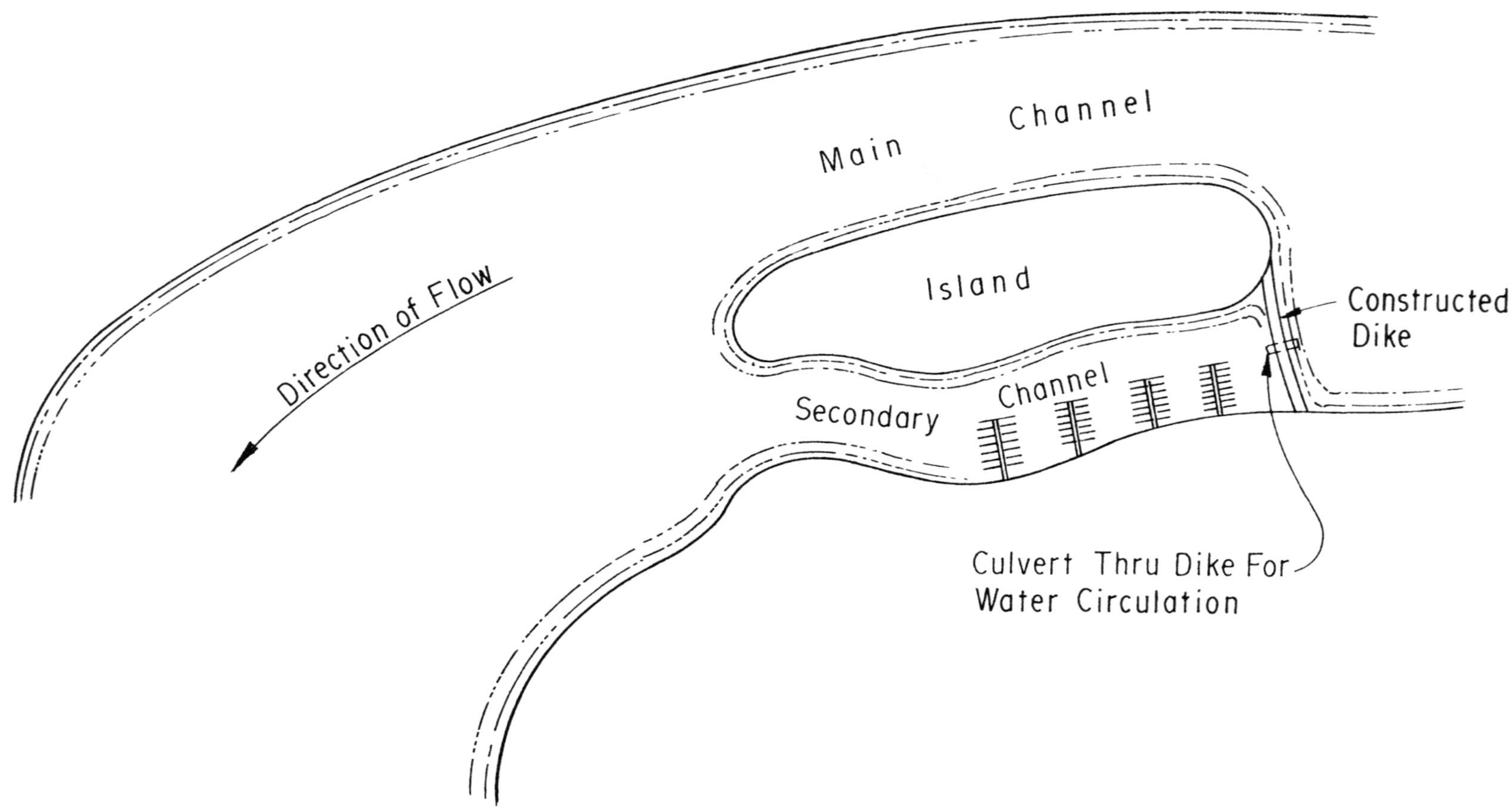

Figure 36. Conversion of river bifurcation into a small-craft harbor.

On rivers or tidal channels that freeze over in winter, ice breakers can be installed upstream from the harbor site to break large ice formations. These are usually timber dolphin-type structures with vertical railroad-rail cutting edges near the water surface, alined in the direction of current flow and bent to form smooth sloping tops. Ice sheets are either broken by the cutting edge or forced up the rail slope until the overhand causes breakage into smaller pieces. Chaney (1961) presents a more detailed discussion of icebreakers.

f. Floating Wave Attenuators. No floating breakwater has yet been devised that will economically and sufficiently attenuate large ocean waves. However, many harbor sites in lakes and partially protected coastal waters (e.g., roadsteads and sounds) meet all the site selection criteria except for lack of an appropriate fixed-breakwater foundation. One reason for this may be that the bottom slopes too steeply from the shoreline so that at the breakwater site the water is too deep. Another reason may be that restrictions in the interest of ecological preservation prohibit fixed-breakwater construction of any kind. In such cases, some type of floating wave attenuator may prove to be the best solution. Although no floating attenuator will eliminate all wave energy, several types are now capable of reducing moderate waves to acceptable proportions.

The simplest type of floating protection is the chained-log boom, used mainly on rivers and small lakes. The log boom is effective only against very short-period waves or "wind chop" and is better for protecting swimming areas or berths for canoes, rowboats, and small sailboats from waves that would not bother most powered small craft. The log boom can be made twice as effective by binding the logs together in units of three; the boom then floats with one log down and two on the surface (Fig. 37).

The next step above the log boom is a stiff membrane skirt extending into the water from a float or barge (Fig. 38). The skirted units must be anchored end-to-end to form a continuous wave attenuator. This principle is sometimes used in skirting the fairway sides of finger floats at the outer ends of floating-slip piers and is partially effective in reducing small boat-wake waves entering the berthing area from the fairway. However, any skirted-float system has little effect on wind waves generated in fetches of a mile or more because of the longer periods of such waves and consequently the greater depths of the water particle orbits that propagate them.

Any solid floating device having small width and depth dimensions in comparison to the wave length is intended to attenuate, and will only "ride" the wave rather than break it. Research has been directed toward the development of devices that break up the orbital continuity of the water particles within a wave or that pit the energy in one part of the wave against another part. Orbital continuity can be destroyed by the controlled release of air bubbles, but the cost increases roughly with the square of the wave height; it would be prohibitive when used in conjunction with a small-craft harbor. This also applies to water jets.

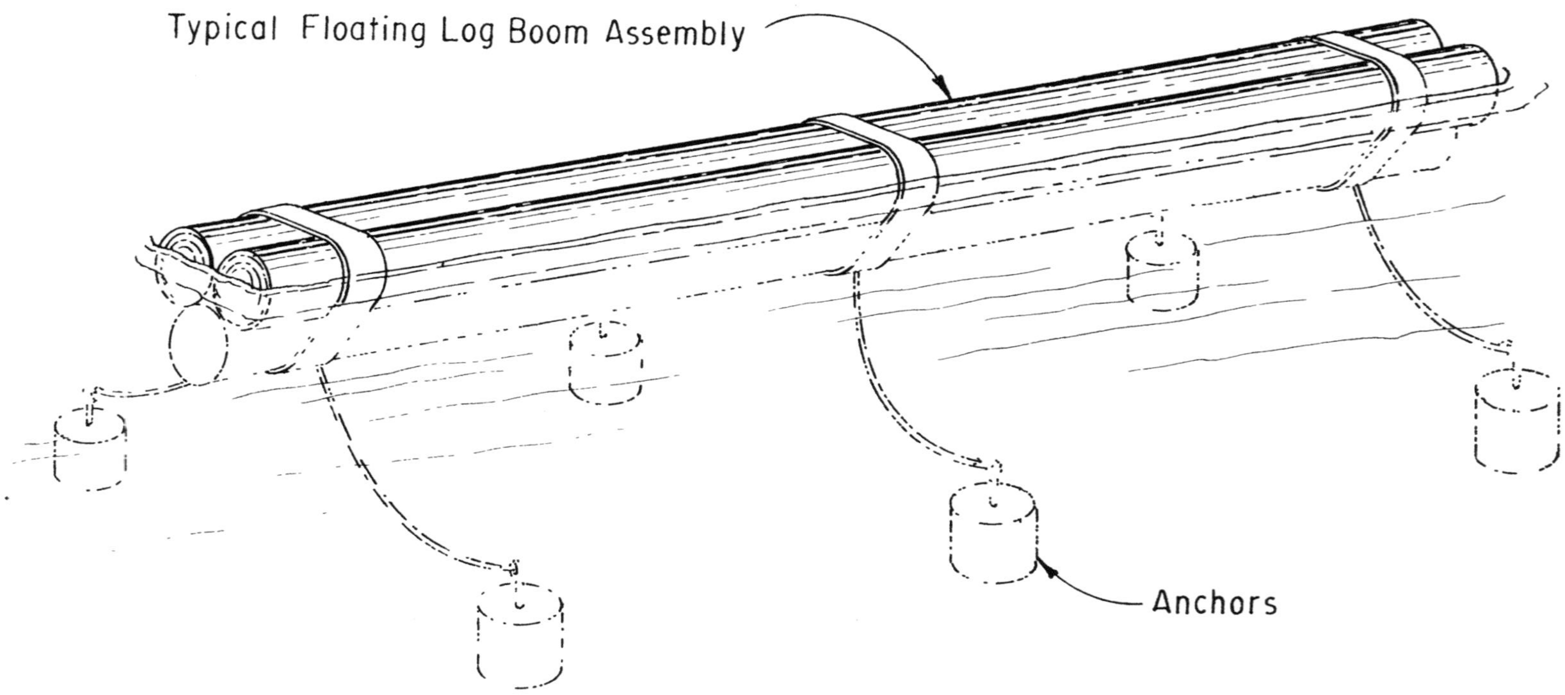

Figure 37. Three-log floating boom for short-period wave attenuation.

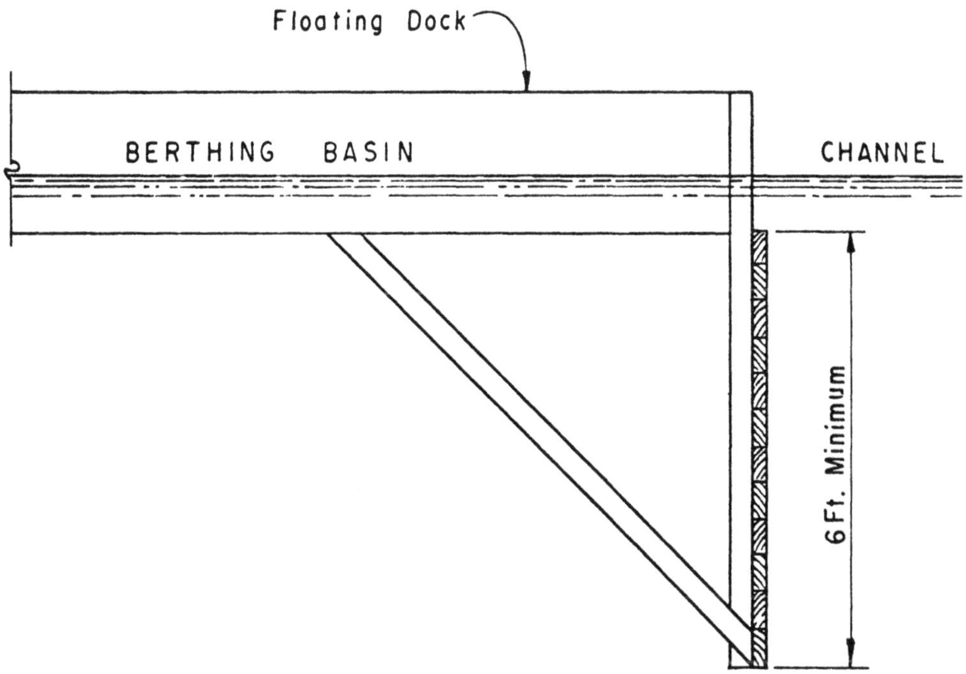

Figure 38. Flank drop-skirt on channel side of floating dock for short-period wave attenuation.

In 1972, a floating breakwater of modular polyolefin pontoon units secured by timber stringers, was installed at Friday Harbor, Washington. The shape of the units when assembled, leave vertical openings occupying 30 percent of the surface area covered between the units. The breakwater is 24 feet wide, 400 feet long, and 5 feet high. It is water-ballasted to float half submerged and is anchored with fore-and-aft chains to submerged piling on the bottom. Scale model tests at the University of Washington have indicated that the coefficient of attenuation should be about 0.2 for the relatively short-period waves that approach the site from the San Juan Channel. Alongside-docking is permitted in calm weather, increasing the harbor capacity by about 60 boats. The structure cost about $320 a linear foot and has performed as expected. The pontoon units have a life expectancy of 25 years; however, it is too soon to indicate how well the assembly will hold up over an extended period of time.

Another approach required a maze of wornout rubber tires secured together to form a continuous porous mass that floats just above the surface (Fig. 39). Test results (Kamel and Davidson, 1968) on various sizes and shapes of tires show that to be effective, the depth of the maze below the surface should be about half the water depth, and the crest width (length in direction of wave propagation) should be about one wave length. Also, the effectiveness of a tire maze is very sensitive to wave length. For example, a maze 10 feet deep and 100 feet wide in 20 feet of water would dissipate about 80 percent of the energy

Figure 39. Tire wave-maze floating breakwaters

in a wave train of 5-foot height and 5-second period. The same maze would dissipate only 20 percent of the energy in a wave train of 5-foot height and 10-second period. The maze would require about 150 to 200 tires per foot of crest length. The maze must be anchored in place, but the forces exerted on the anchor lines would not be excessive. Although the tire maze is not effective against waves with periods in excess of about 7 seconds, except at prohibitive cost, it may be practical for harbor sites not exposed to these longer period waves. If practical, the designer should apply the test results of Kamel and Davidson (1968) to the site conditions to determine the proper dimensions and anchorage plan for an effective tire maze.

Other types of floating breakwaters considered but not tested through long-term use are: (a) an air-filled mattress at the surface or suspended at some distance below the surface to achieve out-of-phase damping, (b) a thin membrane on the surface to achieve viscous damping, (c) random arrangements of horizontal pipes to destroy orbital wave motion, (d) a series of vertical diaphragms to accomplish the same effect, and (e) various types of solid structures with void patterns designed to break up wave action in one way or another (Fig. 40). Although the search continues, none of these types has yet given promise of being a practical solution to long-period wave dissipation.

Figure 40. Floating corrugated metal pipe breakwater, Detroit Harbor, Michigan.

3. Layout Criteria for Marinas.

a. Layout Planning. The proper siting of the various components of a small-craft harbor is a prerequisite for the functional soundness of the overall plan. The following general principles will provide guidance in determining the best allocation of land area within the confines of the harbor boundary (Fig. 41).

For several reasons the larger craft should generally be berthed near the entrance. They are less influenced by residual wave action entering the harbor where action is greater near the entrance. Larger craft require greater maneuver space, which is usually provided in the harbor area near the entrance for the larger volume of traffic. More physical space is needed than for the smaller boats, and if the larger craft are berthed so they need not traverse the inner fairways of the harbor, those fairways may be proportionately narrower. The deeper drafts of the larger craft require a deeper channel and basin; hence, the inner parts of the harbor can be shallower if they are not used by the larger craft. This allowance for a shallower inner harbor is also a factor in initial channel and berthing area excavation.

Commercial craft usually fall in the same category as large private recreational craft with regard to their water area requirements. The berthing areas of commercial and recreational craft should generally be separated because of different adjacent land use requirements. If possible, commercial boats should be located near the entrance in a separate basin or across a fairway from the recreation craft. A commercial fishing fleet will require special hoists and other equipment for moving the fish out of the holds onto perimeter docks and for sorting and preparing the catch for market. In some instances, canneries or freezing plants are located adjacent to the fleet berthing areas (Fig. 42). The general public should be excluded from these working areas for obvious reasons.

Charter boats for sport fishing must have adjacent facilities for selling their services, for controlling the boarding and debarking of clients, and for parking cars. Fish-cleaning stations are often provided for clients to have their catch cleaned by professionals. A viewing area for prospective clients to watch the cleaning operations will help to advertise the charter boat service.

Rental boats should be berthed in the same area and not mixed with private recreational craft. This berthing area should be close to the office where the rentals are handled, with easy access to the harbor entrance. The car parking area for rental boat clients should be separate from the slip rental parking area, but it may be shared with visitors to other facilities in the harbor complex.

Sailboats without auxiliary power should be berthed in slips that open to leeward of the prevailing winds and that can be reached via wide fairways and channels or routes that allow for sailboat tacking with least interference to the powered craft.

Ramps or hoists for launching trailered craft should be separated from the berthing areas as far as possible. The boating habits of the owners of these craft are usually different from those of the berthed craft; conflicts may result if the same fairways are used. If possible, the

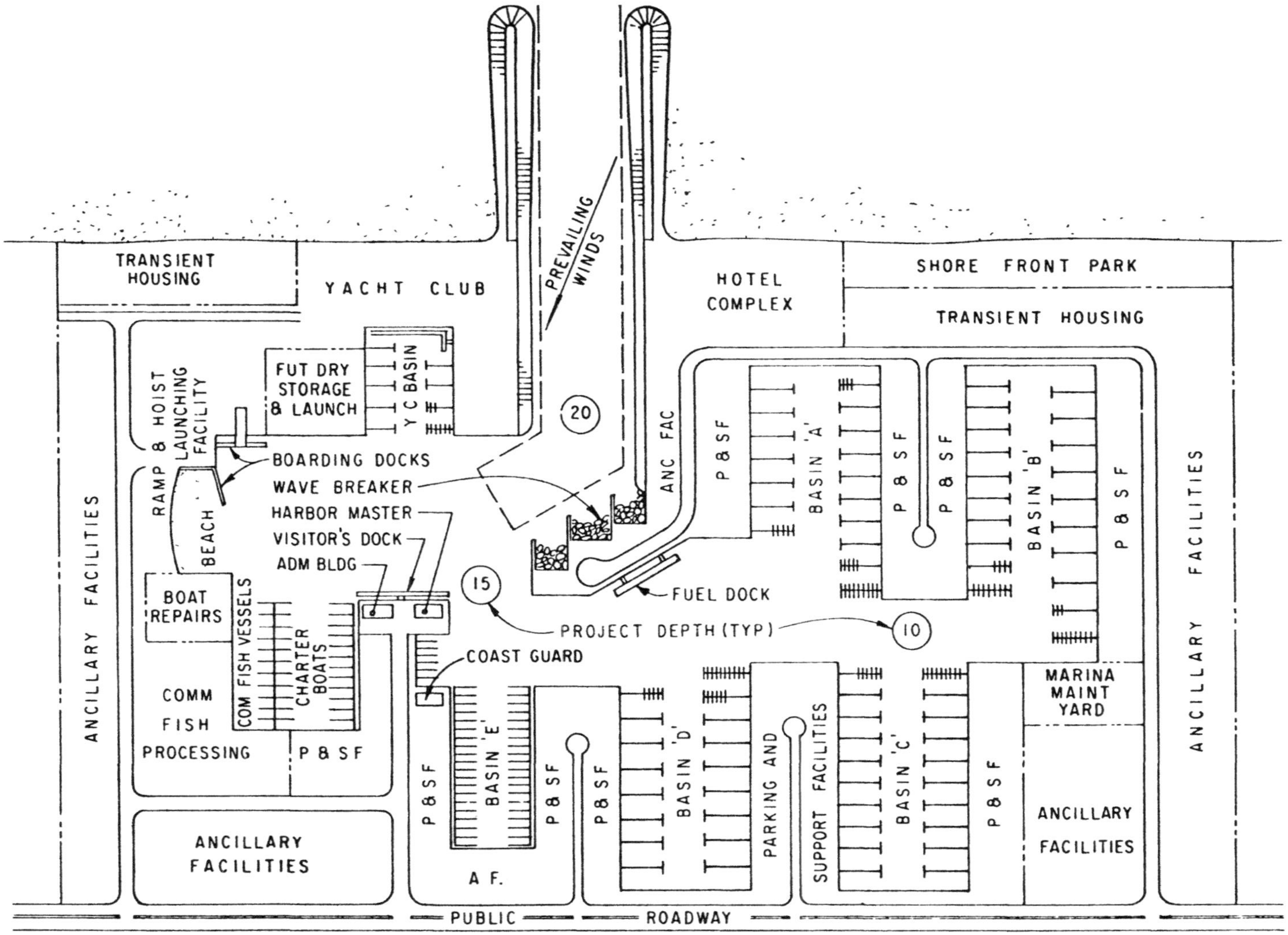

Figure 41. Schematic layout of a marina showing desirable interrelation of facilities.

Figure 42. Commercial area, Westport Marina, Grays Harbor, Washington.

trailered craft should have a separate entrance or be launched directly into the main water body without using the harbor. If trailered craft must use the same protected waters, the launching area should be as near the entrance as possible and physically separated from the berthing areas so that vehicle traffic to berthing areas and trailer traffic to the launching area do not merge. Occasionally, the launching area must be at the inner end of the harbor complex where more parking space is available. The launched craft may then have to share the main channel with the berthed craft, and the channel made wide enough to accommodate traffic from both sources without overcrowding.

The best location for a boat fueling dock is near the entrance in an area that is protected from waves in the entrance channel, thereby causing no interference with the entrance channel traffic. The adjacent land area must be suitable for buried fuel storage tanks and easily accessible for fuel distributing vehicles. The pumpout station should normally be located in the same area and is often on or along the fueling dock so that its operation can be supervised by the station manager. However, it should not be so close to the fuel pumps that a client has to wait for a pumpout operation before he can dock for fueling. It should not be necessary for any boat to go far out of its way for these services. Also, the station should not be in a location where it interferes with traffic flow or constitutes a fire hazard because of its proximity to other harbor facilities or berthed craft.

The harbor administration area should also be located near the entrance and guest docks, where owners of visiting craft can easily come ashore to obtain information. The harbormaster's office should either be a part of or close to the administration complex to provide a good view of boats passing through the entrance. A good view of the berthing areas is recommended, but is not essential and may be impossible in a large harbor.

Vehicle parking lots for the berthing basins should be located so that no parking space in any lot is more than about 500 feet from the head of the pier for the particular lot it is intended to serve. Parking lots for ancillary facilities should be adjacent to parking lots for the berthing basins so that overflow from one lot can be absorbed by the other under time-staggered peak-use conditions. Lots should be well marked for the major area or function they serve.

The boat repair and servicing yards should be located in a remote part of the harbor that has adequate navigation access for the largest craft. If a marine railway, large hoist, drydock, or other device for launching or retrieving large craft is to be provided, it should be included as a part of this operation. The marine traffic generated by the repair yard will be minimal in comparison with the regular entrance channel traffic and need not be a consideration. However, the yard should be readily accessible to large tractor-trailers used for hauling new cruisers or boats to be launched.

If an operational dry storage facility is a part of the harbor complex, it should be located generally in accordance with the criteria that apply to launching ramps. The launchings and retrievals in such a facility are generally accomplished by hoist rather than by ramp, and a well-protected launching basin is necessary for this purpose. If the dry storage facility is for off-season layup only, it should be in a remote area not required for the more important facilities of the harbor; in fact, it need not be in the harbor complex at all. In many places, vehicle parking lots are used in the off-season for dry storage in lieu of providing a separate facility for this purpose.

If a U.S. Coast Guard vessel is to be docked in the harbor complex, it should be located near the entrance where it can move out quickly with the least interference from regular harbor traffic. U.S. Coast Guard personnel should, of course, be consulted and their concurrence obtained as to the exact site.

Boat sales and chandlery facilities should be located along the main access route to the harbor where all vehicular traffic must pass. Restaurants should be located where they have a commanding view of harbor activities but not interfere with or occupy areas that are needed for more vital harbor-support functions. A good restaurant site would be one overlooking the entrance or the main water body just outside the entrance. If transient housing facilities are included in the harbor complex, they should be near the restaurants or adjunct to them. Any recreational facility should be near or readily accessible from the housing facilities. General shopping areas for food supplies, clothing, and drugstore items should be near the harbor's land boundary and away from a water-oriented activity. Yacht clubs should generally be located away from the main activities of the harbor for a more exclusive atmosphere. Higher ground that overlooks the harbor from a nearby viewpoint often makes a good yacht club site. A variation is a yacht club with its own separate basin.

b. Space-Allocation. The total area available to the harbor development often places a restriction on the number of boats that the harbor can accommodate and on the size and scope of the ancillary activity it can support. Several general relationships, found valid for most harbors, may help the planner to make tentative allocations of space, which can later be adjusted to definite dimensions in the final planning stages. Such allocations are important in making adequate allowance for future expansion (Fig. 43).

The average harbor with all-slip moorage can berth about 15 to 20 boats per acre of navigable water area, including main interior channel, fairways, and slip areas, but not the entrance channel. This general rule applies only when the average boat length is 30 to 35 feet and where good basin geometry can be obtained. Because of the wider fingers needed for two-boat slips, they will occupy almost the same area as that required for single-boat slips. When bow-and-stern moorings are used in lieu of slips, about 2 to 4 times as much water area (depending on the water depth) will be required, exclusive of fairways and channels. Single-point moorings require about 6 times the area occupied by the same number of bow-and-stern moorings if full-circle clearance is provided.

For the normal distribution of boats, a minimum of three vehicle spaces in the parking lot will be required for every four boats in the berthing area. About 90 cars can be parked in an acre, so that roughly one-sixth of an acre of parking lot is required for each acre of water area in the harbor. Where the average size of the berthed craft is large—and many are used for social occasions and multi-family cruising—the ratio may have to be increased to a maximum of about three spaces per berth.

An average launching ramp or hoist will launch and retrieve about 50 trailered boats on a peak day, and because of staggered usage, car-trailer parking spaces will be required for only 80 percent of the peak-day ramp or hoist traffic. About 30 car-trailer units can be parked in an acre if pullthrough parking at 45° is provided. This works out to 1.33 acres of parking lot per ramp lane or hoist.

Land area required for harbor service facilities, ancillary facilities, and roads varies from one harbor to another. The minimum requirement is an area roughly equal to the parking area required for berths and operational launchings. This will generally provide enough space only for harbor support facilities and roads. To obtain a good revenue versus cost balance it is usually necessary to supplement slip rentals with leaseholds for ancillary facilities; with the additional parking area required, the minimum leasehold and supplemental parking area needed for the extra services that convert a simple small-craft harbor into a complete marina is about twice the area needed for boatowner parking alone. Thus, once the parking area requirement for slips and launching has been determined, it should be multiplied by 4 to obtain the total minimum land area required for a complete marina. Any additional land that can be obtained may be put to beneficial use later, as a good marina will upgrade its surroundings and attract more revenue-producing ancillary development.

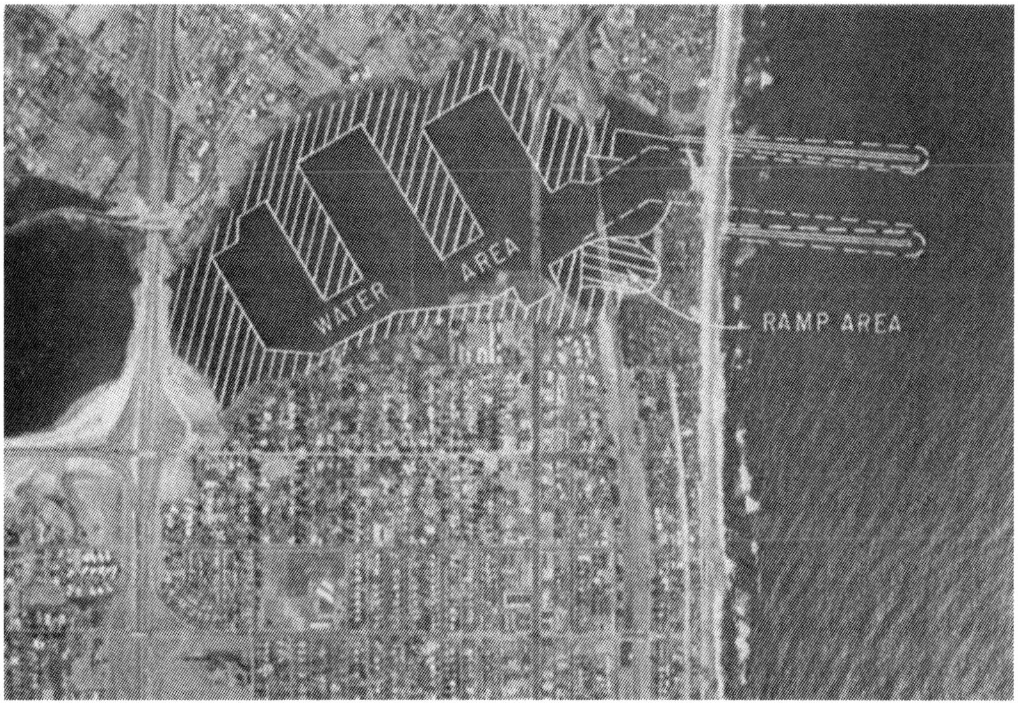

Available Area = 140 Acres (Two Bridges to be Removed)
Want 3-Lane Launching Ramp
Parking = 3 x 1.33 Acres = 4 Acres
Ramp, Road, Wash Area = 1 Acre
Available For Marina – 135 Acres
Let W = Water Area of Berthing Basins & Channels
Then $W + 4\left(\frac{W}{6}\right) = 135$ Acres
$W = 81$ Acres
(This Leaves 54 Acres For Back-up Land to be Filled)
Approx. Berthing Capacity = 81 x 20 = 1620 Boats
Daily Launching Capacity = 3 x 50 = 150 Boats
1770
Entrance Channel Width = 300 + 100 = 400 Ft.

Figure 43. Space allocation for a typical marina.

4. Design Criteria for Basins and Dockage Structures.

a. Water Area Perimeter Stabilization. Within the protected basins of a marina, or along a lake or river shore in front of which boats are to be berthed, some type of bank stabilization is usually desirable. Because the shore areas in these locations are not exposed to large waves or strong currents, the degree of stabilization required against external forces is usually minimal. Many marinas have no bank protection other than that provided by the natural ground cover of the region. Some banks present a rather pleasing appearance if properly cared for; others detract from the esthetics of an otherwise orderly appearing installation.

Armoring is usually required unless a natural bank already is a sand or shingle beach with a flattened slope and adjusted to the action of water exposure. Protection may be needed against scour by boat-wake waves (Sorensen, 1973), impact by floating debris or boats out of control, or gouging by ice. Moreover, the amplitude of periodic water surface fluctuations may make natural ground cover impractical. The most common slope armorings for marinas are riprap, precast concrete block units (usually overlapping or interlocking as shown in Fig. 33), and concrete slope paving. While availability of materials and economic considerations usually dictate the selection, internal forces due to active soil pressures (often intensified by surcharge loading at the top of the bank) must also be considered. These forces usually determine the steepness upon which the slope will safely stand.

Factors in determining the slope to which a bank should be trimmed are: (a) grain size for a beach or an armored slope in uncohesive material, (b) soil tests for cohesive soils, (c) anticipated dry-wet cycles that may affect the slide plane characteristics, such as water level fluctuations and saturation by landscape watering, and (d) surcharge loading at the top of the bank. A sand beach will usually be stable with a 1 on 8 slope in the tide and wave-runup zone and a gravel or shingle beach with a 1 on 6 slope. Finer uncohesive soils such as silty sands should normally be armored in the tide and wave-runup zone and tend to vary in allowable slope steepness above the wave-runup zone from about 1 on 3 to 1 on 2 depending on grading characteristics and median diameter of soil particles. A poorly sorted soil (with a steep grading curve) and small median diameter will require a flatter slope than a well-graded soil (with a flat grading curve) and large median diameter. Cohesive soils such as loams and clays will take steeper slopes but may be subject to plastic distortion and sliding when saturated, especially during extreme low tides and under high surcharge loading. Unless prior experience has already indicated the safe slope limit for the particular soil formation of the site. a soil mechanics engineer should be employed to make the slope limit determination. Below extreme low water where scouring currents (over 3 feet per second) are not present and where wave action is minimal, most soil will be stable on a slope of about 1 on 3 to 1 on 4.

The maximum steepness of an armored slope is often determined by the nature of the armoring material. A small stone riprap of rounded cobbles may not be stable on a slope steeper than about 1 on 2.5, whereas with increasing size and angularity the steepness may be increased to a maximum of about 1 on 1.5 provided the cut slope will stand on this slope until the armor is placed. Extremes of wave action and current flows may determine the minimum stone size of a riprap slope. These possible extremes (sometimes occurring in an off-season when the boats have been removed from their berths) should be determined in project planning and the proposed armoring materials checked for size against applicable formulas for seawall and river revetment design.

Because of the strong pumping action of waves and eddy currents, armor stone ripraps should always have a thickness of twice the average stone dimension, and a filter blanket under this armor layer is a vital requirement. If an appreciable amount of the underlying material should be pumped from beneath the riprap at any point, the armor stones will drop down into the resulting scour pockets, leaving the native soil exposed. Progressive erosion may then destroy large sections of the revetment before the damage can be repaired. Deposition of properly graded filter gravel of adequate thickness (0.5- to 1-foot thick depending on riprap stone size) may prevent such a failure. However, the danger of displacement of this filter material during construction requires careful supervision during riprap laying operations. An alternate filtering device that is gaining wide acceptance is the continuous cloth filter, which ensures continuity and complete filtering action throughout the revetted area (Barrett, 1966). Riprap cross sections specifying these two types of filters are shown in Figure 44.

A slope paved with precast concrete blocks also requires a filter layer as the underlying soil may be pumped through the joints between the blocks. Using filter cloth alone for this purpose may provide inadequate relief for hydrostatic uplift pressures, and may result in individual blocks being lifted out of the section. Unless the underlying soil is exceptionally porous, a gravel layer between the cloth and concrete blocks is recommended to allow quick lateral distribution of locally high pressures to points of relief. The ease with which an entire face of the block paving can slide down a slope dictates a need for an adequate toe-thrust resistor. This usually takes the form of a sheet-pile cutoff wall, or if it can be built in the dry, a poured wall or edge beam.

If the entire slope can be unwatered or can be armored at the extreme low water level, concrete slope paving may be the best solution (Fig. 45). However, a soils engineer should ensure that the soil is not subject to heaving and shrinking, and crack the concrete. Expansion joints should be provided at intervals of 50 to 100 feet depending on the extreme range of temperature and, within the concrete, temperature steel reinforcing bars should be spaced not more than 12 inches apart both laterally and longitudinally. Weep holes connected by French drains should be spaced at about 10-foot intervals down and along the

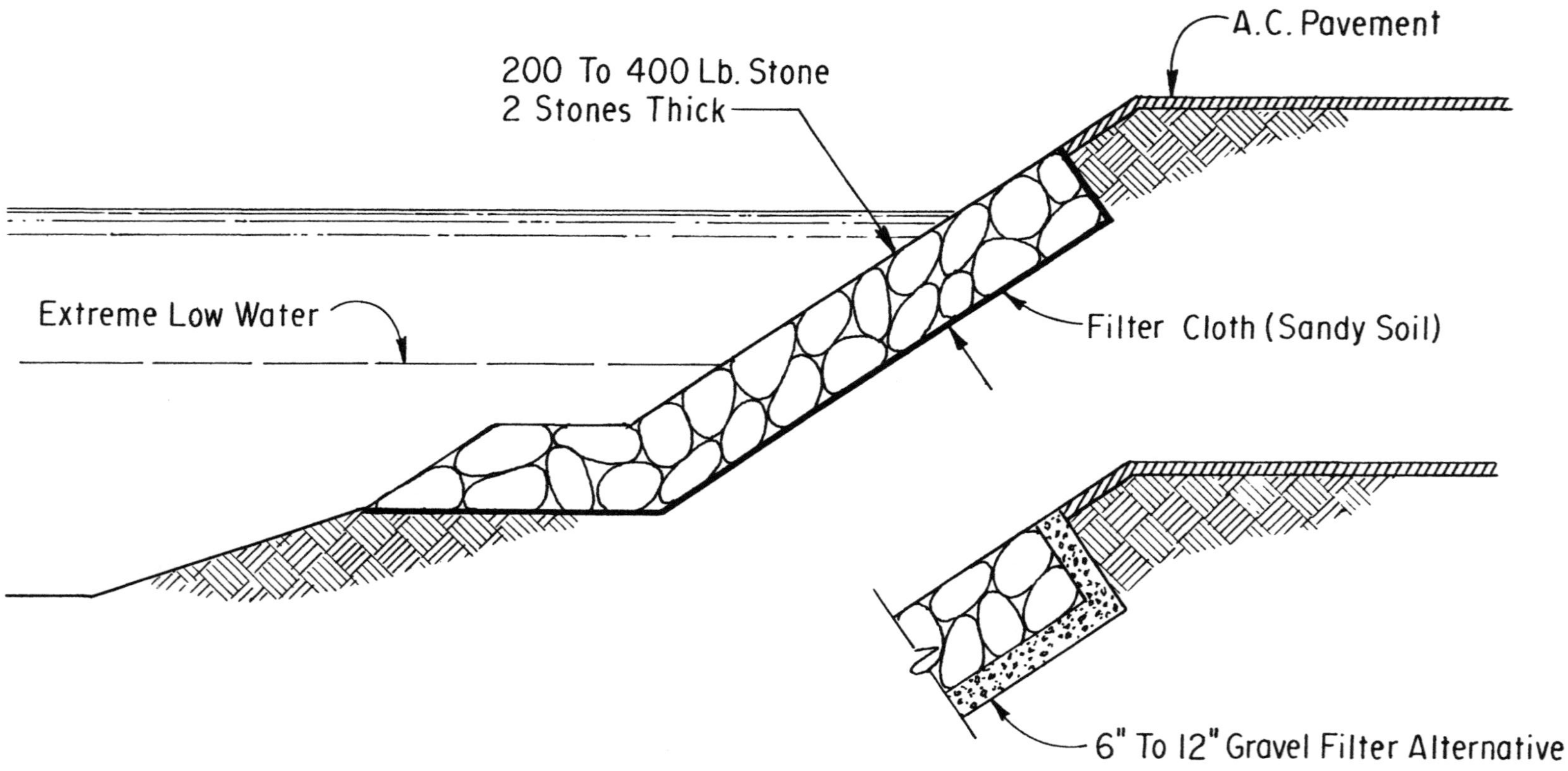

Figure 44. Slope riprap showing two types of filters.

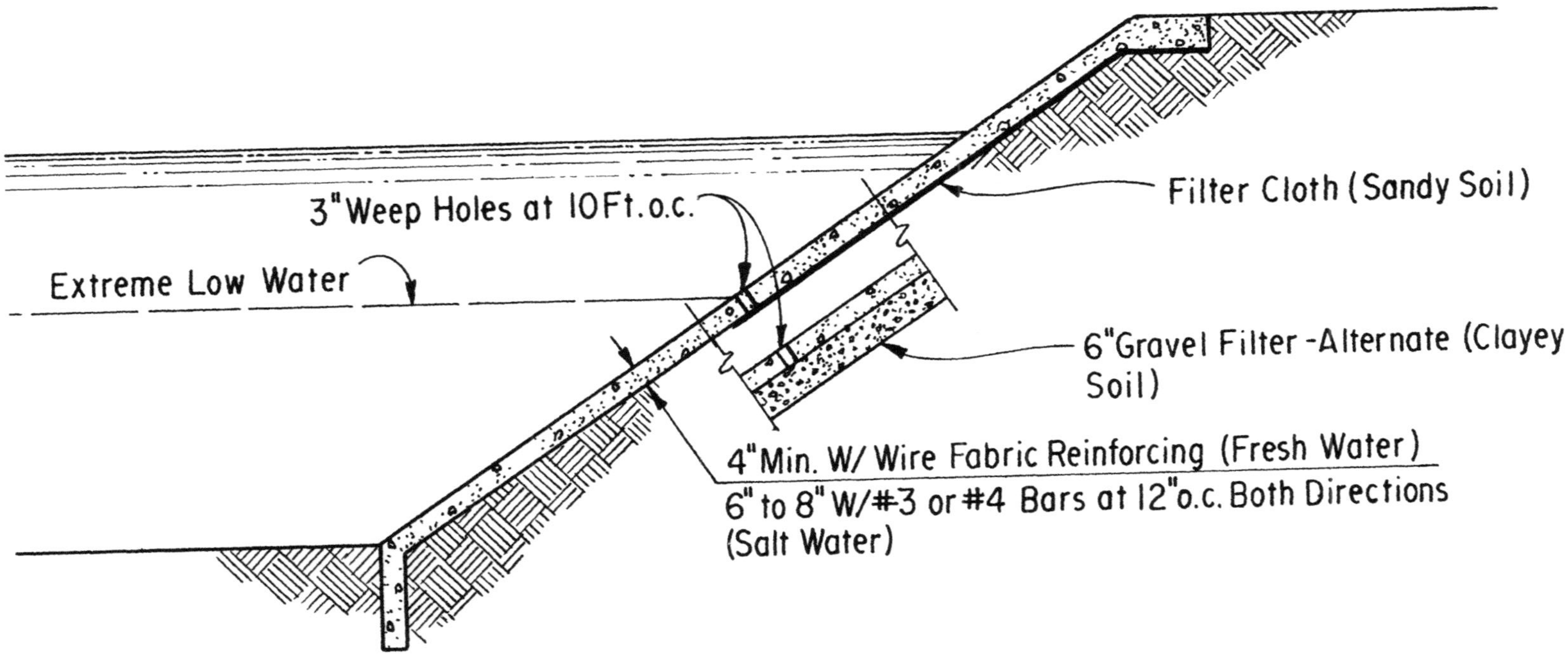

Figure 45. Slope paving showing two types of filters.

slope for relief of hydrostatic pressures, unless the width of the slope is less than about 15 feet and the subsoil is exceptionally porous. Toe thrust should be resisted by a cutoff wall trenched about 2 feet deep along the toe of the slope, and a thickened edge beam at the top of the slope should be provided to resist uneven loading along the shoulder. With careful placement, a good stiff mix may be poured on a slope of 1 on 1.5. For steeper slopes, airblown mortar (Gunite) must be used. Perimeter slopes have been paved with asphaltic concrete, but they usually fail in a few years because of softening in water or uneven settlement of the base material. Asphaltic paving should be used only above the extreme high water level.

An old-fashioned revetment system that permits perimeter slopes to be built steeply at low cost (where adequate small stone is available) is that of rockfilled wire mesh gabions (Smythe, 1961). Normally used only for flowing watercourses and dry bank retaining walls, this construction is also suitable for marina perimeter stabilization. Heavy duty wire mesh baskets with rectangular sides and of convenient size for such construction are manufactured commercially and shipped to the site in collasped form. They come galvanized for freshwater use and galvanized plus polyvinyl-cloride-coated for saltwater use. The filled baskets may be placed vertically on top of each other; for better stability they should be canted back to about a 4 to 1 batter slope. Battering may also be done by setback of successive rows for a terraced appearance. The gabioned slope should be backed by a layer of filter cloth to prevent the pumpout of material from the earthbank behind it. As yet the experience record for such construction in marinas is minimal and its permanence is unknown. The various uses of gabion in marina construction are shown in Figure 46.

Where surging is anticipated, due to partially attenuated wave action reaching the interior basins of a marina, consideration should be given to low reflectivity perimeter stabilization, especially at the inner ends of basins. A beach is the best type of surge absorber, followed by the flat riprapped slope built with large stones. Concrete slopes, either poured-in-place or of block construction, are almost as reflective as vertical walls.

Most vertical bulkheads are more expensive to construct than an armored slope. Where land and water areas are limited or costly, elimination of the waste space occupied by the slope may be an economic requirement.

Of the many types of vertical perimeter walls currently in use, the least costly to build is usually a tied-back pile-type timber bulkhead. Where construction in the dry is possible, the simplest construction usually consists of round or square piling tied to tie piles or deadmen and sheathed on the landward side to a depth well below the seaward toe of the wall. The pile size used for the tieback system and thickness of the sheathing must be determined by engineering analysis. Plastic filter cloth can be used to prevent loss of material through cracks in the sheathing (Fig. 47). Where timber is economical, the tiebacks may be made with timber planks (Fig. 48).

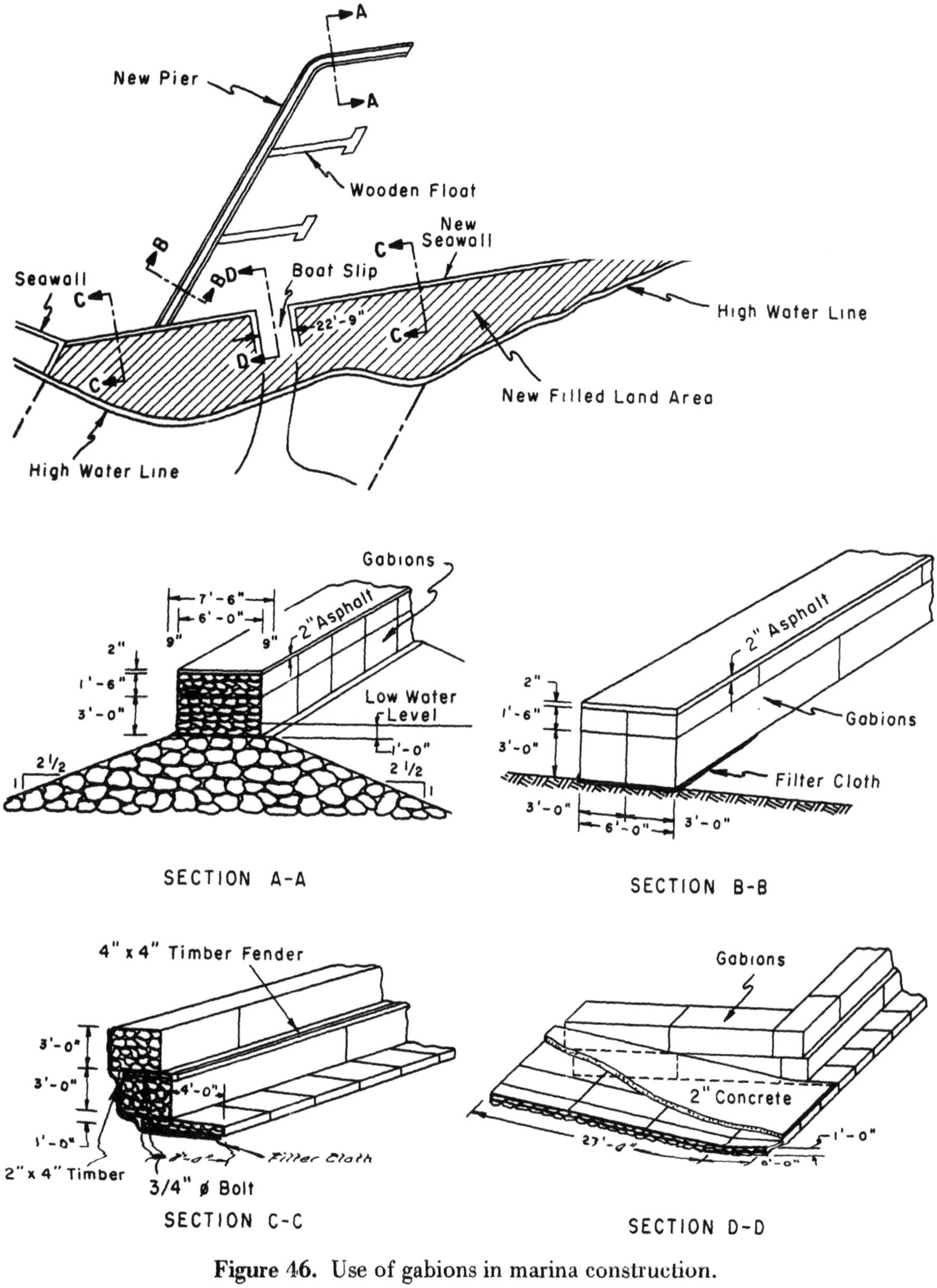

Figure 46. Use of gabions in marina construction.

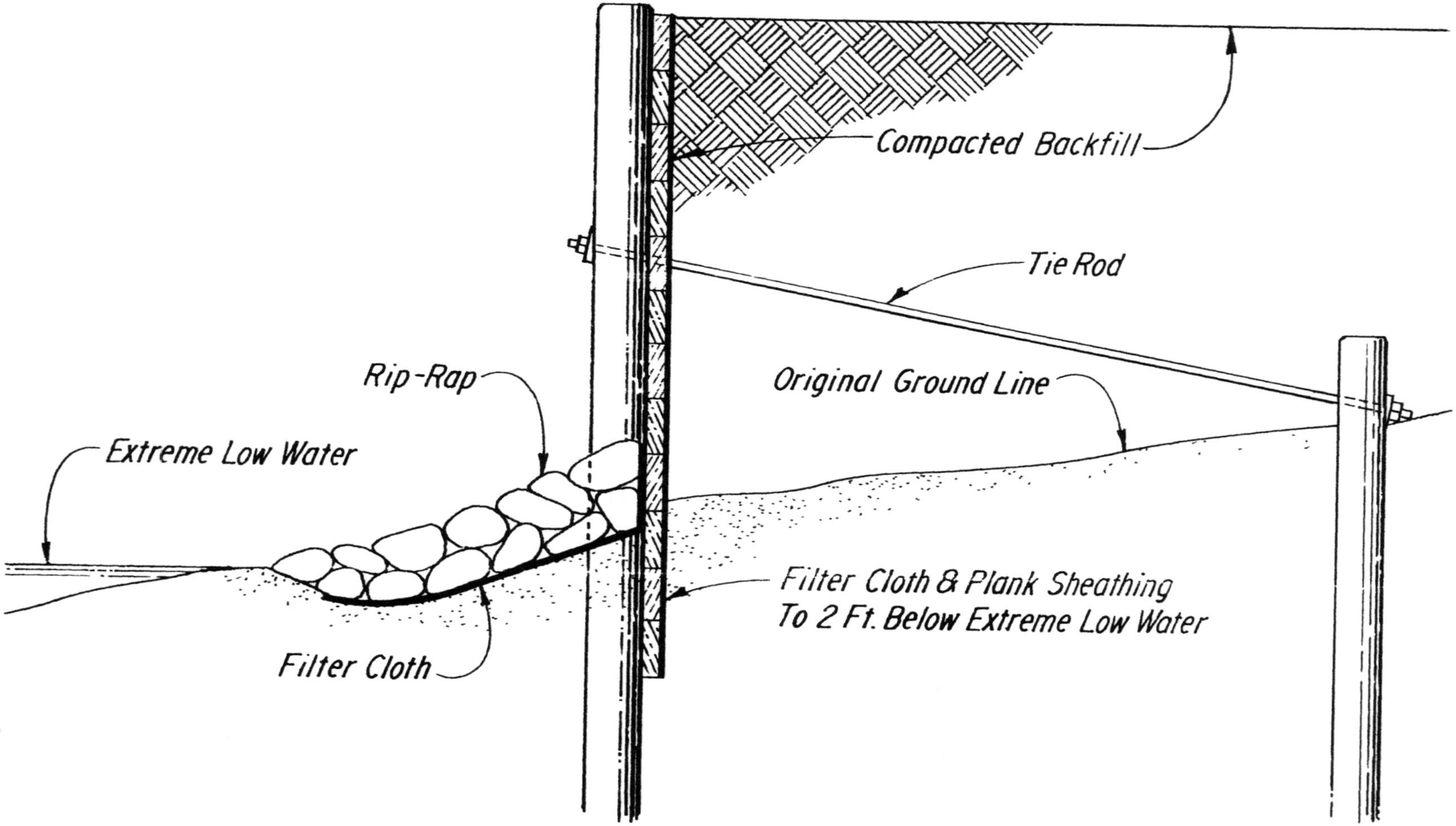

Figure 47. Use of filter cloth to seal timber bulkhead.

Figure 48. Timber bulkhead at Bellingham, Washington. Note projecting planks used for tiebacks.

Where it is difficult or impossible to excavate behind a row of king piles to the sheathing depth required, timber sheet piles may be used in lieu of horizontal sheathing members. An excellent technical guideline on this type of construction titled, "Bulkheads: Design and Construction," was published by the American Wood Preservers Institute (AWPI), in its monthly magazine, *Wood Preserving News*, in 1970. By permission of AWPI, the entire text, tables, and figures of the three-part technical guideline are reproduced in Appendix B.

All timber should be pressure-treated to avoid dryrot and destruction by living organisms. In saltwater, coaltar-creosote treatment in accordance with applicable Federal Specifications or AWPA Standard C18 should be specified. Without proper treatment, timber construction soon falls into unsightly disrepair (Fig. 49), and may be compromised to the point where it cannot fulfill its intended function.

The structural properties of timber piles, wales, struts, and sheathing vary with the types of timber used, with the nature of curing to which it was subjected, with its environmental exposure in the structure, and sometimes by the type of preservative treatment used. These properties, including allowable stresses under various loading conditions and environments, are discussed in the *Timber Construction Manual* published by the American Institute of Timber Construction (AITC), (1966). The standards for various types of lumber are listed in Section 101-65, paragraph 3.5, of the publication; these are changed from time to time, and

Figure 49. Failing timber bulkhead.

only the current issues should be used. The standards are published under the general title of "Standard Grading Rules for – – – ," "Official Grading Rules for – – – ," or "Standard Specifications for Grades of – – – ," with the dashed line being the type of timber covered, such as Douglas Fir, Sitka Spruce, Southern Pine, and Tidewater Red Cypress. Some of these standards are repeated in the Uniform Building Code (1970), or in the most current edition. Wherever a timber bulkhead of more than minimal height or length is required, the designs discussed in the AWPI Technical Guidlines (App. B) should be checked against the latest grading standards for the type of timber used.

Timber bulkheads often fail because of corrosion, abrasion, or fatigue of metal connectors, or because of abrasion of the wood by loose connectors and not as a result of deterioration of the wood members. For marine exposure, all hardware should be galvanized and the following minimum sizes used:

In or Below Splash Zone

Bolts .... 1-inch diameter
Plates .... 0.5-inch thickness
Washers .... Ogee (standard size to fit bolt)

Above Splash Zone

Bolts .... 0.75-inch diameter
Plates .... 0.375-inch thickness
Washers .... 0.25-inch plate (ogee operational)

In general, the exposure of bolts to the atmosphere should be reduced to a minimum by using only one washer or plate per unit. Adjacent timbers should be in contact with the bolt to prevent exposure. Boltholes should not exceed the diameter of the bolt by more than 0.062-inch. Drift bolts or spiral bolts should have a driving fit. Washers should bear evenly and fully on the timber, and where the axis of the bolt is not perpendicular to the face of the timber, beveled plates or washers used.

Steel or aluminum sheet-pile bulkheading with an exposed face of not more than about 6 feet can be constructed quickly and at fairly low cost with special lightweight sheet-pile sections (Fig. 50). Matching cap beams and deadman anchors of the same metal are available for some of these lightweight sheet-pile systems. Their manufacturers claim that with proper control of alloy content in the base metal, or application of preservative coatings, these sheets can be resistant to seawater and a wide range of hydrogen-ion (Ph) values in freshwater. The primary advantages of the lightweight sheets is that they require only the simplest equipment for handling and driving, and yet provide a sandtight wall with considerable tensile strength at interlocking joints. Manufacturers usually provide design details that require only the simplest engineering analysis. Asbestos-cement sheet piles have been used successfully for low bulkhead walls to avoid the corrosion problem. Because of the low allowable fiber stress of the material, the active earth pressures should be checked

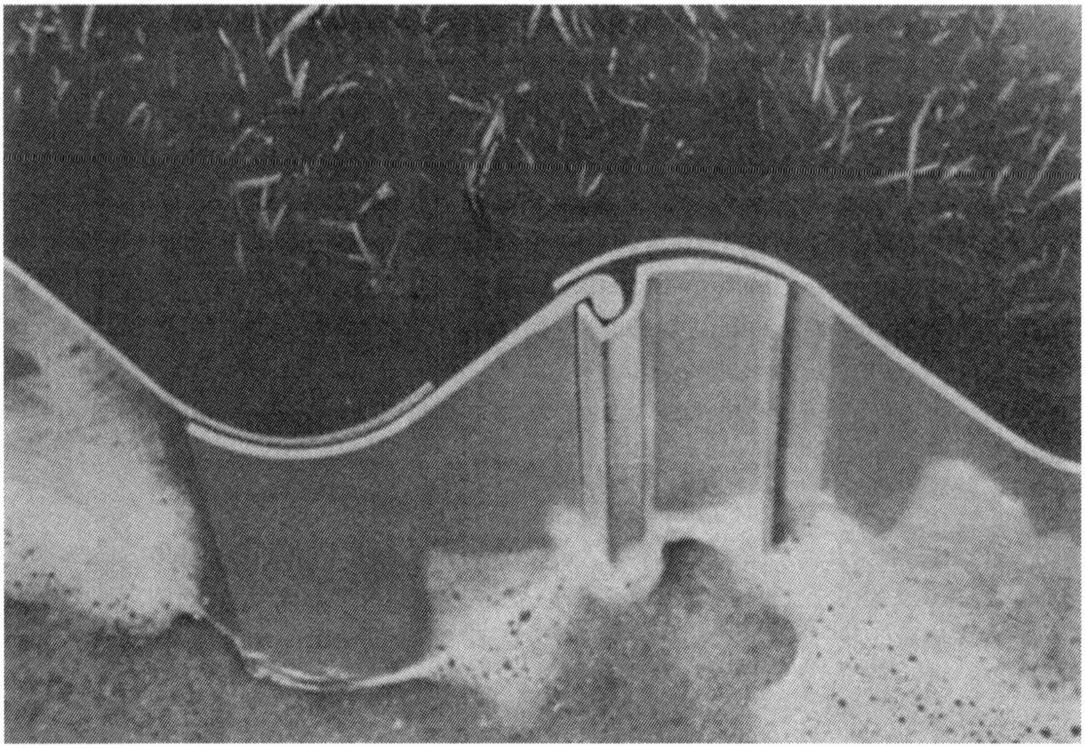

Figure 50. Low-height, free-standing aluminum bulkhead
(Courtesy of Kaiser Aluminum & Chemical Sales, Inc.).

carefully against the section modulus to make sure the sheets will not be overloaded. The sheets are brittle and should only be used where they can be jetted into place. Some cements in asbestos-board tend to deteriorate after long periods of immersion; therefore, only sheets that are certified by the manufacturer for marine applications should be used.

Where the exposed face is more than about 6 feet in height, the section modulus provided by the lightweight sheets is usually insufficient to resist the active soil pressures, and careful engineering analysis is required to determine design limits. Some of the heavy gage steels may prove satisfactory to about 8 feet of exposed wall, but beyond that, a stronger section is needed. Standard rolled steel sections are generally used for steel sheet-pile walls with an exposed face more than 8 feet high. These walls can be designed as freestanding or cantilever-type structures, if the space just behind the wall is required to be kept free of any obstructions such as a tieback system. The extra pile length for such construction is usually costly, however, and tiebacks with either a single or multiple-wale system are more economical. Detailed engineering analysis of the soil mechanics involved and structural analysis of the steel sections used will ensure a safe and most economical design.

The design principles discussed in the AWPI Technical Guidelines (App. B) are generally applicable to steel sheet-pile construction by substituting the structural properties of the steel members used for those of timber construction. Chaney (1961) also provides a treatise on bulkhead design for steel and other materials.

Several corrosion-retarding procedures have been devised for steel components in a water environment, including galvanizing, aluminizing, flame-spray metal coatings, various types of brush coatings, and cathodic protection. An excellent, economical coating for large members is coal-tar epoxy, shop-applied to *white metal* steel surfaces and, before installation, carefully examined for flaws with approved electrical equipment. Some coatings are available for application in a wet environment, but experience records are too short for reliable determination of their effectiveness. A 15-year experiment was started in 1967 (Watkins, 1971) to test the performance of 32 different systems of steel-pile protection in the Atlantic Ocean off the Virginia coast. When completed, this test should provide some fairly reliable data on protective systems in a marine environment. Meanwhile, a good practice is to allow for the loss of 0.003 inch of metal per year from each surface of an uncoated steel section of standard alloy content where the average water temperature ranges between 50° and 70°F. in colder coastal waters (0.002 inch in freshwater). This factor may be decreased slightly in colder waters, but must be increased significantly for higher temperatures. A good preservative coating (applied to white metal) should double the effective life of the section. An efficiently operating cathodic-protection system may extend the life of the steel structure indefinitely, but it should be installed by an expert.

Special precautions should be taken to avoid the use of different metals or even different grades of the same metal in close proximity in seawater because of the galvanic action thus induced. The low conductivity of freshwater does not present a galvanic-action problem, but dissolved oxygen content and splash zone oxidation causes rusting. The Ph factor of the water and the soil into which the piles will be driven should be checked and the recommendations of a corrosion expert obtained wherever a metal sheet-pile system is to be installed in freshwater. Uhlig (1953) presents a detailed treatise on corrosion theory and methods of retarding or eliminating corrosion. The compatibility of various metallic fasteners with certain base metals in seawater is shown in Table l.

Concrete bulkhead walls come in a variety of sections and in combinations with other types of construction. The most commonly used combination in marina construction is a wall with a vertical or slightly battered face extending to about extreme low water level, and a slope extending from that level down to the project depth of the marina. The slope should be armored with riprap from the wall to about 5 feet below extreme low water to prevent wave or eddy current scour at low water levels. Because of the danger of boats grounding on the riprap slope, such construction should not be used where the bulkhead is marginal to a navigation fairway unless the limits of the deepwater channel are marked with buoys, or where berthing facilities or other structures prevent inadvertent navigation close to the wall. This combination of concrete structure and revetted slope is often the most economical solution.

Table 1. Galvanic Compatibility of Fasteners in Seawater

Base Metal	Aluminum*	Carbon Steel	Silicon Bronz	Nickel	Nickel-Chromium Alloys	Type 304 SS	Nickel-Copper Alloy 400	Type 316 SS
Aluminum	Neutral	C	Unsatisfactory†	C†	C	C	C†	C
Steel and Cast Iron	NC	Neutral	C	C	C	C	C	C
Austenitic Nickel Cast Iron	NC	NC	C	C	C	C	C	C
Copper	NC	NC	C	C	C	C	C	C
70/30 Copper-Nickel Alloy	NC	NC	NC	C	C	C	C	C
Nickel	NC	NC	NC	Neutral	C‡	C‡	C	C
Type 304	NC	NC	NC	NC	May Vary §	Neutral‡	C	C
Nickel-Copper Alloy 400	NC	NC	NC	NC	May Vary §	May Vary §	Neutral	May Vary §
Type 316	NC	NC	NC	NC	May Vary §	May Vary §	May Vary §	Neutral §

* Anodizing would change ratings as fastener.

† Fasteners are compatible and protected but may lead to enlargement of bolthole in aluminum plate.

‡ Cathodic protection afforded fastener by the base metal may not be enough to prevent crevice corrosion of fastener particularly under head of bolt fasteners.

§ May suffer crevice corrosion under head of bolt fasteners.

NOTE: C = compatible, Protected. NC = Not Compatible, Preferentially Corroded.

The concrete section may either be of unreinforced gravity construction, or a reinforced section. Where construction in the dry is possible, the reinforced concrete L-wall is usually the most economical for exposed faces not exceeding 15 feet in height (Fig. 51). For greater exposures, counterforted walls or cellular construction may be more economical. Design details for monolithic wall sections are too complex for inclusion in a manual of this scope. The design effort falls into two general categories: analysis of the external forces and structural design of the section to resist these forces. A treatise on external forces is available in Bowles (1968), and Navy Bureau of Yards and Docks (current DM-7). Bowles (1968) presents enough design data and examples of internal design to cover most sections used in marinas.

Where the wall site cannot be conveniently unwatered, concrete sheet piling is usually the simplest alternative, but tiebacks are usually required. The most difficult part of concrete sheet-pile construction is the achievement of sand tightness in the joints. Joints are usually of the tongue-and-groove type that are not provided with interlocks for horizontal tensile strength. A common method of achieving the necessary joint tightness is the casting

Figure 51. Concrete L-wall under construction, Dana Point, California. Note filter cloth taped to concrete at weepholes and expansion joints preparatory to backfilling.

of opposing grooves on the edges of the piles above the mudline (Fig. 52). After all the sheets have been driven, joint grooves are jetted clean and grout is pumped into them by starting from the bottom. To contain the grout if serious leaks occur, elongated plastic bags can be inserted into the joint along with the grouting pipe. The grouting pipe is lifted out as the grout is pumped in, and the bag remains in the groove preventing escape of the grout through cracks along the opposing edges. Although the plastic material prevents bonding of the grout to the concrete, the joint is closed enough for sand tightness. Sometimes it is possible to place a strip of filter cloth behind each joint, thus obviating the need of joint grouting for sand tightness.

Concrete sheet piles must be reinforced and, in a saltwater environment, the reinforcing steel must be kept 3 inches behind the concrete saltwater interface. Sheet piles with a single plane of reinforcing can be used for walls with less than about an 8-foot exposure; two planes are required to provide the necessary section modulus for walls of greater height. Since the backfilled face of the wall will not be visible, the sheet piles can be cast directly on the ground provided the bed is first carefully rolled and graded to a perfect plane. To achieve a smoother face, a thin polyethelene sheet may be spread over the bed before setting the reinforcing cage and edge forms.

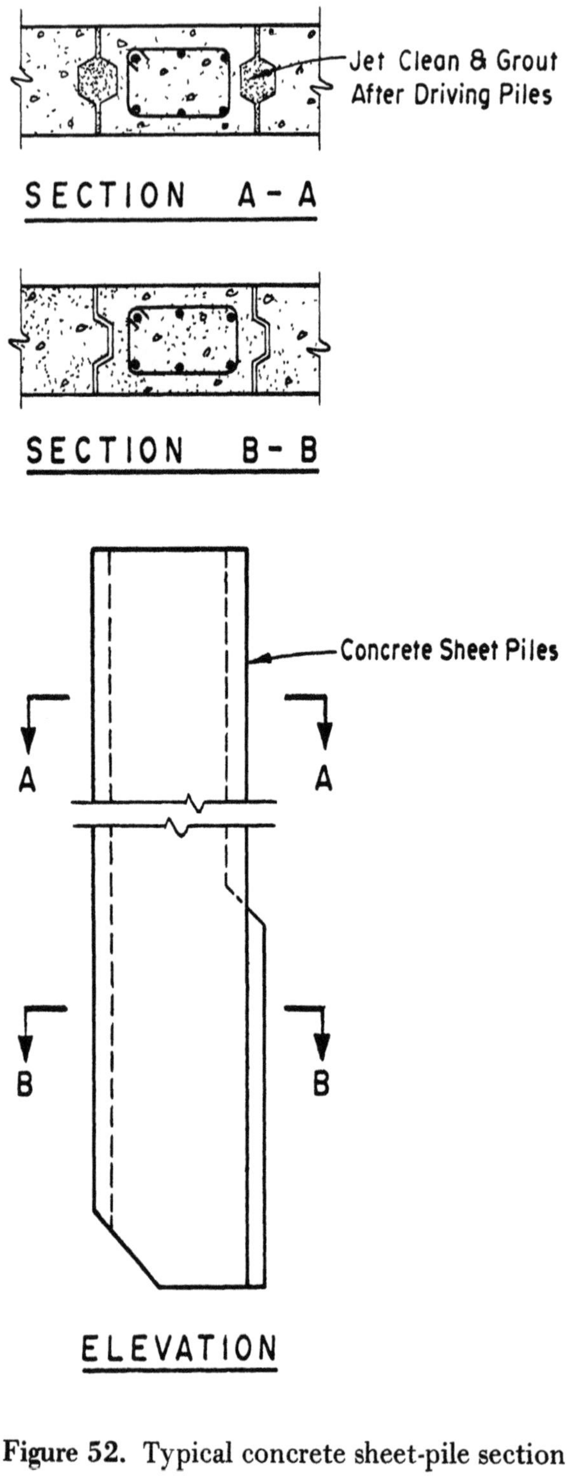

Figure 52. Typical concrete sheet-pile section with grooves for grout-sealing joints

The pile tip is mitered to keep it closed into the adjacent pile while driving or jetting. The upper part of the pile is held against the adjacent pile with a rope sling, and a temporary timber backing frame is used to keep the wall on line as the work progresses. In sandy soils, jetting alone is used to seat the piles, but in loam or clayey soils, they must be driven into place to avoid loose seating. The tieback system differs from that of metal or timber sheet piling in that the tie rods are usually embedded in a cap beam poured around the top edge of the completed wall (Fig. 53).

Except for width of the pile, the proper dimensions for the concrete sheet piles must be determined by engineering analysis in the same manner as for timber or metal sheet piles. The pile width (incremental length of progression along the wall per pile) may be varied to suit the placement conditions of the site. In sandy soils where jetting to grade is permissible, the width of the pile is limited (in combination with the thickness and length) only by the capacity of the handling equipment and certain convenience considerations. These factors usually limit the width to about 4 or 5 feet. Where the piles must be driven in tight soils, the width must be decreased to prevent damage in driving. In tight soils, a good practice is to make the width about twice the pile thickness, and the reinforcing cage must be strengthened with more looping near the top to prevent shattering of the concrete under hammer impact.

Concrete perimeter bulkheads are the most durable if properly designed and constructed. In a marine environment, it is important that the ingredients of the concrete, mixing and placing in the structure be carefully controlled. Chaney (1961) discusses in detail the types of cements that are suitable for various conditions, the selection of aggregates, placement, and resultant structural properties of the cured concrete in the structure. For ready reference, the ACI Standard 211-1-74, *Recommended Practice for Selecting Proportions for Normal and Heavyweight Concrete*, published by the American Concrete Institute is reproduced in Appendix C.

Guidance for selecting types of concrete mixes for special environments is published by the California Division of Highways, Bridge Department, in a *Bridge Planning and Design Manual*, 1969. The part on "Corrosion Protection for Concrete" is reproduced in Appendix D.

Marinas are often built in areas where the substrata have low bearing capacities, and a heavy wall may settle unless adequately supported on bearing piles (Fig. 54). The tendency of silty materials and soft clays to flow out from under walls or to push the entire wall toward the basin under saturated conditions requires careful analysis. Proper drainage through French drains or weep holes under the base of the wall and provision of batter piles will usually counteract these destructive tendencies. A soils analysis will reveal the required extent of such drainage requirements. All concrete construction must be designed in

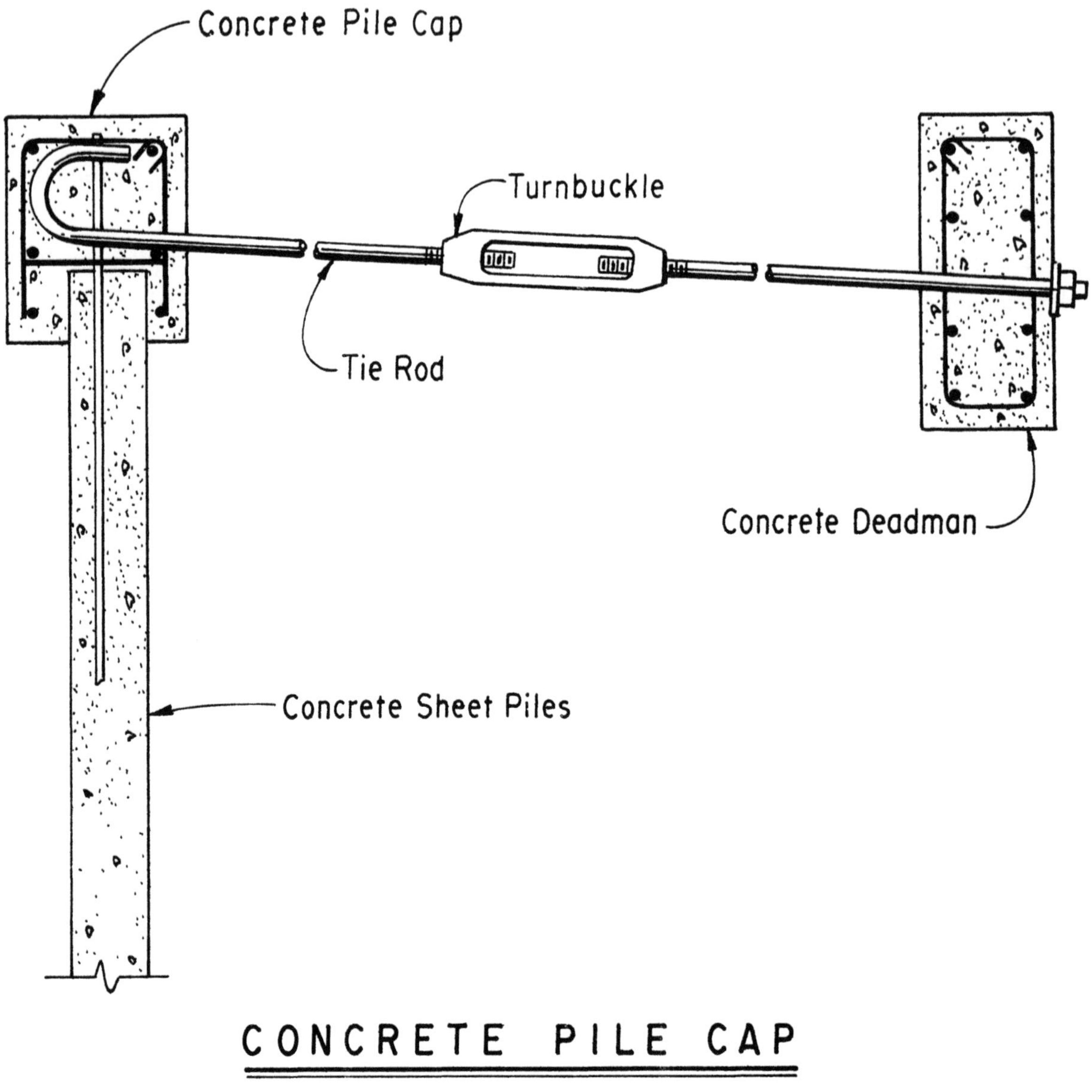

Figure 53. Cap and tie rod system for concrete sheet-pile wall.

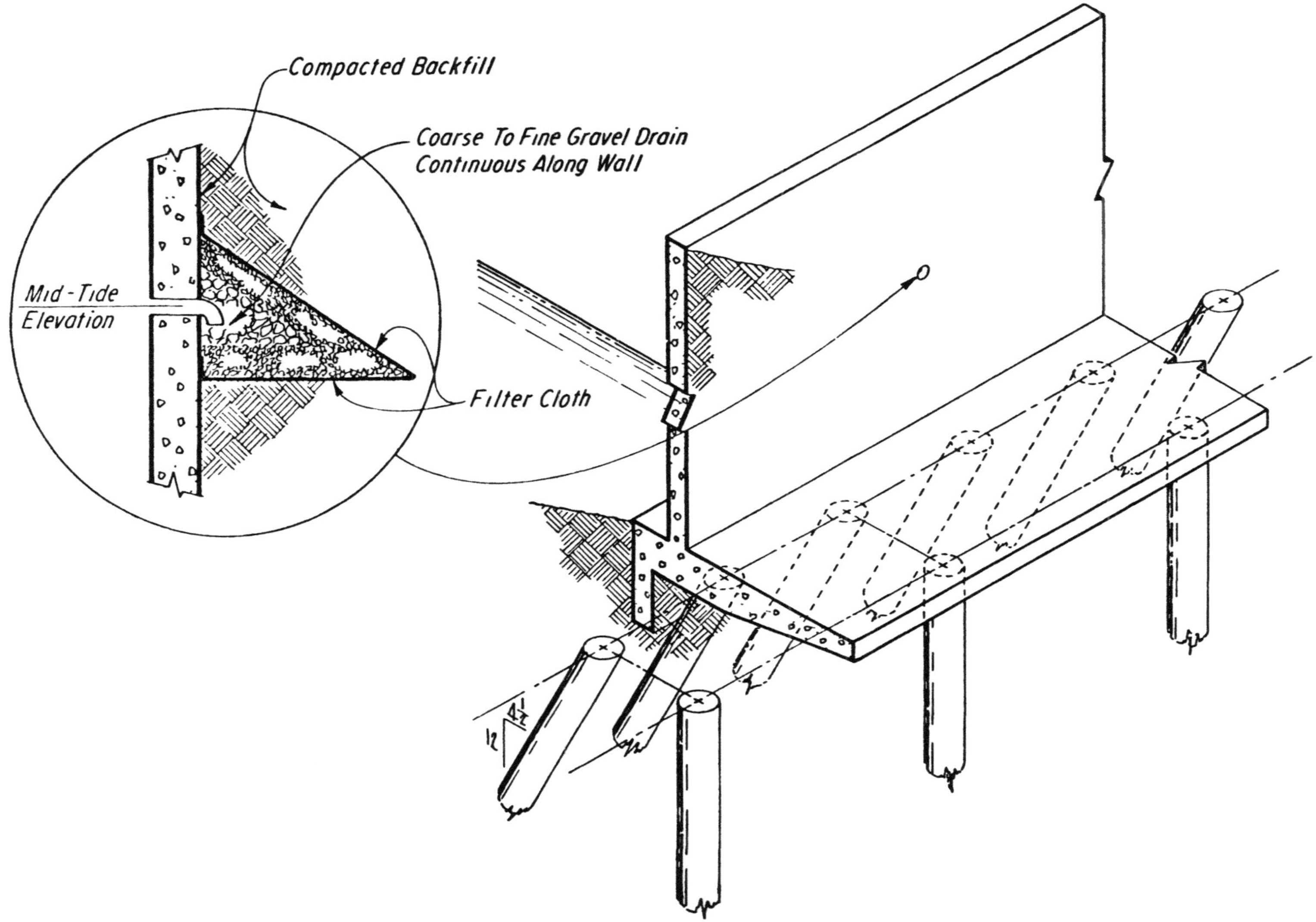

Figure 54. Typical pile-supported L-wall for weak soils.

monolithic units separated by construction and expansion joints. Unless these joints are made and spaced properly, cracking, bulging, or separation of units may occur. Adequate imbedment of reinforcing steel in all concrete construction is vital to the prevention of spalling. Here again, proper design and control of construction procedures are necessary to prevent failures (Fig. 55).

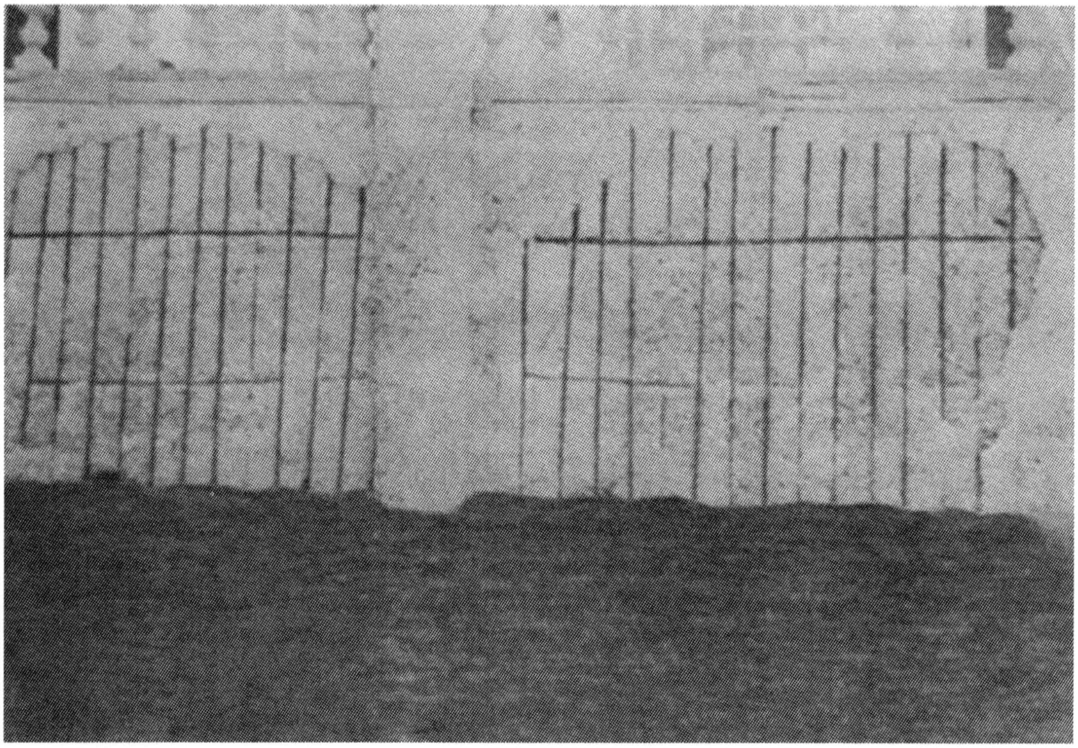

Figure 55. Spalled face of concrete wall due to inadequate rebar imbedment.

If necessary to cross the basin perimeter to gain access to the berthing area, provision must be made for a pier landing. When a landing is directly on a perimeter wall or top of the slope, adequate structural detailing for each landing must be incorporated in the perimeter structure. Because this landing usually places an extra load on the wall or top of the slope, additional engineering analysis is required to ensure: (a) the adequacy of this perimeter structure to receive the load, and (b) the capability of the substrata to support it. The load may require extra bearing piles under a wall or a special load-distributing seat or abutment at the top of a slope. Whatever the requirement, it should be considered and be planned as an integral part of the perimeter treatment and not deferred to a later phase of the design effort.

All bulkhead walls and perimeter slopes steeper than about 1 on 1.5 should be protected by handrails if the adjacent water is more than about 2 feet deep or if at low water level the drop along the edge exceeds about 6 feet. Municipal or county building and safety agencies often provide design criteria for handrails. In the absence of handrail criteria, it is customary to make the top rail 42 inches above the adjacent ground level and to design the top rail and supporting posts for a minimum lateral thrust of 30 pounds per foot of rail. One secondary rail should be provided about midway between ground level and the top rail. Some agencies require a handrail system with no openings larger than 6 inches, necessitating insertion of wire mesh panels or other means of compliance.

The selection of materials for the handrail will depend on the type of bulkhead construction. With timber bulkhead construction, the handrail is usually of timber. Galvanized pipe handrails, are commonly used and provide long life except where subjected to salt spray. One of the best handrail materials is aluminum with a hard anodize finish 1.5-millimeters thick. Some aluminum handrails have an experience record of over 15 years without maintenance in a salt-air environment. They are available commercially, are adaptable to almost any foundation or face-fastening situation, and have corner-turning and termination systems to meet a variety of requirements.

One type of perimeter treatment has been developed that produces a natural rustic appearance and may be suitable for the channels and water areas adjacent to a marina where bank erosion may be a problem. It consists of special molds for the exposed face and use of colored concrete to blend with the native soil and, when completed, is similar to those produced by natural weathering. Although the concrete must be placed in the dry, this treatment could be the answer to many river- and lake-connecting channel problems where preservation of esthetic quality is a requirement (Fig. 56).

b. Basin Depths. The interior basin requirements of a small-craft harbor are determined generally by the same criteria applied to entrance and basin approach channels. An additional factor that must be considered, however, is the effect of bottom depth on structures of the berthing system, such as fixed-pier supports, floating pier guide piles and dolphins, and interior wave and surge baffles. The cost of each structure increases roughly with the square of the depth; hence, the basins should be no deeper than navigational considerations dictate. For this reason, as well as for space utilization economy, it is customary to berth the larger and deeper-hulled craft near the entrance and to decrease basin depths in step increments toward basins or parts of basins away from the entrance. Thus, the narrower fairway and smaller turning area requirements of the small craft decreases the amount of water area required for nonberthing use. Also, the shallower back basin areas reduce pier and slip construction costs. In the berthing basins, assuming that the maximum depression of a boat below the stillwater surface will be about 2 feet (due to wave action and scend), the depth should be at least 2 feet below the keel of the deepest-craft boat at extreme low water (see Sec. V, para. 4d, Fig. 63, for values of deepest draft).

Figure 56. Molded concrete bank simulating natural conditions (Courtesy of Mr. Harlan Glenn).

In rivers, natural lakes and manmade reservoirs, bottom depths must often be accepted as they are, sometimes well below a higher level indicated to be adequate by navigation criteria. Then, less conventional anchorages must be devised for floating systems, and floating wave baffles may be needed. In rivers and off-river basins where sediment deposition becomes troublesome, initial project depths may have to be increased to allow for bottom buildup between periodic maintenance dredging activities. Also, during initial planning, consideration must be given to the interference problems of later dredging. Fixed structures must be kept to a minimum in areas where accretion is anticipated. Where possible, provision should be made for periodic removal of shoals by means of land-based equipment operating from strategically placed moles or platforms.

c. *Interior Wave Barriers.* Within a small-craft harbor, troublesome waves are often generated by boat wakes (Sorensen, 1973) from a busy fairway or strong winds blowing from a long interior watercourse. The best way to keep these waves out of a berthing area is to place a solid sheet barrier between the source and the berths (Fig. 57). Where a solid barrier may prevent the necessary water circulation needed for water quality control, a baffle made of alternating panels on either side of a pier framework, may be a better barrier (Fig. 58). A single diaphragm with windows will seldom give satisfactory results; too much

North Side

South Side

Figure 57. Shilshole Bay Marina, Washington. Interior steel pile-supported timber wave barrier.

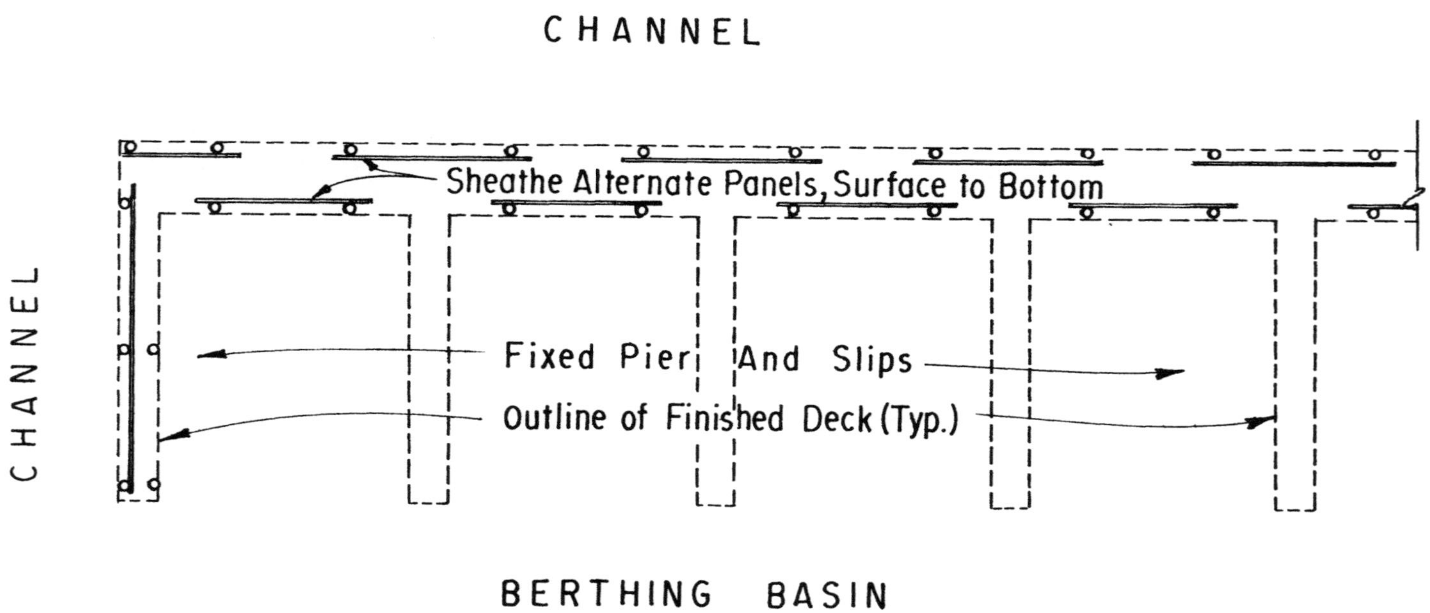

Figure 58. Alternating wave baffles protect a berthing area. Panels stop most of the surface wave action but openings allow water circulation.

window space is required for the necessary circulation and wave transmission through the windows defeat the purpose of the barrier. Where only a single diaphragm is used, the best means of achieving some water circulation without increasing wave transmission to unacceptable limits is to carry the solid sheathing about three-fourths of the way to the bottom and leave the bottom one-fourth completely open except for supports. This will effect a high degree of attenuation of short-period waves, and although it will still allow the longer-period waves to regenerate, a satisfactory fraction of the circulatory currents will pass under the barrier.

If a fixed barrier is impractical because of depth problems or lack of space, some degree of wave protection can be achieved by a continuous drop panel on the outside of a floating walkway, along the margin of the fairway from which protection is sought. Where a series of fixed or floating piers extends out to a main channel, a frequently used alternative is to attach drop panels to the end fingers (Fig. 38). To be reasonably effective, the panels should extend at least 6 feet below the water surface.

A good way to ensure that troublesome boat-wake waves will not enter the berthing area is to design the harbor so that heavily traveled fairways or channels are separate from the berthing areas (Sorensen, 1973). Separation of basins from channels can be done by geometric arragement of the interior water areas. The layout of Marina Del Rey, discussed in Section VIII, is a classic example of basin separation by moles.

d. Berthing Facility Arrangements. The oldest methods of securing unattended small craft are anchor moorings and beachings. Any type of anchor mooring, from single-point free-swinging to multiple-anchorage ties, results not only in the inefficient use of space, but the need for shoreboats in transferring personnel between the moored craft and shore. Beachings are not only difficult for larger craft, but take up considerable perimeter area. For these and perhaps other reasons, neither anchor moorings or beachings appear compatible with modern marina operations except in a few isolated cases.

When a visiting craft arrives at a small-craft harbor it is normally docked alongside a float or fixed pier near the harbormaster's office, and the owner proceeds to make his arrangements for temporary berthing or for a longer-term slip rental. The first type of facility, the landing dock, requires alongside breasting of the craft, usually with ties to mooring cleats or curb timbers on the dock foreand-aft of the boat. This type of docking is not suitable for long periods of unattended berthing; the constant differential movement of boat with the dock may cause hull abrasion or may accelerate wear on a fendering system. It is used, therefore, only for short stays, like at a fueling dock. One exception is the trailing floating slips sometimes used in flowing rivers (Fig. 7), which can only be used satisfactorily in conjunction with alongside dockage.

In a boat slip, the craft may be tied away from the dock structure, usually with fore-and-aft ties on both sides. In a single-boat slip, the craft is flanked on each side by a finger pier. In a double-boat slip, various methods have been used to avoid breasting against a finger, including a tie pile centered between the finger ends, three-point ties, steel whips, and cooperative switch-tie systems agreed upon by two-slip occupants (Fig. 59). A compromise scheme places a narrow tie finger (usually half-length) between the two boarding fingers (Fig. 60). The full-length, single-slip system is the most desirable from the boatman's standpoint, but is more costly than a double-slip or any of the compromises. The system selected will depend largely on what the boat owners of the local area are accustomed to or the amount of slip rental fees they are willing to pay.

Small boats in relatively quiet waters may be berthed to a dock with stern hooks or bow clamps at a considerable savings in both structural costs and water space, but boarding is more difficult. When not in use, lightweight sailboats are sometimes pulled onto a floating dock equipped with special launching rollers and cradles to keep the hulls dry (Fig. 61). Charter boats are sometimes tied to a system of tie piles with their sterns close to a crew boarding dock (Fig. 62). When the boat is scheduled for a cruise, access to the stern is readily available for tending the mooring lines and subsequent movement of the craft to the passenger boarding dock.

Berthing facilities for commercial fishing fleets are more utilitarian than those serving the recreational fleets. Because the hulls are generally more rugged, breasting against a dock is common, and two or more boats under single ownership or cooperative arrangement are often clustered beam-to-beam at a single dock.

Slip arrangements vary, usually for best conformance to the shape of a basin. The most common is a series of piers or headwalks extending perpendicular to the bulkhead to a pierhead line, with finger piers extending at right angles from the headwalk on either side. For power craft, widths of fairways between finger ends are usually 1.75 to 2 times the length of the longest slips served, while for sailboats the width is 2 to 2.5 times the slip length. Slip widths have been increased in recent years because of the wider beams of some craft. A graph from the boat manufacturers' current catalog data and showing average beam width and maximum depths of keels for various lengths of craft is shown in Figure 63. The graph also shows suggested average widths for right-angle slips where the actual dimensions of craft to be berthed are not known. Where basin configuration or curtailment of water space dictates a need for skewed slips, the slip spacing must be calculated, allowing about 1.5 feet of clearance on each side between hull and finger for boats up to 35 feet in length and 2 feet of clearance for longer craft. With the increase of houseboats and multihulled craft, it may be advisable to provide a number of extra-wide slips, depending on past experience record and projected needs of the area (Fig. 64).

An interesting proprietary system arranges modular floating slip units in star-shaped clusters (Fig. 65). Access to a cluster is either by shoreboat or by a star-to-shore extension of one of the fingers.

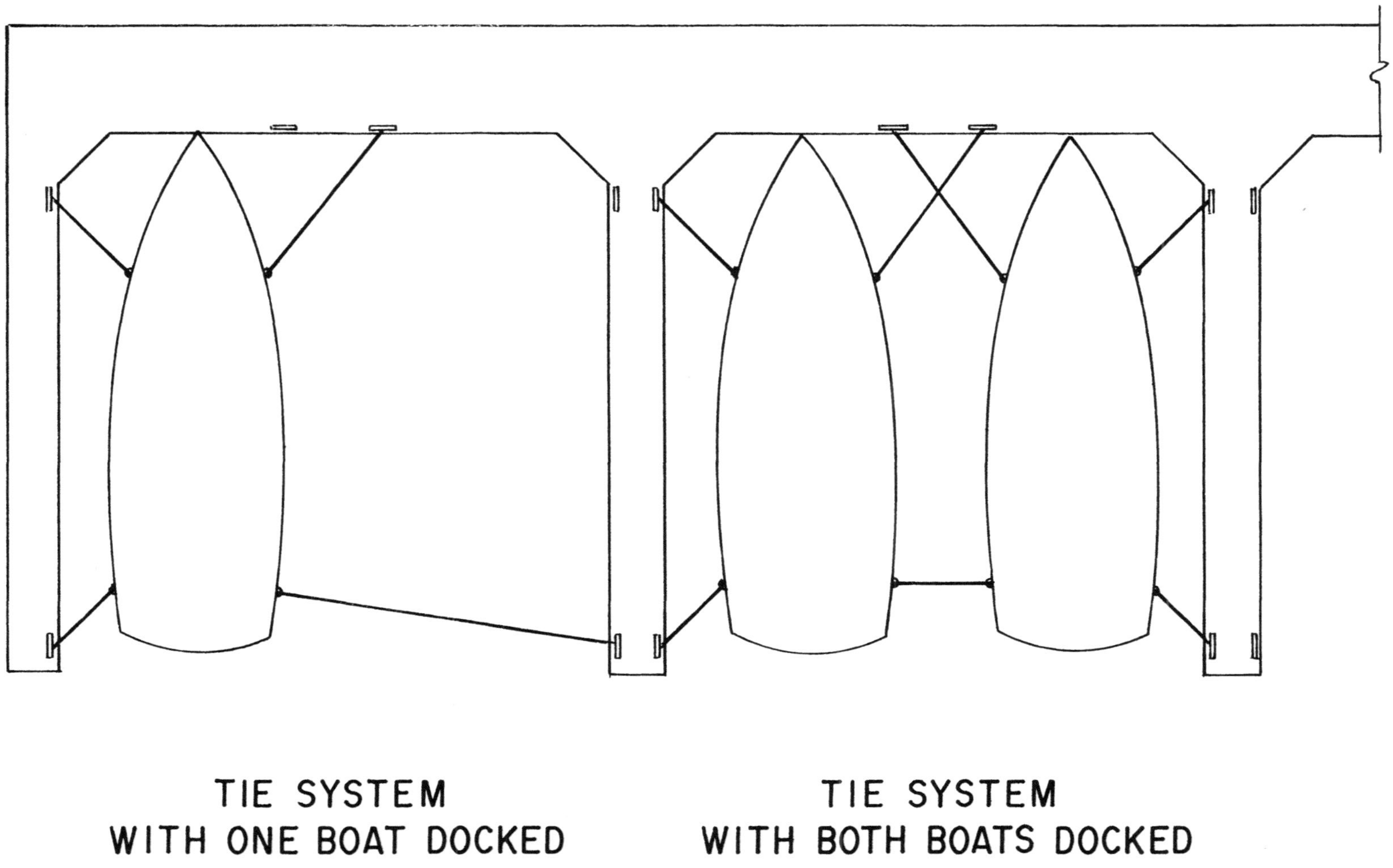

Figure 59. Cooperative switch-tie system for a two-boat slip.

Figure 60. Two-boat slip compartmented by a narrow tie finger

Figure 61. Lightweight sailboats stored on floating docks.

Figure 62. Charter boats tied stern-to-dock, Miamarina, Florida
Note fish cleaning facilities along dock.

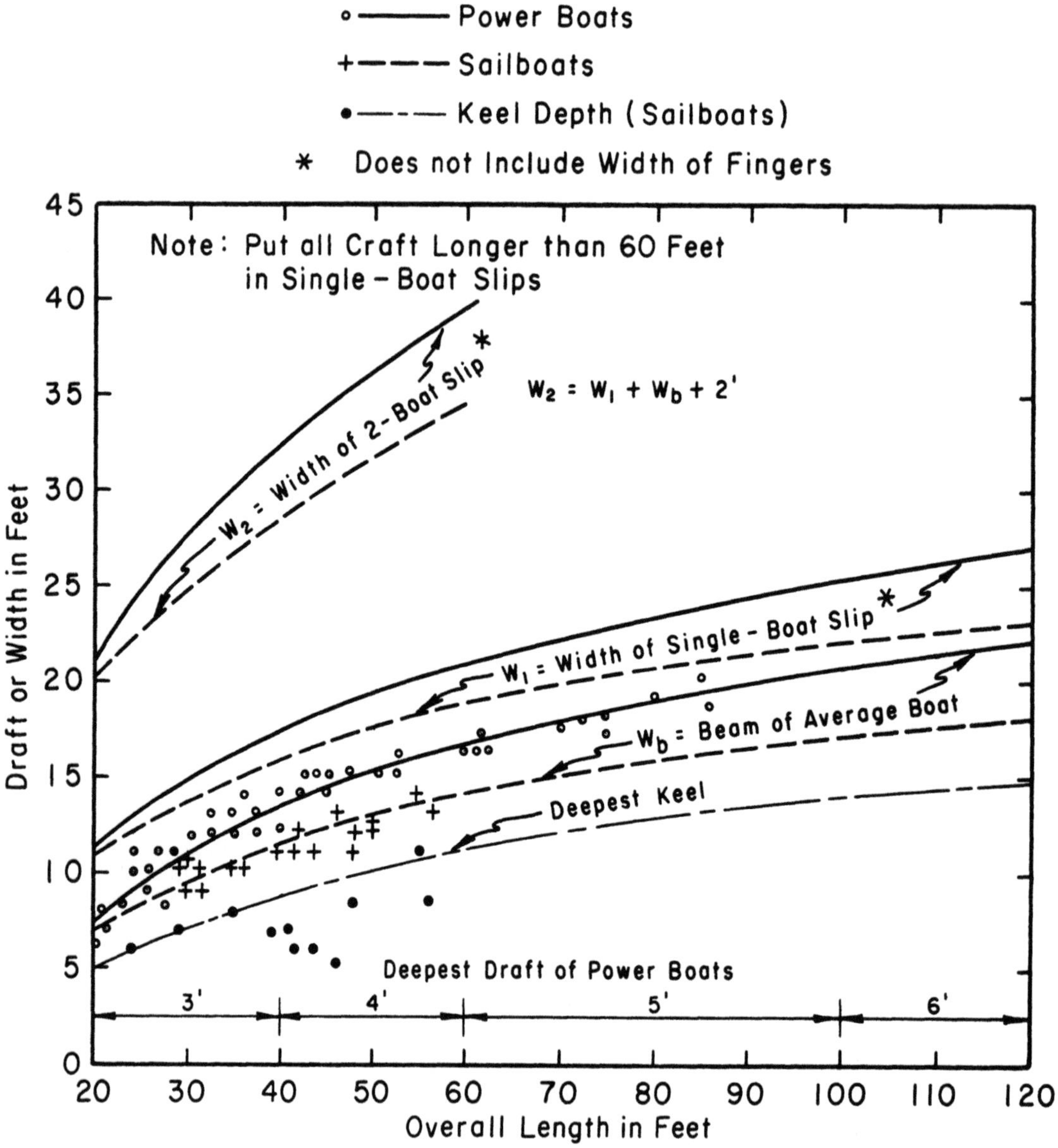

Figure 63. Dimensional criteria for berthed craft.

Figure 64. Tri-maran berthed at West Palm Beach Marina, Florida. Consideration should be given to the berthing problems of extra-wide-beamed craft in future marinas.

Figure 65. Modular star-cluster floating slips (Courtesy of Harbor Host Corporation).

Widths of headwalks and finger piers vary from one region to another, although cities, counties and other local agencies have begun to set minimum limits. The average headwalk width is about 8 feet, with a range of about 5 to 16 feet. The wider headwalks usually have some width for bearing-pile risers, locker boxes, firefighting equipment, and utility lines (Fig. 66). The narrower piers often have all obstructions moved to knees at the junctions of finger piers. Extra-wide headwalks are usually in fixed-pier installations because of the higher cost of floating construction. Long, fixed headwalks can also serve as roadways for service vehicles.

Boarding fingers for single-boat slips are usually about 3 feet wide, normally the minimum allowed for floating construction because of the instability of narrower floats. For this reason, floating fingers longer than 35 feet are usually 4 feet wide. In double-boat slip construction, a finger width of 4 feet is common for all slip lengths. A compromise system of alternating full width and narrow fingers (about 12 inches) is used in some areas. An interesting use of short cantilevered fingers supported only by headwalk piles, and used for boarding (not mooring ties), is shown in Figure 67. A reverse variation of the cantilevered finger is the above-deck-level finger, with its outboard end supported on a pile and the inboard end resting on the headwalk (Fig. 68). A hanging ladder system used for boarding from high level fixed fingers is shown in Figure 69.

e. Fixed-Pier Structures. Structural design criteria for fixed-timber headwalks and fingers are presented in Chaney (1961). All timber used for construction should be treated to avoid damage by dry rot and living organisms. Both Chaney and the American Society of Civil Engineers (1969) present data on timber treatment. The American Wood Preserver's Association (1971), Standard C-1, describes the treatment processes, materials used, and results of preservative treatment for wood products by various preservatives applied by the pressure process. AWPA Standard C-18 extends the coverage of Standard C-1 to include the specific requirements for pressure-treated piles and timbers used in marine construction. These standards are updated periodically to include the latest techniques and materials; hence, only the current editions should be used. Because of possible failure due to undetected weakness in the wood, all deck planking should be of nominal 2-inch thickness and not less than 6 inches in width. Galvanized nails and hardware should be used. Deck nailing should penetrate the supporting timbers at least 3 inches so the nails will not pull up under repeated flexure caused by passing traffic. Creosoted piles that project above deck level should be protected with battens (Fig. 70) or some protective sheathing.

Metal framework berthing structures are generally too costly to fabricate commercially available basic components, but several systems have been developed that use factory-built components for easy field installation. Most are of tubular and pressed-steel construction with either stamped metal or timber plank decks. Bottom conditions in the berthing basin must be checked to determine if the anticipated loading will cause settlement or if the depths in the basin are too great for the system. Most prefabricated systems are for small individual docks along a lakefront or riverbank and are not normally suitable for large installations.

Figure 66. Locker boxes, firefighting equipment and utilities along a wide main walk at Dinner Key Marina, Florida.

Figure 67. Cantilevered boarding fingers at Occoquan Marina, Virginia

Figure 68. Raised-level finger pier junctions with main walk. This could be a stumbling hazard.

Figure 69. Hanging boarding ladders, Old Town Yacht Haven Alexandria, Virginia.

Figure 70. Batten-protected creosoted-pile tops, Lighthouse Point Marina, Pompano Beach, Florida.

In areas where timber is scarce or costly, reinforced concrete construction is frequently used for fixed-level berthing systems. The structural design criteria are similar to timber construction except that connections and fastening devices are different and the dead load to be supported is greater. However, lightweight concretes are sometimes used in stringers and decking to reduce the dead load. If enough concrete cover on the reinforcing steel is not provided, cracking and spalling from rust swells may result. As for bulkhead construction, all concrete in a saltwater environment should have 3 inches of concrete covering all steel reinforcement. This amount of cover may be insufficient if care is not exercised in placing and vibrating the concrete. With a properly designed mix and careful placement, a good dense concrete can be obtained to outlast almost any type of construction.

The lasting quality of concrete was demonstrated in an experiment started in 1905 by the U.S. Army, Corps of Engineers. Eighteen large test blocks of concrete were placed on a shelf just below the waterline on the seaward side of the Los Angeles Harbor breakwater at San Pedro, California. The concrete was varied from block to block both in the mix and type of cement used (several commercial brands were available at the time). Seventeen blocks were recovered, core-drilled and returned to the shelf in 1932; six were again recovered, core-drilled, and replaced in 1972. Cores from the 1932 operation tested from 2,180 to 6,010 pounds per square inch and those from the 1972 operation tested from

3,060 to 6,020 pounds per square inch. Original form marks were still visible and edges were still sharp. Three blocks had gained strength since 1932, two had remained the same, and one had decreased slightly.

Although prestressed concrete construction has been used successfully in many other types of marine structures, it probably has not been used for pier decks in small-craft harbors. One reason is the high cost of special prestressing beds needed for this construction. If two standard width precast deck sections could be agreed upon by the construction industry, one for headwalks and one for fingers, the sections could be turned out in long mass-produced beds at prices competitive with other systems, especially with the use of lightweight aggregates. A sketch of the section and details of possible seating arrangements on pile supports is shown in Figure 71.

One type of construction that uses a combination of concrete and timber in a fixed-pier structure is illustrated in Figure 72. The concrete piles (when properly treated) have a reasonably long life. They can be readily replaced when worn or damaged, either piecemeal or in their entirety. A timber superstructure is desirable because of the ease with which attachments may be made after final construction.

The mainwalks and finger piers of a small-craft harbor generally have no handrails. Public safety has sometimes been an issue, but the prevailing logic seems to be that the danger of a person falling is not sufficiently great to justify the cost and inconvenience of handrails on either fixed or floating dockage systems. Where small children are likely to venture out on the docks unattended, locked gates can be installed at the basin perimeter or on the approach pier leading from the perimeter (Fig. 73). If a fixed pier is too high above the water surface during low water levels to constitute a hazard, a floating system is required. This is important for easier boarding of boats.

f. Floating Pier Structures. Where water levels do not fluctuate more than 2 feet, the berthing docks and slips are almost universally of fixed construction. If the normal range is between 2 and 5 feet, the use of a floating system is optional. Where water levels fluctuate more than 5 feet, a floating system is mandatory. The cost of a floating system is usually greater than a fixed system, but the difficulty in keeping boats properly tied and the inconvenience of boarding or leaving boats during extreme low water often justify the choice of a floating system.

A successful floating dock system has the best possible combination of flotation units and structural system. Numerous floats have been used, and most are described in Chaney (1961), and American Society of Civil Engineers (1969). The most successful are the foams, such as extruded polystyrene (Styrofoam), expanded-pellet polystyrene, and foamed polyurethane. Although foam floats have been used extensively without any surface protection, they attract marine growth and living organism. Large accumulations of aquatic plants on the floats can be unsightly, and may foul propellers. Also, birds and sea animals have been reported to tear out pieces of foam while seeking marine life burrowed on the underside of the foam surface. For these reasons some external protection is now usually

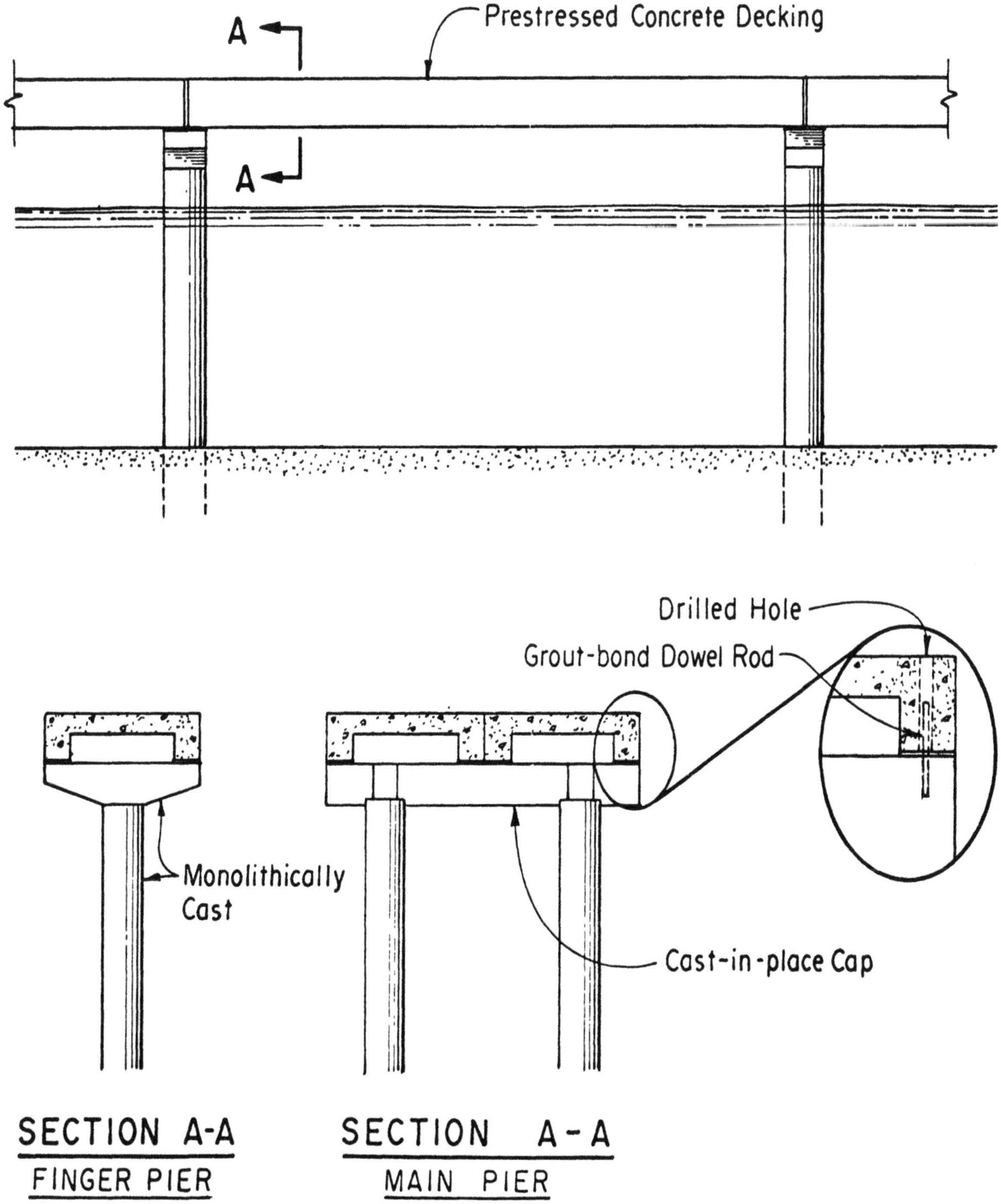

Figure 71. Proposed use of prestressed deck slabs for fixed pier construction.

Figure 72. Concrete-supported timber deck, Dinner Key Marina, Florida Note cantilevered support for locker boxes.

Figure 73. Pier gate, Coyote Point Marina, San Francisco Bay. Note trash collection station in foreground.

applied to all foam floats, especially in seawater. This protection may be a brush of spray coat of polyvinyl-acetate emulsion or dense polyurethane, a fiberglass and resin application, a plaster coating, or concrete encasement of the foam.

Extruded polystyrene is often preferred to the heat-spanded pellet product (beadboard) because it is of uniform quality and is completely impervious to water. However, unless specially treated, it can be damaged by petroleum product spills and must be protected by a hydrocarbon-resistant coating where this is a potential hazard. The expanded-pellet product can also be made hydrocarbon-resistant and is usually less costly than extruded polystyrene. If improperly made, it can be penetrated by water and may lack the desired amount of cohesiveness. To reduce costs, some manufacturers have used a less than minumum amount of material and have overexpanded the beads, producing a friable foam with high void content. Unequal heating during manufacture will also produce a foam with various flaws. Adequate density specification and proper heat-expansion techniques will correct both problems. Because of the widespread use of polystyrene for flotation, the following specifications, recommended by the California Department of Navigation and Ocean Development (State of California, 1971), are quoted for the guidance of prospective purchasers of polystyrene for flotation of structures in marinas:

"(1) Materials: Cellular polystyrene may be formed by the expansion of high density beads or granules in a mold or directly from the base resin by extrusion. The material shall be firm in composition and essentially unicellular. No reprocessed materials shall be used.

(2) Dimensions: Unless otherwise specified, the manufacturers' standard size will be acceptable if incorporated into the design with a minimum of field cutting. The tolerance in each dimension shall be plus 1 inch or minus 0.5 inch.

(3) Color: As normally supplied by the manufacturers for the particular type of polystyrene. Variation in color indicative of damage or deterioration will not be accepted.

(4) Surface Finish: Surface shall be stressed, polished, free from pits, blisters, cracks, dents, waviness, heat marks, or deep scratches.

(5) Odor: The material shall be free from any objectionable odor.

(6) Exterior Coating: In all locations where the waterfront is subject to infestation by marine borers which damage polystyrene, the flotation material shall be protected with an adequate material capable of resisting any anticipated attack by marine organisms.

(7) Physical Properties: Specimens from polystyrene planks shall conform to the requirements stated below:

(a) Density: 1.5 pounds per cubic foot (minimum).

(b) Compressive Strength: 20 pounds per cubic inch minimum at 5 percent deflection.

(c) Tensile Strength: 40 pounds per square inch minimum at break.

(d) Shear Strength: 25 pounds per square inch minimum at break.

(8) Moisture Absorption: The maximum water absorption shall be 0.12 pounds per square foot of skinless or rindless surface when tested by immersion method in accordance with U.S. Department of Defense, Military Specifications Mil-P-40619 (3 April 1962) 4.5.7.

(9) Hydrocarbon Resistance: Polystyrene planks to be used in the vicinity of gas docks or other areas subject to petroleum products floating on water shall be hydrocarbon resistant. The materials shall show no apparent softening or swelling when tested by the immersion method specified in the U.S. Department of Defense, Military Specifications MIL-P-40619 (3 April 1962) 4.5.10.

(10) Shape: Surfaces of the finished planks shall lie in normal planes so that the plank, when installed in final position in the floating dock, shall lie in a true horizontal plane with the water. Edges formed by molding or cut sections may be either rounded or square."

Polyurethane foam is more costly than polystyrene foam, but is sometimes preferred because it can easily be "foamed" into a mold without expensive processing. Also, it is naturally hydrocarbon-resistant. Two types are available, but only one is nonabsorbent—the monocellular variety, which should always be specified. Like polystyrene, it should have protective covering for marina use.

Care must be used in selecting coatings to ensure compatibility with the base foam. Polyester resins cannot be used with polystyrene, but will bond well with polyurethane. Polyvinyl-acetate emulsion and dense polyurethane may be applied directly to polystyrene foam which makes a fairly tough coating. Epoxy glues should be used to bond separate boards of polystyrene foam; epoxy-bonded protective coatings may be used with either foam. If the additional protection of a fiberglass and resin application is desired over polystyrene, an epoxy coating compatible with the resin must be first applied. Where wrap is used, it should be lapped well over the top shoulder of the foam and not carried to the top edge, as the wrap may peel away from the foam (Fig. 74).

The two most common methods used to secure foam floats to deck structure are: (a) hardware dowels driven on a skew into the foam through holes in the bearing boards, and (b) 2-inch nylon strapping. The strapping method is preferred because it is more secure and easily removed if the float has to be replaced or repaired. Bearing boards must have enough bearing surface to prevent crushing the foam under maximum loading. The foam can be loaded to about 5 pounds per square inch; however, it is weak in bending strength, and bearing contact areas should be spaced not more than 2 feet apart, and continuous along each edge.

Most of the hollow shell floats are now being replaced with foam-filled shells. Few shell-type floats are being manufactured without a foam core of some kind. Problems with leakage, internal condensation of moisture, and vandalism (mainly bullet punctures) are the

Figure 74. Fiberglass wrap peeling away from foam float

reasons for this change. Fiberglass shells are still the most common of the thin shell types, but foam fillers are often provided. Tubular metal floats for freshwater use are now nearly all foam-filled. They are particularly serviceable in areas where ice formation and heavy floating debris are problems. Monolithic concrete shell-and-deck units are now almost universally cast around foam cores. Nearly all the metal box floats (Fig. 75) are also foam-filled.

Figure 75. Metal box floating pier system. Note collection of surface debris due to continuity of flotation elements, with no gaps for debris to flow through.

One type of floating construction that appears very promising is the fiberglass or plastic-coated shell with a molded foam core over which a reinforced concrete deck is poured (Fig. 76). The edge beam and the crossbeam and tie rod system in this construction make the units exceptionally strong, if the concrete is properly mixed and placed. Although topheavy, there is no danger for a headwalk with finger piers of this construction to turn over. The extra weight and stiffness of the concrete deck add an element of inertia under pedestrian traffic that makes it approximately equal to the monolithic concrete shell-and-deck float in stability. However, the extra weight of concrete or part-concrete floating units places a severe stress on the unit connecting stringers and on the finger headwalk connections under moderate wave or surge conditions. For this reason, heavy floating systems should be installed only in well protected basins.

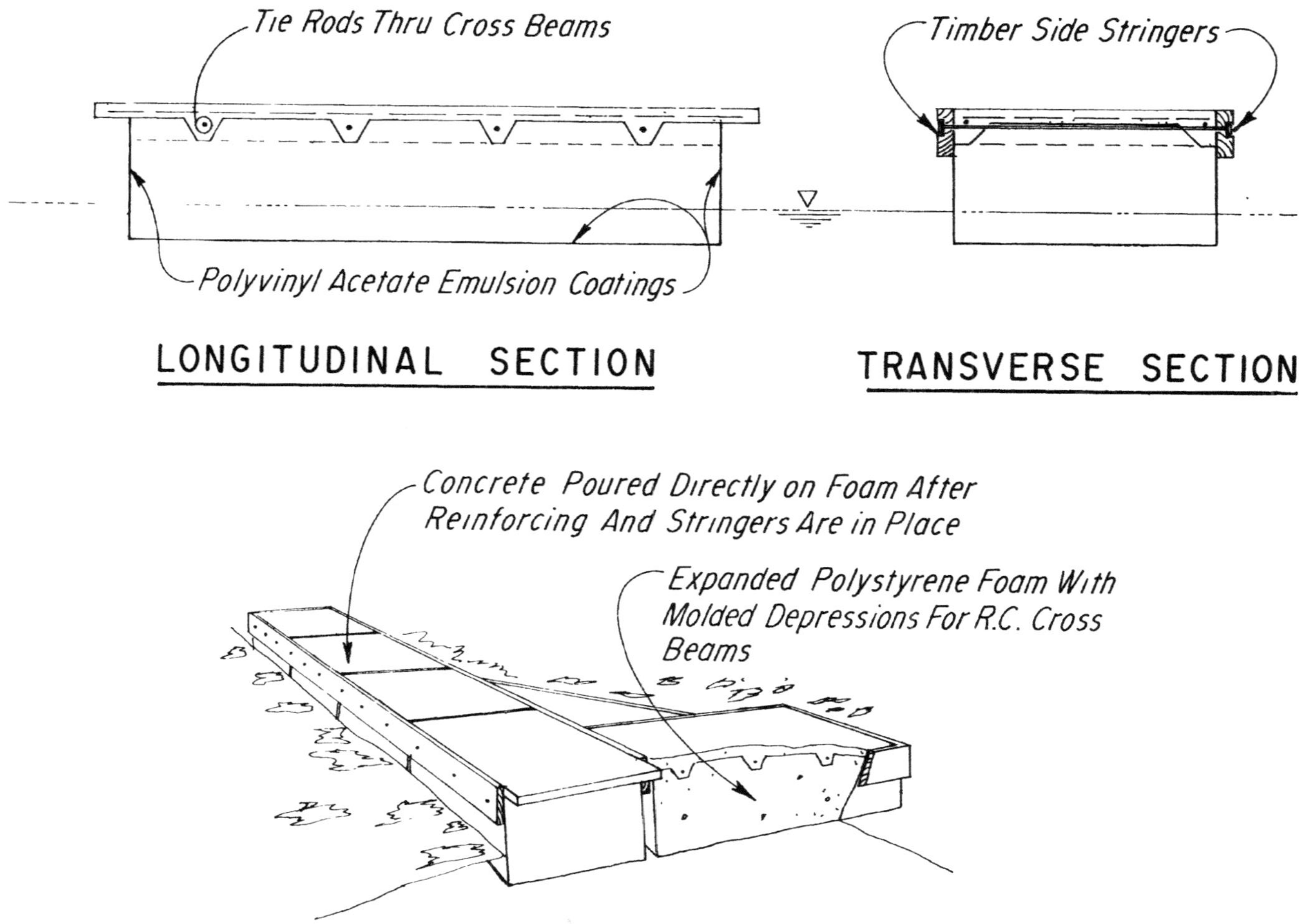

Figure 76. Foam float with concrete deck (courtesy of Floating Dock Systems).

PATENT NO 3,488,709 LICENSED THROUGH HARDWICK ENGINEERING & ASSOCIATES

A difficulty with concrete decks built over foam cores is the occasional formation of a hazardous ice sheet over the decks in cold weather. Hollow concrete floats are not as susceptible to this problem, probably because the trapped air conducted heat from the warmer water to the deck slab and kept it above freezing temperature. The foam acts as an insulator preventing this heat transfer. No low-cost solution has been found for this problem that may be faced in certain regions.

Lightweight floating docks tend to be "bouncy" and, for this reason are often rejected in favor of the heavier types. One thin shelled float deliberately leaves a pocket of unfilled space below the foam core. After launching, these pockets fill with water through small holes punched in the bottom of the shells. The trapped water moves with the float, adding measurably to its inertia without increasing the load on the supporting foam. The result is less bounciness with no increase in the deadweight of the floating components before launching (Fig. 77).

g. *Vertical Loading and Deck Levels.* Minimum deck loading criteria for fixed structures are usually specified by a building and safety agency of the area from whom a construction permit must be obtained. Without specific design requirements, fixed structures built over the water should be designed for a deck loading of not less than 50 pounds per square foot for fingers and 100 pounds per square foot for main walks and building floors. Where vehicles are to be allowed on main walks, the design loading should be increased accordingly. The structural integrity of any floating system requires careful analysis to ensure its capability of supporting the design loads as well as resisting windloads, currents, and impact stresses. Floating slip systems are normally designed with flotation adequate to support the dead load plus a live load of 20 pounds per square foot of deck space. The height at which the deck rides above the water surface under dead loading only is determined to some extent by the sizes and types of boats to be berthed. The height usually ranges from 15 to 20 inches, but the level selected must be constant throughout the system. In this respect, gangways often add considerable extra dead load in the landing area of the floating system, requiring enough additional flotation to maintain this prescribed level. Some agencies specify the allowable deck level limits for dead loading plus a further requirement that the system will not settle more than 8 or 9 inches under full live loading. This effectively places a lower limit on the percentage of the deck area that must have flotation. Figure 78 shows the minimum percentage of deck area requiring flotation for various minimum freeboard requirements with various dead and live loads and the resulting submergence of the floats under the assumptions noted, provided the floats are square cut (vertical sides).

If the design dead load deck level is other than 18 inches, subtract the number of inches the design deck lies above 18 inches from (or add the number of inches the design deck lies below 18 inches to) the required minimum $F_{(D+L)}$ and enter the left side of the graph at that point (Fig. 78). The correct values of P, S_D and $S_{(D+L)}$ will then be given by the procedure shown in the following example problems:

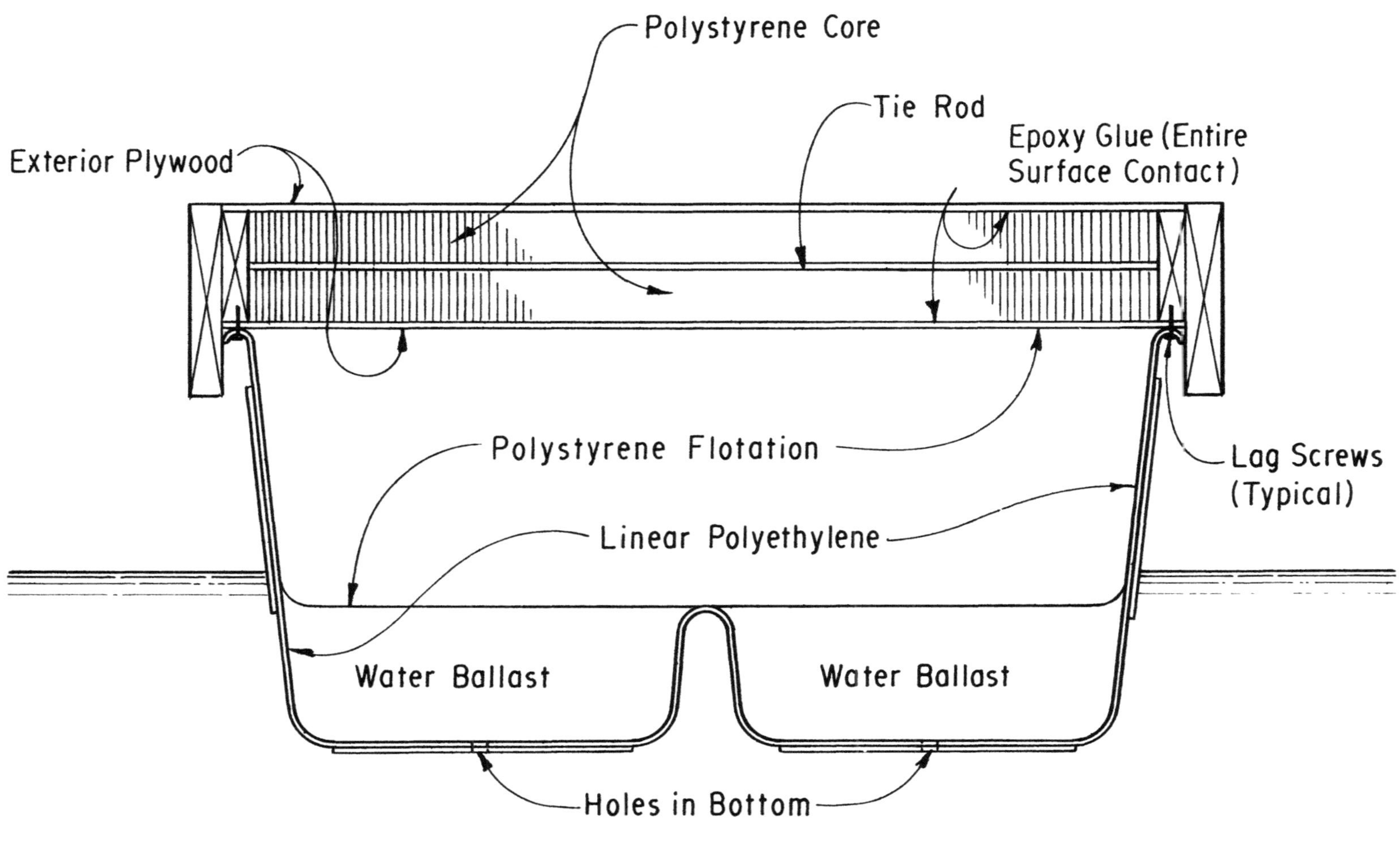

Figure 77. Water-ballasted floating dock (Courtesy of Marina Associates).

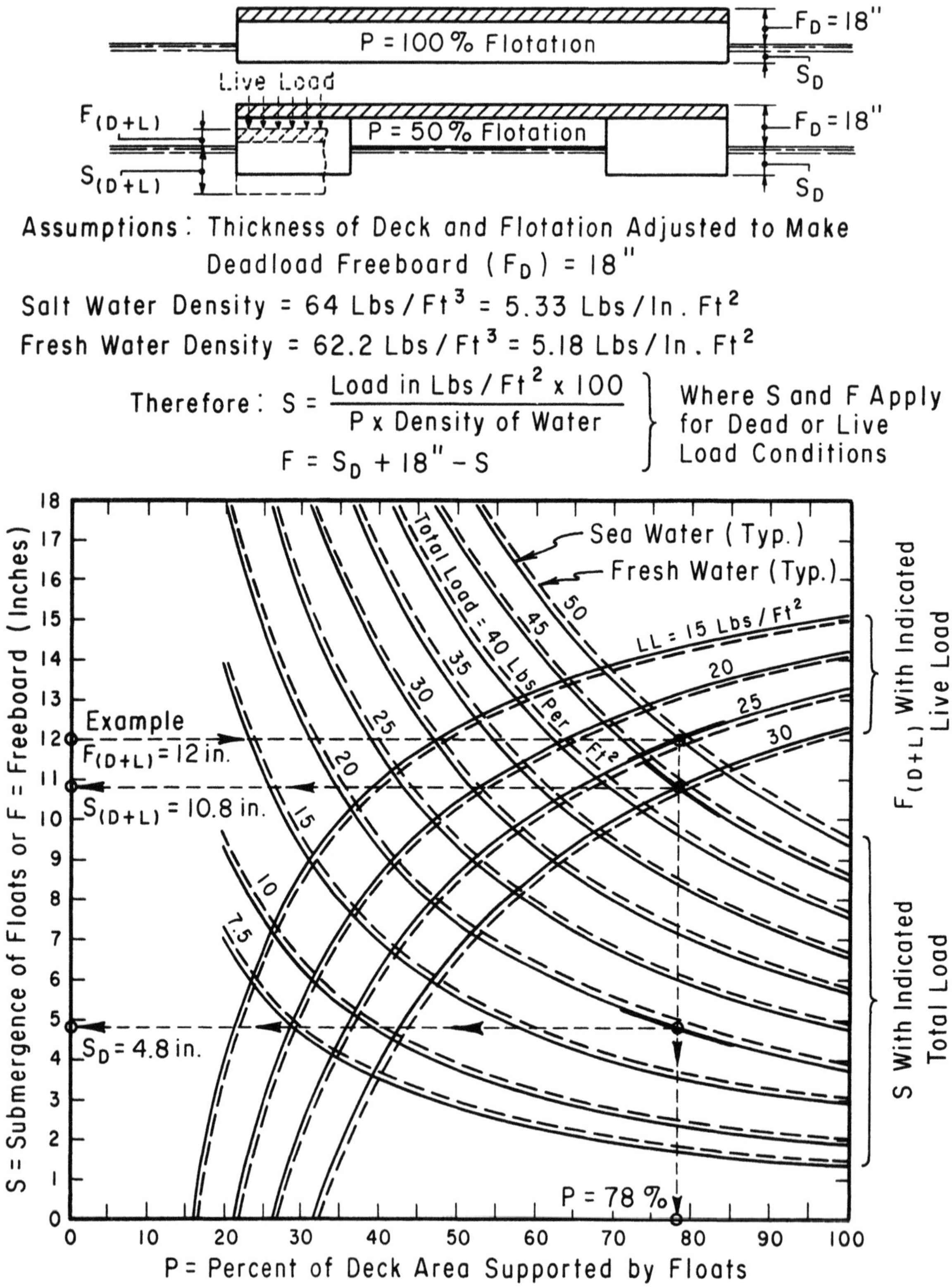

Figure 78. Reduction of freeboard due to concentration of flotation elements for F_D = 18 inches.

* *

EXAMPLE PROBLEMS

Example 1 (for $F_D = 18$ inches)

Given: DL (dead load) = 20 lbs./ft.²

LL (live load) = 25 lbs./ft.²

$F_D = 18$ in.

$F_{(D+L)} = 12$ in. (in seawater)

Find: S_D

$S_{(D+L)}$

P

Solution: Enter left side of Figure 78 at $F_{(D+L)} = 12$ in. and intersect with curve LL = 25 lbs./ft.²

Read P on bottom scale (P = 78 percent)

From intersect with LL = 25 lbs./ft.², move vertically to intersect with Total Load curves to:

DL + LL = 45 lbs./ft.² and read $S_{(D+L)} = 10.8$ in. on left scale.

DL = 20 lbs./ft.² and read $S_D = 4.8$ in. on left scale.

Answers: P = 78 percent; $S_{(D+L)} = 10.8$ in.; $S_D = 4.8$ in.

Example 2 (for $F_D \neq 18$ inches)

Given: DL = 20 lbs./ft.²

LL = 25 lbs./ft.²

$F_D = 20$ in.

$F_{(D+L)} = 12$ in. (in seawater)

Find: S_D

$S_{(D+L)}$

P

Solution: $F_{(D+L)}$ must first be adjusted because dead load deck level $(F_D) \neq 18$ in. (See text)

$F'_{(D+L)} = F_{(D+L)} - [F_D - 18] = 12 \text{ in.} - [20 \text{ in.} - 2 \text{ ft.}]$

$F_{(D+L)} = 10$ in.

Using $F'_{(D+L)}$, find S_D, $S_{(D+L)}$, and P in a similar manner to Example 1.

Answers: P = 58 percent; $S_{(D+L)} = 14.4$ in.; $S_D = 6.4$ in.

* *

If the float sides are not vertical, the depth of submersion of the entire system can be readily computed for any given deck load from a table showing the load supported for each inch of submersion of the float. This table is normally provided by the pontoon manufacturer. Entrapment of floating debris is a major defect in the use of 100 percent flotation (Fig. 75).

A major consideration in stringer or side skirt design for the floating system is to determine at what level the stringer or the attached fender stripping will make contact with the hull of any boat that might be berthed against or near it. Some boats have a low rub strake that may catch under the bottom edge of a stringer if it is too high off the water. The stringer then should be wide enough to extend down to within about 8 inches of the water surface under dead load only. However, some boats have high gunwales topping strongly battered hull sides that allow the deck to protrude over a low level finger pier. For this reason, some low level floating fingers are provided with vertical fender posts that extend upward from each side a few feet above deck level at intervals of about 8 feet (Fig. 79). The level at which the deck floats above the water surface and the method of fendering the slips will be determined largely by the characteristics of the using craft, the local custom, and the desires of the patrons.

Figure 79. Floating slips with vertical fender posts (Courtesy of MEECO Marinas, Inc.).

The fixed-pier approaches and hinged gangways leading from the basin perimeter to a floating dockage are designed to carry more live loading than the specified flotation loading. This occurs from a concentration of loading in an area of the fixed part of the system or on the gangways at one time without the knowledge of the persons causing the overloading. Once on the floating system, however, the same persons will become aware of any overloading because of the canting of a deck surface or the partial submergence of a float. Then they will instinctively move back or spread out to regain a safe load situation. For this same reason the deck system and all interfloat stringers should also be designed for fixed system loading; this is usually 40 to 50 pounds per square foot, but it may be specified otherwise by the controlling agency. A common additional specification for deck loading is that a concentrated load (e.g., 500 pounds) can be placed anywhere on the deck surface without stressing the framing members beyond their design capacity and without tilting the deck more than about 6° from the horizontal.

h. Lateral Loading. The maximum lateral loading of a fixed or floating system is usually produced by strong winds blowing against the structure and berthed craft. Such loading usually exceeds normal docking impact loads or current drag. The design lateral load is normally specified in terms of a given wind speed acting on the above water profile of the system and craft. This loading in pounds per square foot for wind speeds up to 120 miles per hour is shown in Figure 80. This is for steady-state winds neglecting gusts. The higher loads of gust winds are seldom transmitted to a dock because of the inertia of the boats and the flexibility of the tielines and piling. Design winds are often specified by the local building and safety agency; if not, they should be determined by analysis of local wind records.

In analyzing the system for lateral loading it is customary to check two directions: parallel and perpendicular to the main walk. If the anchorage or lateral bracing system is adequate for these directions, it will normally be adequate for all other directions. Some agencies simplify the calculations by specifying that the average profile height for boats in open berths be 15 percent of the slip length. This assumption is conservative, and if not specified, a lower design profile height can be obtained from Figure 81, which shows actual average profiles for most craft in marinas. In calculating the parallel windload on a line of boats, all shielded boats may be assumed to experience only 20 percent of the windloading that is applied to the first (unshielded) boat. In calculating the perpendicular windloading on a system, the total projected area on which the wind acts is obtained by multiplying the average craft profile height by the slip width and that product by the total number of slips; then, add to the result the above water areas of the finger pier ends exposed to the wind. Where slips are provided on both sides of the main walk, the area calculation should include the side that berths the set of boats with the largest average profile height. This value should then be multiplied by 115 percent to account for the wind force on the sheltered or leeward boat row. A sample calculation for lateral loading on a typical open pier system is given in Figure 82.

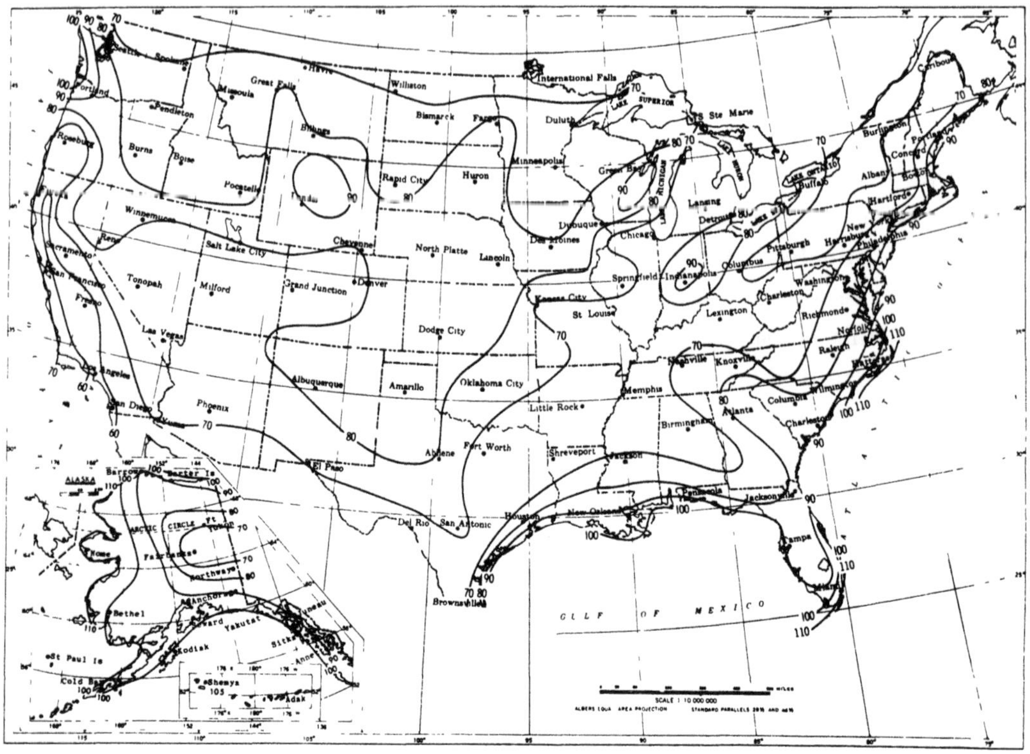

Figure 80-A. Isotachs showing fastest mile of wind (in miles per hour) 30 feet above ground, 50-year period of recurrence (from ASCE Paper 6038, Journal of Structural Division, July 1968).

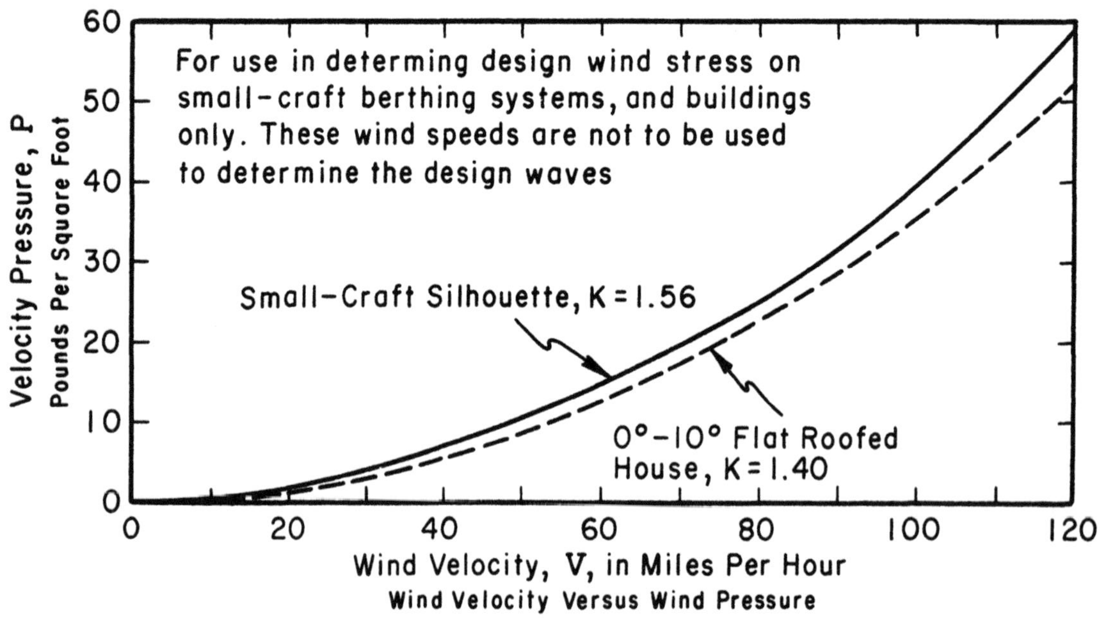

Figure 80-B. Windloading against a vertical face.

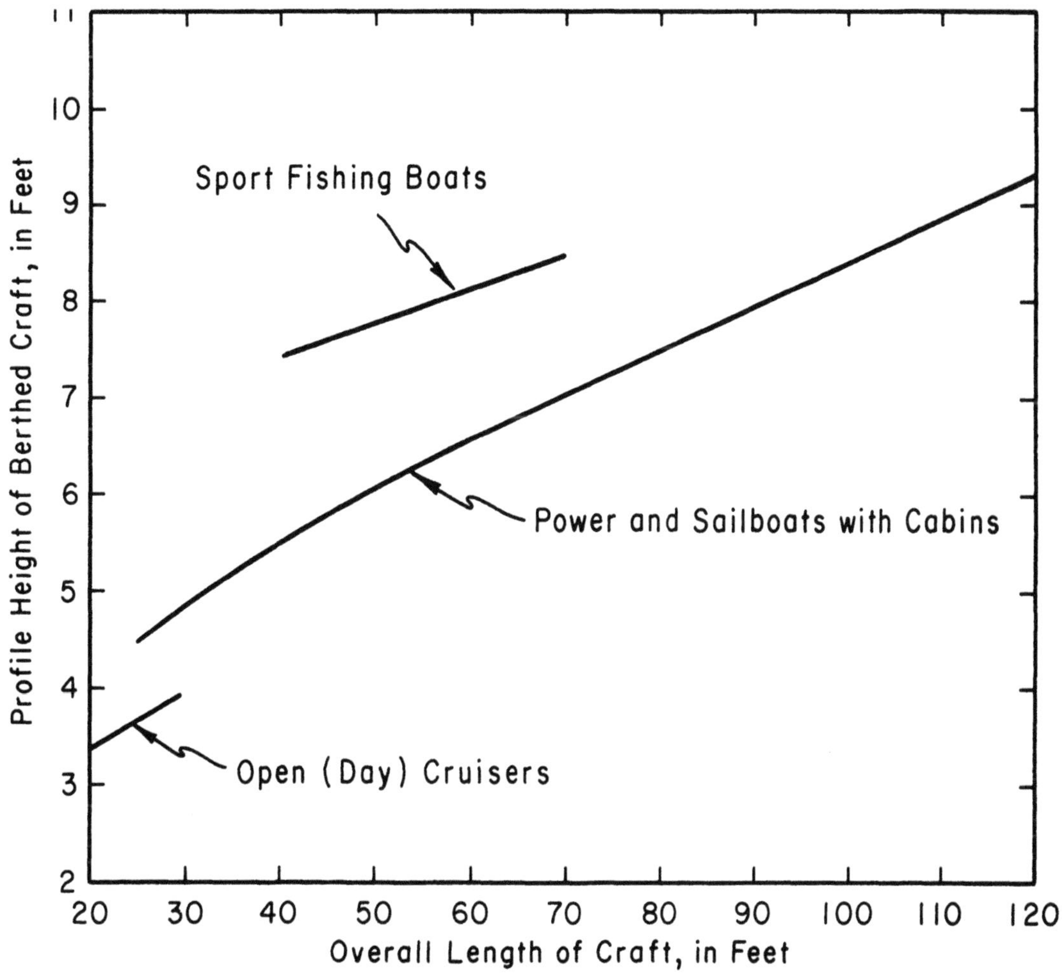

Figure 81. Average profile height versus length of craft.

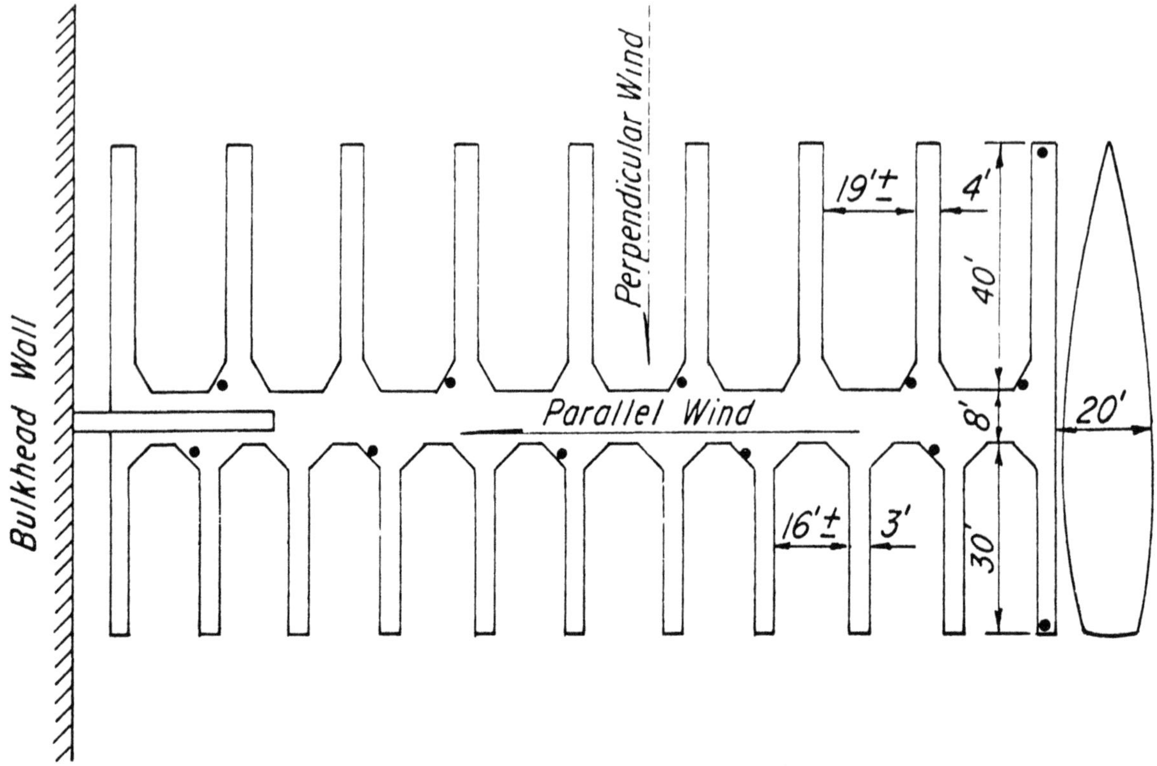

Design Wind Load: 15 Lbs/Sq. Ft.

Perpendicular Wind

40 Ft. Boats: 8 Each x 19 Ft. Beam x 5.5 Ft. Height x 15 Lbs. = 12540
Fingers: 9 Each x 4 Ft. Width x 1.5 Ft. Height x 15 Lbs. = 810
Multiply By 1.15 For Row of 30 Ft. Boats = 1.15 x 13350 = 15352.5
78 Ft. Boat: 20 Ft. Beam x 7.4 Ft. Height x 15 Lbs. = 2220.0
Total Perpendicular Load = 17572.5

Parallel Wind

78 Ft. Boat: 78 Ft. x 7.4 Ft. Height x 15 Lbs = 8658
40 Ft. Boats: 8 Each 40 Ft. x 5.5 Ft. Height x 3 Lbs. = 5280
30 Ft. Boats: 10 Each 30 Ft. x 4.8 Ft. Height x 3 Lbs. = 4320
Total Parallel Load = 18258
Single Pile Capacity = 1530

$\frac{18258}{1530} = 11.933$

Use 12 Piles

Figure 82. Sample calculation of windloading on a floating slip system.

The bearing load and cross bracing requirements for most fixed open pier systems will normally result in adequate structural stability to resist design windloading; but, if there is any doubt, the system should be analyzed. However, floating systems must always be analyzed to determine the number of guide piles required, or with a cable-moored system, to determine the adequacy of the lines and anchors. Any covered system must be analyzed for windloading against the building itself and not the berthed boats. This results in exceptionally high lateral loading in most areas, often requiring dolphins or subsurface cables and anchors rather than an array of single guide piles.

Guide piles are usually the least costly anchorage system. A typical 14-inch diameter prestressed concrete pile with anaylsis for design moment calculation is shown in Figure 83. A sample analysis of the lateral load resistance of the same pile in a typical bottom formation with the load applied at a given distance above the bottom is shown in Figure 84. Where guide piles of other types and materials are used, the allowable bending moment must be computed in accordance with standard structural design procedures. The pile length and resistance to lateral loading may be computed by the method shown for the typical case. This requires an analysis of soil samples taken from the basin substructure to determine the slide angle, the dry density W_d, and the saturated moisture content V. If strata of different properties are penetrated by the piles, a more complicated calculation will be required and the services of a soil-mechanics specialist should be obtained.

The spacing of guide piles in a floating pier system is largely a matter of judgment. It must be assumed that a floating main walk with 40 or 50 feet between adjacent piles is fairly rigid in a horizontal plane and will take bending loads due to wind stresses, exhibiting almost no flexure. Most guide piles will flex in cantilever bending from the bottom allowing windloads to be distributed evenly throughout the pile system, provided the guides do not have more than 3 or 4 inches of play, and the center of resistance is fairly concentric with the center of loading. Each pile guide should be designed to resist the design load of the pile in any direction and to transmit load safety to the structural members of the main walk, either directly, or through fingers. Several commonly used types of pile guides are shown in Figure 85. Guides that do not completely surround the pile may fail to transmit the load in certain directions and thus throw inordinately high loads on other piles.

In marina sites subjected to frequent or prevailing moderate to strong winds or currents, certain guides of a guide-pile system take all the lateral load most of the time and therefore wear out quickly. Some guides have lateral adjustment devices that permit adjustment of the driven pile and will distribute the load equally to all piles in the system. This is seldom the case, however, and it is important that the harder working guides be made easily replaceable or strongly wear-resistant. This may be done by fitting the guide with easily replaceable hardwood blocks or with heavy-duty rollers (Fig. 86).

Pile Penetration   14"φ Prestressed Pile

Soil Properties

Data From Soils Report:

$\theta = 35°$

$WD = 115$ p.c.f.

$V = 35\%$

$W_s = 115 - \left[\frac{100-35}{100}(64)\right] = 115 - 42 = 73^{\#}/Ft.^3$

$P_p = 73 \tan^2(45 + \frac{35}{2}) = 73(3.69) = 269$

$P_A = 73 \tan^2(45 - \frac{35}{2}) = 73(.272) = 19$

$P_p - P_A = 250^{\#}/Ft.$

For 14"φ Pile 117(250) $= 292^{\#}/Ft$

Penetration: Reference Anderson, Paul "Substructure Analysis And Design" 2nd Edition, 1956, Ronald Press, New York - Ref: p.47, Eq (2-4):

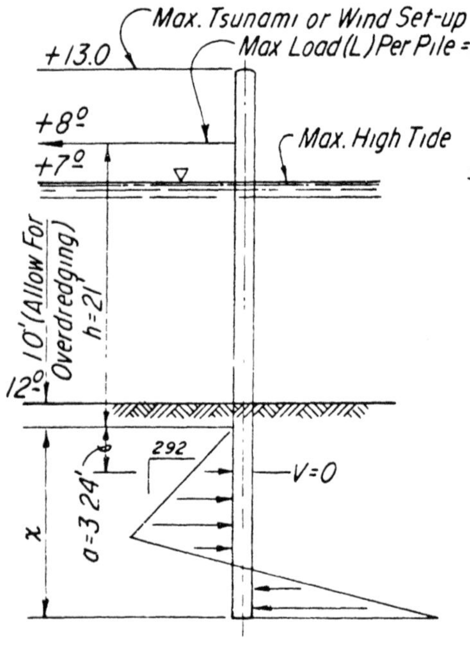

$$x^4 - \frac{8L}{P_p - P_A}x^2 - \frac{12Lh}{P_p - P_A}x - \left[\frac{2L}{P_p - P_A}\right]^2 = 0$$

$$x^4 - \frac{8(1530)}{292}x^2 - \frac{12(1530)(21)}{292}x - \left[\frac{2(1530)}{292}\right]^2 = 0$$

x	x^4	-4.18x	-1320x	-110	=0
Use → 12	+20700	-6020	-15830	-110	-1260
13	+28550	-7070	-17150	-110	+4220

Total Length 12
21
5
38 ← Length

Check Moment:

Find Where V=0: $a = \sqrt{\frac{2(1530)}{292}} = 3.24'$

$M = 1.53(21 + 3.24) - \frac{1}{6}(292)(3.24)^3$
$= 37.1 - 1.7$
$= 35.4^{'K} = 425^{''K} = 413^{''K}$
(3% Over)
Say O.K.

Figure 83. Determination of required guide pile penetration

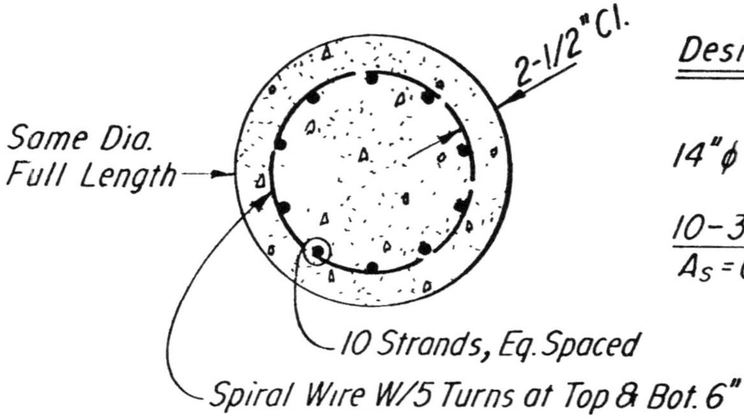

PRESTRESSED CONCRETE PILE

(14" DIAMETER)

Design Moment Calculation

14" ϕ Prestressed $\quad I_c = \dfrac{\pi r^4}{4} = \dfrac{\pi (7)^4}{4} = 1{,}890\ in.^4$

10–3/8" o Strands $\quad A_c = \pi r^2 = \pi (7)^2 = 154\ in.^2$

$A_s = 0.08\ in^2/Strand$

$$\text{Prestress} = \frac{10 \times 11\,200}{154} = 727\ psi$$

$$\text{Modulus of Rupture} = f_{ct} = 600\ psi$$

$$\text{Mom. at Cracking} = \frac{(.727 \times .600)(1890)}{7} = 358''^{K}$$

$$\text{Ult. Mom.} = 0.37' \times 10 \times .08 \times 250 \times 14 = 1034^{K}$$

$$S.F. = \frac{1034}{358} = 2.9 \text{ Higher Than Nec. For Wind Load}$$

$$\text{Use } S.F. = 2.5 \text{ For Bending Only}$$

$$\text{Design Mom.} = \frac{1034}{2.5} = 413^{K}$$

Figure 84. Determination of allowable bending moment for a prestressed concrete pile.

Figure 85. Examples of pile guides

Figure 86. Pile guide with heavy duty rollers. This guide is needed only for "hard-working" piles.

In most installations pile handling and driving equipment are so large as to preclude placement of the piles after the floating system is installed. It is then necessary to drive each pile precisely in its predetermined location by means of shore control and then move the floating system into place and build the guides around the piles. This requires careful planning and may necessitate the hookup of some floating elements after the main walk is in place. A good location for a guide pile is in a knee at the junction of the main walk and a finger. Usually the two end fingers that form the T at the end of the main pier may make an ideal location for breast-docking large cruisers of twice the length of the slips. For this reason, end fingers are usually wider than the others and secured with guide piles at their ends. Where end-of-pier breast-docking is permitted as a permanent feature of the system, a large vessel must be considered in windloading calculations.

As previously stated, finger piers longer than about 35 feet are normally wider than the shorter fingers intended for boarding access and other reasons. In a floating system, the width of the finger is often controlled by the need to withstand a large bending moment caused by lateral windloads wherever they are applied in a horizontal plane. The full design windload against a single slip-length boat is assumed to apply at the center of the finger, or half the load is assumed to apply at the outer end of the finger, whichever creates the greatest structural design problem. Either type of loading to about 40 feet can be resisted by building adequate strength into the stringers and main walk attachments to accommodate the cantilever bending forces. It is usually necessary to place a pile at the end of each finger for longer slips. This procedure will require more guide piles to resist the windloading on the whole system. However, the internal strength of each finger and connections to the main walk must still be considered in the structural analysis.

Some floating systems are being marketed that rely for lateral support on pipe piles of rather small diameter driven through pipe sleeves after the system is in place (Fig. 87). The

Figure 87. Close fitting pipe-sleeve pile guide. Piles are driven through sleeves after floating system is in place.

analysis of such systems is made complex by the flexibility of both the pipe and the floating components. The resistance of the bottom materials to cantilever bending of the piles from bottom is usually small, and it is necessary to consider the probability that a reverse cantilever action occurs in which the pile may be considered fixed within the sleeve. A simple reaction force must then resist the pile tip at some point below the mud line. The sleeve and attachments to the floating framework, and the framework itself must all be adequate for cantilever action in the pile. A sample calculation of the forces involved in this type of guide pile is shown in Figure 88.

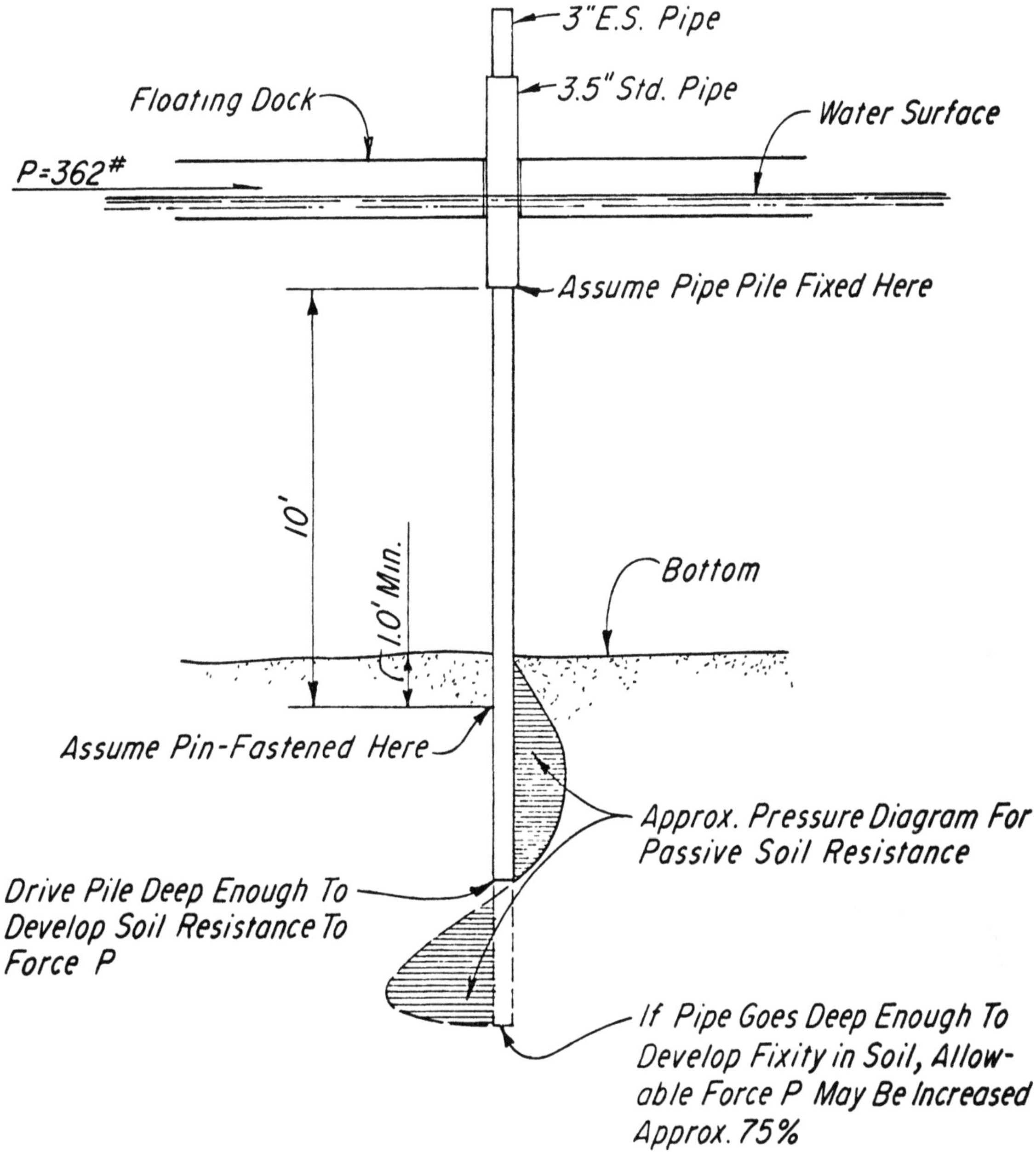

For 3" E.S. Pipe, $S = 2.23\ In.^3$

Allowable Moment, $M = fs = 20{,}000 \times 2.23 =$

$44{,}600\ In.Lbs = 3620\ Ft.Lbs.$

Allowable Force, $P = \frac{M}{L} = 362\ Lbs.$

Figure 88. Determination of allowable loading of sleeve-guided pipe pile.

The anchorage of a covered floating system is usually required to be much stronger than that of an open system. In sheltered sites an ordinary guide-pile system may be adequate if the number of piles required is not prohibitive. It is more common to use some other anchorage system. Dolphins that develop their strength on the A-frame principle rather than through the combined cantilever bending strength of their component pile members are often in this situation (Fig. 89). Considerable internal strength can be built into the

Figure 89. A-frame dolphins guide a floating pier at Shilshole Bay Marina, Washington.

structural framework of the cover so that points of lateral support may be widely separated. A single dolphin may develop 10 to 20 times the strength of one guide pile, and few may be needed. If the water is deep enough, submerged structural ties across finger ends well below the keel depth of the berthed craft are sometimes used to strengthen a covered floating system (Fig. 90).

Since a large amount of flotation is required for covered slips and high windloading must be resisted by the anchorage system, a separate berthing system and cover may be desirable. At sites where the bottom is not too deep and the water level does not fluctuate more than a few feet, the cover may be a pole-supported shed. Long poles driven into the bottom support the roof and sides of the sheltering structure. The floating slips are then placed within the covered area and the supporting piles of the shed are spaced so that they can be

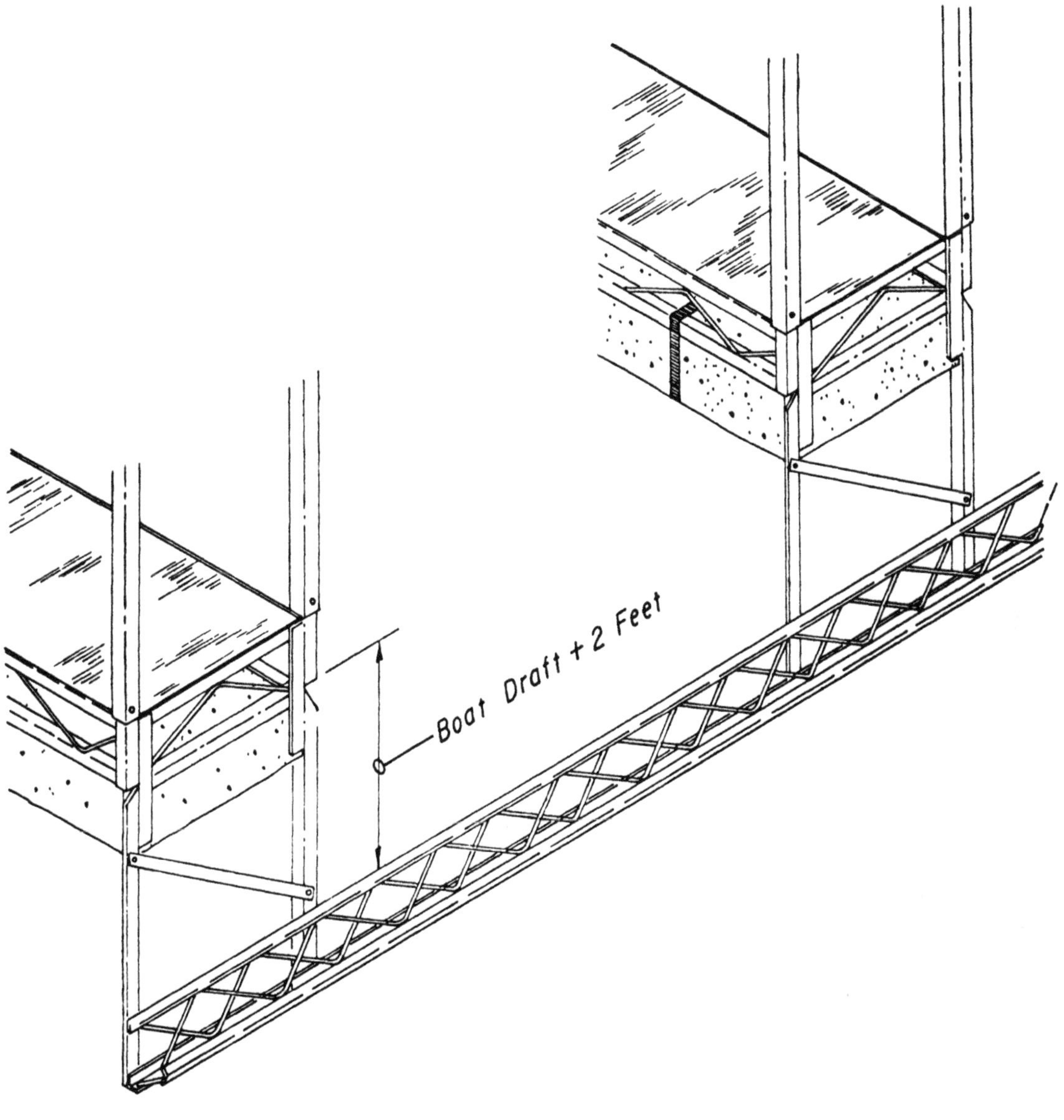

Figure 90. Submerged crossties strengthen floating slip system.

used as finger-end guide piles. The eaves of the roof must be high enough to clear the superstructures and masts of all the berthed craft at the highest water level.

At some sites, the water may be too deep or the water level may fluctuate with too great an amplitude to permit guide-pile anchorage. A cable and bottom anchor to moor the floating system is then necessary. Such a system is usually more costly than a guide pile or dolphin system. To maintain proper distance from the shore, horizontal displacements of the entire installation will occur as the water level changes. Chaney (1961) contains a description of one such system. Another scheme that may be used where the bottom is more rugged, as in mountain reservoirs, is shown in Figure 91. Where this anchorage system can be installed in the dry before the reservoir is flooded, the work will be greatly simplified. Anchors that can be installed in deep water include massive concrete weights, large boulders with drilled-and-grouted eyebolts, pulled-in ship anchors, and shot-in anchors.

Where the surface level fluctuation amplitude is small, a system of long lateral lines with sinkers or counterweights may provide for automatic adjustment to meet the specific requirements of lateral positioning at different levels. In cases where adjustment is done by winches, it is important to devise a system that is simple to operate and more foolproof. Systems that require several adjustments to be made concurrently should be avoided. To devise a system that requires only one winch is usually impossible, but with ingenuity a two-winch system can usually be designed for all the necessary adjustments. The two-winch system should be planned by a rigging expert. Steel cable manufacturers often have such personnel or can recommend a qualified rigger.

A floating dock system is common along the bank of a river or canal where a unidirectional current prevails all or most of the time. Several river anchorages are possible, but all are based on the trailing slip principle shown in Figure 7. The differences are usually based on the varying methods for anchoring the system against the pull of the current. The American Society of Civil Engineers (1969) shows a river anchorage system in which a dolphin is used for one of the anchors. Each boat is moved into a berth against the current from a downstream access. When leaving, the mooring lines are unfastened and the boat is guided carefully by hand to clear all moored boats and other obstructions until it drifts out of the berthing area.

In locations where the bottom drops steeply from the bank, or where the bottom should remain free of piles, anchors, and cables, a floating system may be held away from the bank by double-hinged struts. Struts are generally used only for small docks and slips because of the large stresses that can be transferred to the struts and shore anchorages by small lateral loads on the floating system. If the gangway can be attached to the bank at the same level as that of the strut system, it may substitute for one of the struts. The hinges must be larger than the ordinary gangway hinge to resist the higher tensile and compressive stresses to which the struts are subjected. Because of the limited use of these systems they are not discussed in detail, but included as a possible solution to a specific problem.

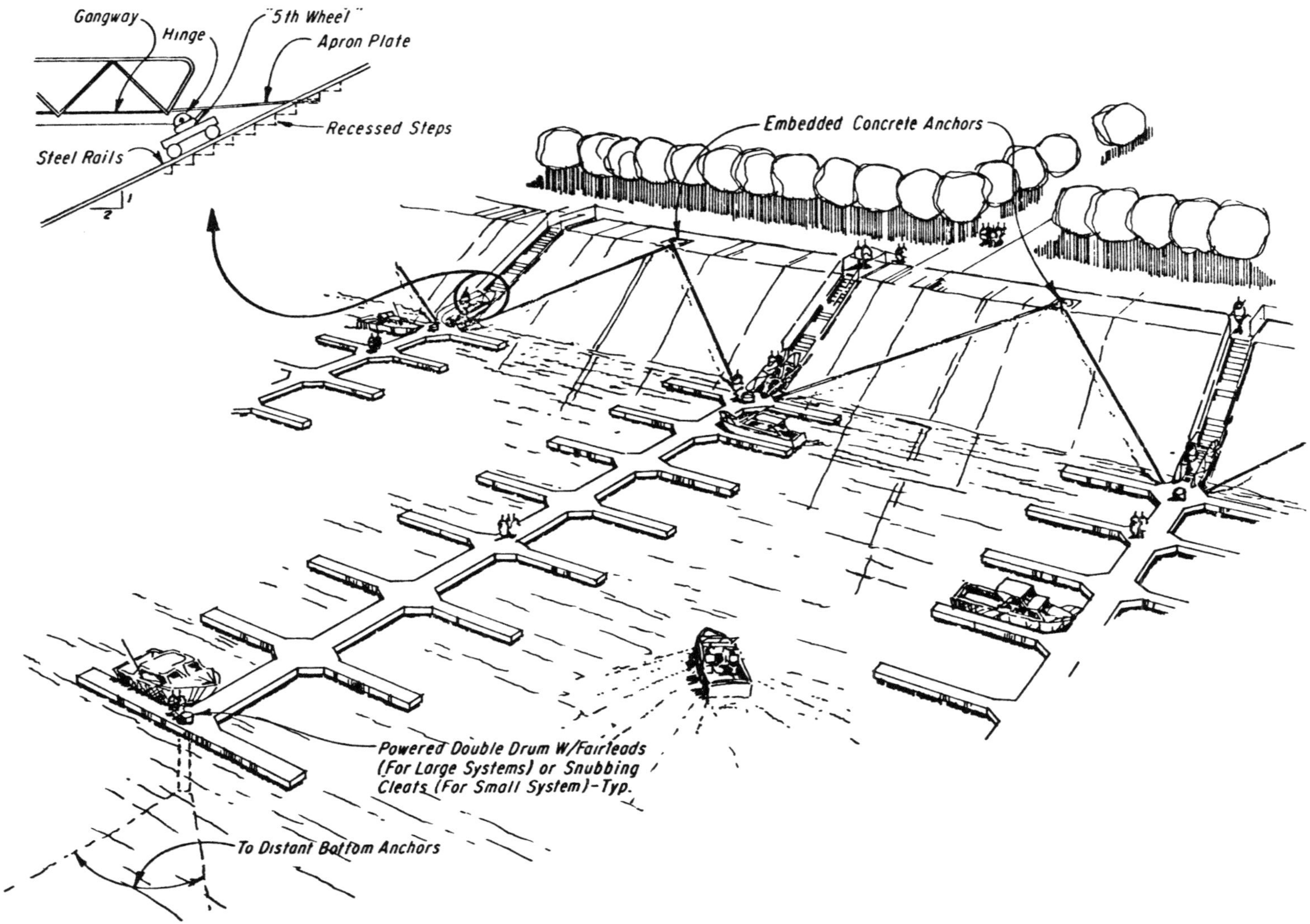

Figure 91. Deepwater anchorage of floating pier with drawdown adjustment. Schematic is but one method of drawdown adjustable anchorage; each situation requires special engineering and detailing.

The forces imposed by windloading on a fixed berthing facility can be resisted easily by almost any adequately designed structural system. It is only necessary to check the resistance of the array of supporting bearing piles to determine if they can take the lateral loading in cantilever bending or if cross bracing is needed.

i. Interconnection of Floating Components. A floating system requires considerable analysis of the internal forces imposed by the various types of loading to which it may be subjected. As has been pointed out, each floating finger must be considered separately, and the design lateral load in many locations will be great enough to require considerable strengthening of the commonly used stringer and crosstie deck systems and the connections of fingers to headwalks.

Pull tests made at Marina del Rey, California, in the early 1960's on certain proprietary flotation systems that had been accepted without structural check elsewhere revealed weaknesses that required major modification of structural systems. One problem involved buckling of fingers due to the eccentricity of shear-stressed members with compression and tension members under lateral loading. Corrective action included increasing the size of the stringers to resist torsional forces and the addition of a system of cross struts and tie rods. Some of the clip angles attaching fingers to headwalks failed in internal bending and had to be strengthened with gusset plates. Other clip angles held but caused failure of the headwalk stringers to which they were attached. Some bolted splice-joints failed for lack of enough bolts or insufficient end distance in the joining members. An important lesson learned was that measurable strengthening can be effected in any flotation system tied together with timber stringers by the addition of cross struts and tie rods under the deck at frequent intervals (Fig. 92).

The wave-produced vertical deflections of any floating system can be dealt with in either of two ways. One is to join fairly short, rigid components with hinged connections. Unless the hinges are quite massive and kept well greased, this procedure usually results in rapid wear and ultimate failure of the hinges by a peening action, which progressively elongates the holes of the hinged connections. This peening action caused by repetitious reversal of stress becomes noisy as the situation worsens and may be a source of irritation to people in the harbor. The other method is to join all components semirigidly by continuous headwalk stringers and bolted finger connections; a modification is to join the fingers to the main walk by extending the finger stringers under the main walk, joining its components in a crosslocked system (Fig. 93).

The semirigidity of any extensive floating system without hinges sets up some rather severe vertical bending stresses in the connecting stringers when waves pass under the structure, especially in heavy systems such as those with concrete floats or decks. In a basin of 10-foot depth, e.g., a typical local wind wave of, say, 5-second period, will have a crest-to-crest length of about 80 feet. If this wave is 2 feet high (about the maximum allowable for any floating system) and approaches along the axis of a headwalk, the

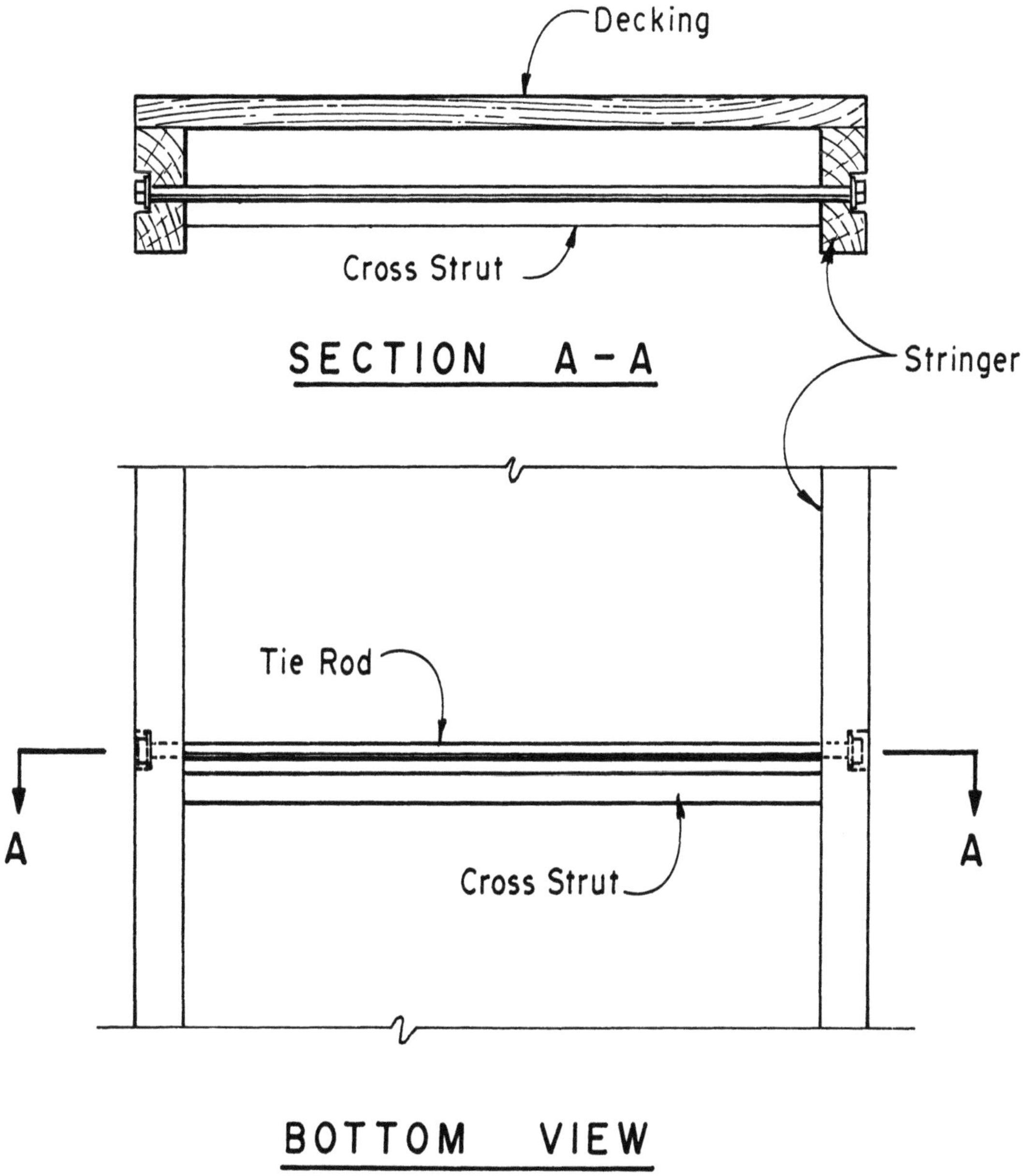

Figure 92. Crossties and struts used to strengthen floating pier decks.

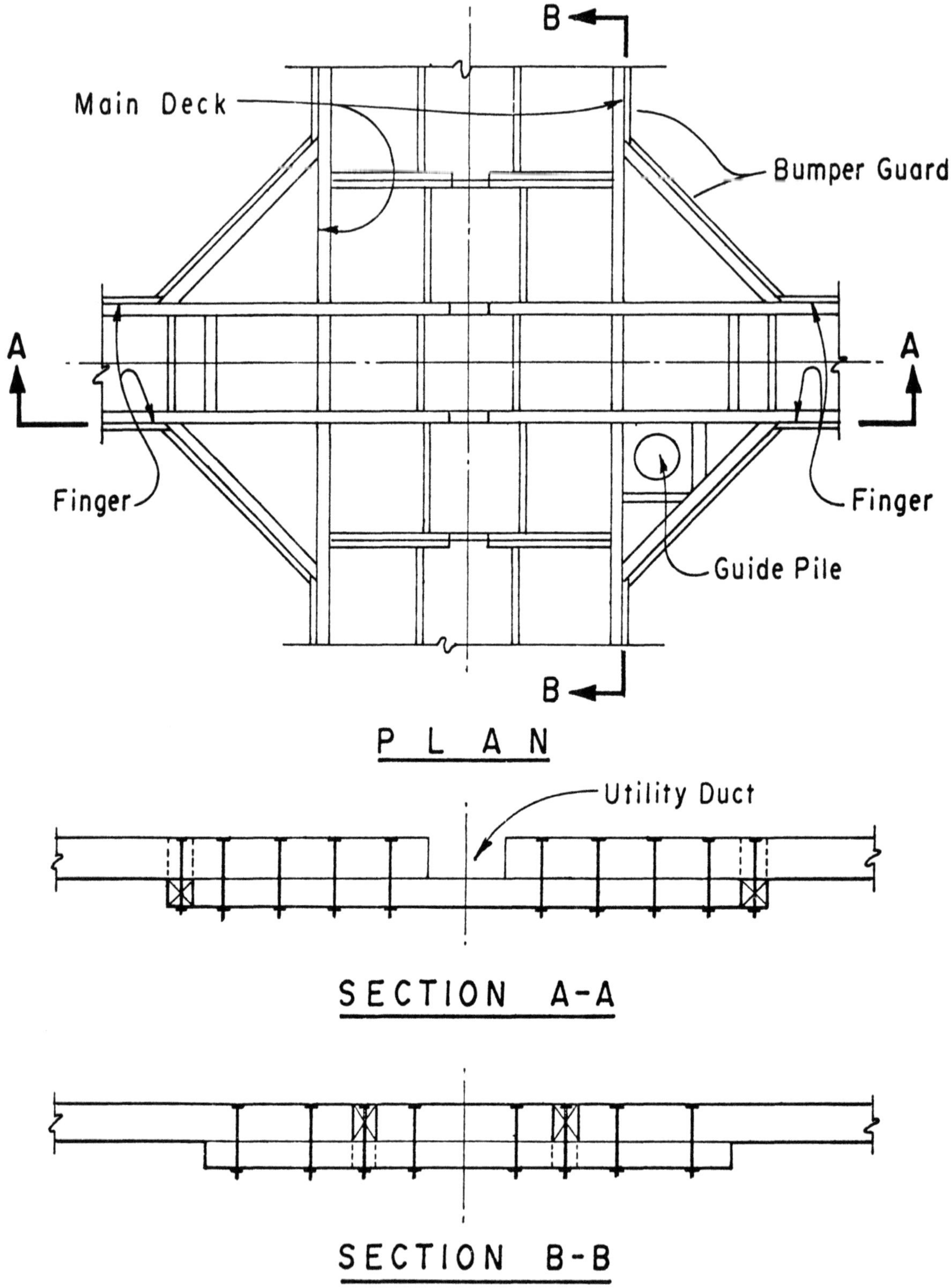

Figure 93. Crosslocked connection of finger and headwalk stringers.

headwalk deck and supporting floats will conform more or less to the wave profile, as the ordinary stringer system will not support even the dead load in bending with supporting points 80 feet apart. The floats in the crest area may be depressed a few inches below, and those in the trough area raised a few inches above their free-floating position, but the resultant deflection will still be about 1.5 feet in 45 feet. Either the stringers must flex this amount or their splice-joints must loosen and allow a hinging action that will permit this deflection. Once this type of hinging begins, the structural integrity of the system starts to deteriorate, and major repairs may be necessary if the loose joints are not tightened immediately.

With proper splice-jointing, timber stringers will flex enough between joints and float attachments to accommodate all reasonable wave action in a berthing basin for an indefinite period. No evidence of pure fatigue failure in a timber stringer system has yet been reported. Most damage from interior wave action has resulted either from loose bolts in the stringer splices or from loose connections between the floating elements and the stringers. Much of this damage could have been prevented by the use of larger bolts, larger washers, and better positioning of splice points.

In many floating berth marinas the fingers are hinged to the headwalks because of the above-mentioned differential movements caused by wave action. Where necessary, the hinges should be of rugged construction with fairly large pins. Special cast-steel hinges used for this purpose in the Shilshole Bay Marina, State of Washington, have performed effectively, even with the heavy concrete floating system used and the proximity of the basin to occasional high waves coming from Puget Sound to the area just outside the two entrances. In most cases, however, hinging of finger connections is unnecessary if strong, rigid junctions are used.

In some proprietary steel-frame systems the floating components are interconnected entirely by bolting. They are intended only for freshwater lakes and rivers, as the galvanizing or other coatings will not provide adequate protection in a saltwater environment. The quieter waters of most freshwater locations will not overstress these systems, but where higher waves from a large lake may penetrate into the berthing area, these systems should be checked for structural adequacy under wave action of the type previously discussed.

j. Deck Materials and Surfacing. Many wood-plank decks without a coating have given many years of satisfactory life without maintenance. In most larger installations, however, some coating is provided for improved appearance and to minimize splintering. These coatings usually have a roughened surface texture to give a nonskid quality. Because of the shrinkage and expansion of wood planking and the tendency to curl as the planking loosens and gains moisture, the planks should not be more than about 10 inches wide and spaced about a quarter of an inch apart. Diagonal planking is sometimes used for floating docks to provide cross-bracing strength (Fig. 94).

Figure 94. Diagonal decking strengthens a floating slip structure

All commercial grades of exterior plywood are now made with waterproof glues of excellent quality and may safely be used for exposed decking. A slight crowning of the plywood is sometimes induced by crown-trimming the framework to which it is fastened (usually by lag-bolting) to guard against a *bird-bath* effect in wet weather. Some plywoods have a special plastic surface that is waterproof and provides an excellent base for painting. Some floating dock manufacturers provide a heat- and pressure-bonded silicone-base synthetic coating with an embossed pattern that provides excellent nonskid and long-wearing quality (Fig. 95).

The laminated plank described in the American Society of Civil Engineers (1969) is another type of timber deck that has many favorable qualities, including a pleasing appearance and a high degree of stiffness. It is usually made of nominal 2- by 3-inch or 2- by 4-inch cedar sticks glued together side-by-side in a special endless-belt press that moves the plank slowly through a curing oven. As the plank emerges from the press it passes through a planner that finishes the sides and edges.

The sandwich deck has a polystyrene core glued to plywood top and bottom faces and edged with 2-inch lumber to achieve a degree of stiffness equal to or exceeding that of the laminated deck. It is lighter in weight than most decking, and by increasing or decreasing the thickness, the height of the deck above the water surface can be adjusted as desired. Cost of the sandwich deck compares favorably with the laminated deck.

Concrete slab decking provides both stiffness and durability, but unless the quality of the mix is carefully controlled, it may crack or spall shortly after installation, and repair is difficult. Before engaging the services of any concrete deck manufacturer, examples of similar installations should be examined to judge the quality of the product.

Metal decks of various types are available in modular units for use in a freshwater environment. Modern coating techniques have been used for good weathering and wearing quality. However, all parts must be factory built, and field modification is difficult if they must be tailored to meet special dimensional requirements. Once bent, or pulled apart at their connections, repairs may be difficult.

All fiberglass deck panels and other synthetic decks are available with certain flotation systems now on the market. Most of these panels have little resistance to torsional bending forces and must rely on the framework for stiffness and prevention of damage by racking action in rough water. Some panels have excellent wearing qualities. but are rather brittle and may break or chip if twisted or if a heavy object is dropped on them. Each must be analyzed for compatibility with the flotation system and the physical environment of the site.

k. Approach Piers and Gangways. Access to the docks and slips from the basin perimeter is usually accomplished by a fixed-pier approach to a fixed-berthing system and a hinged gangway to a floating system. The fixed-pier approach is usually an extension of the

Figure 95. Deck panels with nonskid surface

headwalk to a landing on the bulkhead wall or an abutment at the shoulder of a sloping bank. In a floating system, one end of the gangway must be supported on floats; hence it should be lightweight and only long enough to result in some predetermined maximum slope at extreme low water level. The slope is usually specified by the controlling agency; a slope of 1 on 3 is about the maximum that can be traversed safely without the use of cross cleats. If a steeper slope must be used, a hinged staircase with self-leveling steps should be provided.

In the interest of safety, all gangways should have handrails designed to the same standards as those on the perimeter walls. Gangways are normally narrower than fixed-pier approaches because of the need to save weight. A 3-foot clear width between handrails is usually the minimum, and 4 to 5 feet may be specified if the pier served has many berths or if the gangway traffic is expected to be heavy. Gangways are normally hinged at the top inside edge of a perimeter bulkhead wall. If the perimeter is a sloping bank, a short fixed-approach pier is usually required to reduce the length and weight of the gangway.

The hinge at the upper end of the gangway should be of fairly rugged design. An example of a gangway that has been standardized for Marina Del Rey, California, is shown in Figure 96. If the gap between the end of the gangway and perimeter deck is greater than

Figure 96. Rugged gangway "pin-in cradle" hinge. Note loose fit to accommodate lateral movement of gangway.

1 inch at any level of the water surface, a cover plate should be provided. The gangway may be of any design that meets the structural requirements. Manufactured gangways that meet most design criteria are available in sizes for most marina requirements (Fig. 97). Before gangways are designed for an installation, cost data should be obtained from a manufacturer; they can usually be purchased at less cost than would be required to custom-build a gangway.

Where the bottom end of the gangway rests on a floating deck that is restrained in horizontal movement by guide piles or cables, it must either slide or roll with changes in water level. Because of the tendency of this rolling or sliding end to creep sideways and severely stress the hinges at the upper end, guides are usually provided to prevent any lateral movement (Fig. 98). Sliding ends are used for gangways weighing less than about 500 pounds. Heavier gangways should always be provided with rollers and aprons. The apron plates should be long enough to clear the ends of the roller guides at low water level and attached to the bottom end of the gangway with a pipe-and-rod hinge (Fig. 99).

5. Design Criteria for Support Facilities.

a. Utilities, Locker Boxes, and Fire Equipment. Freshwater from a municipal water system or a local water supply development is normally piped out along the main walks of a marina, mainly for use by boaters to wash craft and the adjacent dock area, but can be used for fire protection. Where no fire hydrants are provided, a 0.75-inch line will serve about 20 berths and a 1-inch line about 40 berths. Where provided, fire hydrants are spaced at about 100-foot intervals and serve standard 75-foot, 1.5-inch-diameter canvas hose racks in a red colored firehose cabinet (Fig. 100). Local regulations will usually establish the size of the supply line, normally not less than 2 inches. In a saltwater environment, the lines should be either plastic or copper. In freshwater harbors, galvanized iron pipe may be used if permitted by local regulations, and the owner is willing to accept the probable shorter trouble-free lifespan of the material. Bronze hose bibbs, usually 0.75-inch diameter, are provided for every one or two boats on each side of but not on the main walk. Some backflow-prevention system acceptable to local authorities should also be provided. Unprotected risers should not extend higher than 1 foot above the deck, and preferably attached to a supporting pile, a locker box, or other dock fixture. Because the valves may eventually break or wear, they should be attached to the piping by a threaded adapter at or near the point of support; replacement can then be made without disturbing a soldered or a plastic joint. All risers should be brass or copper pipe rather than plastic. Hose racks will help to achieve an orderly appearance of the docking area (Fig. 101). A shutoff valve should be provided to each pier served.

The landside water main serving a berthing system should be protected with a backflow preventer. A positive neat appearing junction of buried and exposed lines should be

Figure 97. Typical large and small gangways

Figure 98. Simple metal guides for a light gangway

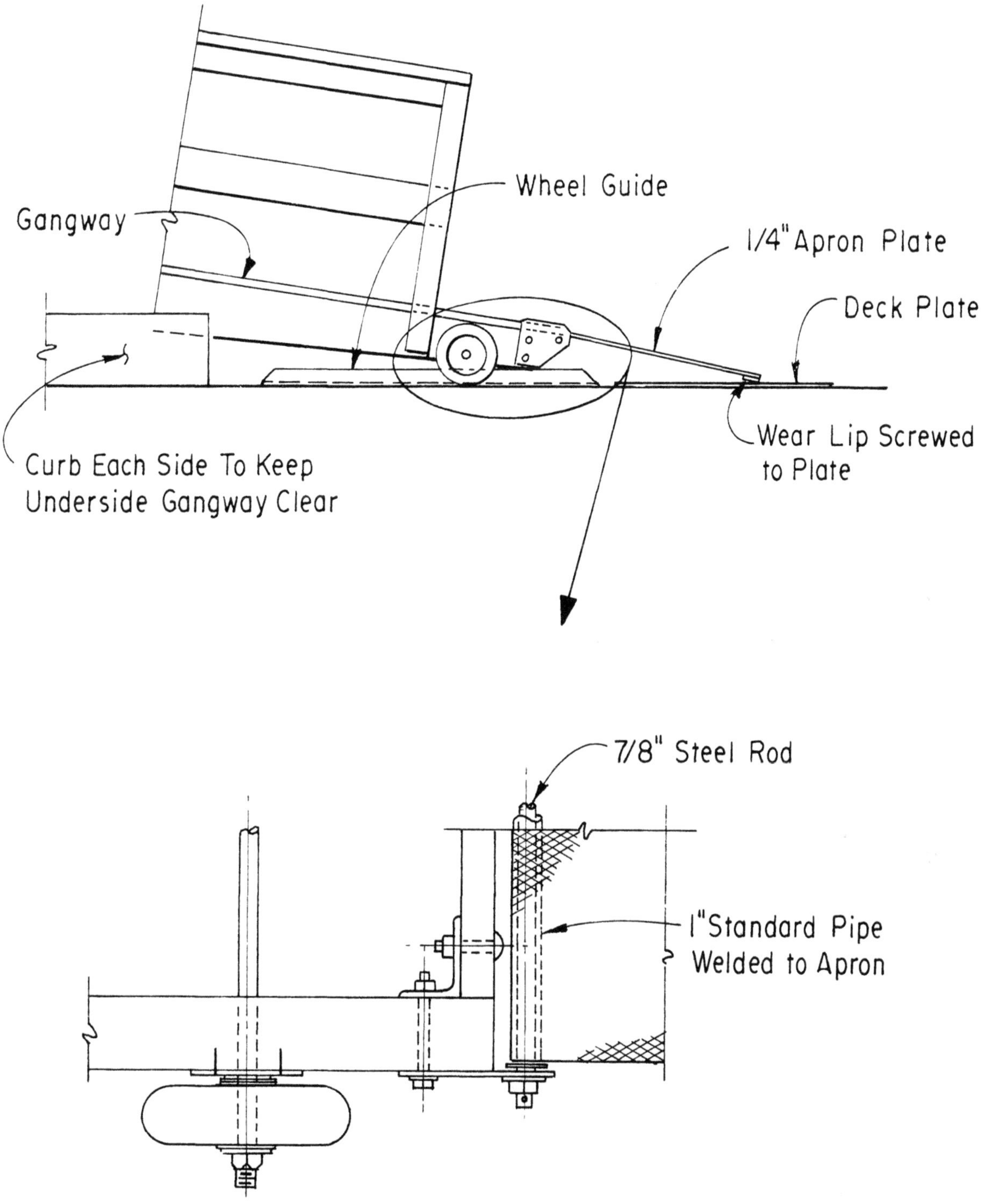

FRAMING AND HINGE DETAIL

Figure 99. Pipe-and-rod apron plate hinge for heavy gangway.

Figure 100. Firehose cabinet on a floating deck.

Figure 101. Rack for garden hose on pile extension

provided where the buried lines emerge from the basin perimeter. If the docks are floating, flexible hose connections are needed at each end of the gangway. The hose should always attach to a downward facing pipe end and hang in a U-shaped loop to prevent early fatigue failure (Fig. 102). In galvanically active soils, buried pipe requires either a corrosion resistant wrap or cathodic protection. It is customary to provide a free water supply to the slip renters; hence, only one water meter is normally required for the marina. Ancillary facilities may be metered separately.

An electric power outlet is normally provided at each slip for boats over 20 feet in length, with breaker switches at each outlet rather than just at the pier landing. Capacities will vary from 20 to 50 amperes, depending on the amount of electrical gear carried by the various berthed craft. Most craft require 120-volt service, but many larger cruisers also need 240-volt service. In a few installations, service is provided to the docks by overhead wiring (Fig. 103), but a below-deck-level conduit is preferred. Special outside outlets are available that are virtually foolproof and completely weathertight. Some are placed in riser racks just above deck level (Fig. 104). Others are supported on locker boxes or other convenient deck facility. Local agencies sometimes specify that only twist locks be used, but the outlet selected should be of a noncorrosive outdoor-type, and well protected from the elements (Fig. 105). It is important that outlets be located and oriented for convenient plug-in. Some outlet boxes have been manufactured with such a penchant for weatherproofing that the outlet plugs are difficult to reach and have become a source of irritation to harbor patrons.

Calculation of distribution circuits should always be done by a qualified electrical engineer to avoid inadequate sizing of lines. A 25 to 50 percent demand factor may be applied on each pier in estimating the connected load, but judgment must be based on local experience. Conduits should be large enough for wiring to be pulled without difficulty. Electrical conduits are often placed alongside the waterlines and telephone or public address system conduits. Grounding wires should always be returned to land and not to grounding plates in the water.

The electrical system for any large marina should be designed by an electrical engineer, but for the small operator who may desire to design and install his own facilities, the following data will be helpful. Where required, an owner's construction permit can be obtained from a local inspection authority. A copy of the local electrical codes should also be obtained.

Service voltage to the utility outlets may be a 120-volt, 60-hertz, 2-wire system or a 120/240-volt, 3-wire system with the neutral (white or green) wire grounded for craft requiring 240 volts.

The measure of power is termed *watts*. The information on watts is obtained from equipment nameplate data or may be computed if the values for amperes and volts are

Figure 102. Proper connection of hose and pipe at gangways.

Figure 103. Overhead dock wiring creates a cluttered appearance

Figure 104. Power outlet racks for a fixed concrete pier

Figure 105. Well protected outlets may thwart quick plug-ins. See also Figure 108.

known. If any two of the three variables (watts, amperes, volts) are known, the third can be computed as follows:

To find watts when volts and amperes are known, multiply the variables.

15 amperes X 120 volts = 1,800 watts

To find amperes when watts and volts are known, divide watts by volts.

1,800 watts ÷ 120 volts = 15 amperes

To find volts when watts and amperes are known, divide watts by amperes.

1,800 watts ÷ 15 amperes = 120 volts

The longer the circuit for a constant load and conductor size, the greater the power loss in the circuit due to resistance of the wire. The resistance in the wire dissipates some of the power by converting it to heat. For example, the input voltage may be 120 volts, but the voltage may be only 114 volts at the output end, which is a drop of 6 volts or 5 percent. To reduce this loss to 3 percent, a larger wire or cable with a larger cross-sectional area (and therefore a lower resistance) would be required. To determine the cable or wire to be used for a 3-percent voltage drop instead of 5 percent, the following method may be used. This method is based on the *International-Ohm-Mil-Foot* (I-O-M-F), which has been established as a standard constant of 10.8 ohms per foot of wire length. The ohm is defined as a unit of resistance. An example of this method follows:

* *

EXAMPLE PROBLEM

Problem: What size wire should be used to serve a load of 50 amperes on a single-phase 2-wire circuit 300 feet from the power source? The supply voltage is 120 volts and the resistance loss is to be kept to about 3 percent.

Solution: Find 3 percent of 120 volts = 120 volts X 0.03 = 3.6 volts; I-O-M-F constant for two wires = 2 X 10.8 = 21.6; Distance = 300 feet, one way; Current = 50 amperes.

Find wire size in circular-mils:

$$\text{Circular mils} = \frac{\text{constant} \times \text{length} \times \text{ampheres}}{\text{acceptable voltage drop}} = \frac{21.6 \times 300 \times 50}{3.6\text{-volt drop}}$$

$$= \frac{324{,}000}{3.6} = 90{,}000 \text{ circular mils}\,.$$

From a wire table in the National Electrical Code (1971), page 487, find the following:

Size No. 1 AWG (American Wire Gauge) is 83,690 circular-mils (too small)

Size No. 0 of 1/0 is 211,600 circular-mils (too large)

Since there is no size between these two, use the wire closest to the requirement, the 83,690 circular-mils wire, and tolerate the slight difference in drop. In using the No. 1 wire, note that the voltage drop is about the 3 percent required and is close enough for practical purposes:

$$\text{Volts drop} = \frac{21.6 \times 300 \times 50}{83{,}690 \text{ circular-mils}} = 3.87 \text{ volts}$$

$$\text{Percent drop} = \frac{3.87 \text{ volts}}{120 \text{ volts}} \times 100 = 3.2 \text{ percent}$$

* *

Outlets should be ampere-rated—20 amperes for 0 to 20 amperes; 30 amperes for 0 to 30 amperes; and 50 amperes for 0 to 50 amperes. In addition, outlets must contain a separate grounding pole for each white-green grounding conductor.

Circuit breakers (used as circuit protective devices instead of fuses) provide quick restoration of service after an outage. Maximum circuit loads should not exceed 75 to 80 percent of the breaker-trip rating to compensate for higher than normal ambient temperature conditions.

Experience has indicated that maximum demand on circuits is between 20 and 50 percent of the connected load outlets. However, each case should be individually estimated.

Lighting circuits are rated 100 percent of installed lamp-load in watts plus 25 percent, and demand is 100 percent. For best results, lighting facilities should not be supplied from outlet receptacle circuits.

All equipment must be weatherproof or raintight and safely installed, and not be a hazard either electrically or mechanically to persons, craft, or navigation.

Dock lighting methods vary considerably throughout the United States. Many older marinas were illuminated with a few floodlights around the basin perimeter. This method of lighting proved inadequate, as light failed to reach some of the outermost slips and the floodlights caused excessive glare, both directly and by water reflection (Fig. 106). Most floodlights have been replaced with a system of smaller lights set between 8 to 12 feet over the main walks of the marina at frequent intervals (Fig. 107). Complaints about glare have led to the small, well-shielded lights set at about 30 inches above deck level that are now prescribed by somc harbor administration agencies (Fig. 108). However, one marina

Figure 106. High perimeter floodlighting may cause excessive glare.

Figure 107. Intermediate-height, shielded dock lights typify most marina lighting systems.

Figure 108. Low level dock lights reduce glare but may not provide adequate illumination (Courtesy of Harvey Hubbell Corporation).

reported that about 75 percent of the low-level lights were broken off or stolen during the first year. The manager felt that these lights failed to illuminate the area sufficiently for the necessary security at night. Most managers and many boaters still prefer the medium-height standards.

An intercommunication system is desirable at any harbor berthing more than a few dozen boats. Some marinas merely provide pay telephone stations at convenient intervals; others have phone jacks for every berth. Some have only a public address system for paging; others combine this system with a telephone at the shore end of each main walk. The system selected should best meet the anticipated requirements of the slip renters. A two-way public address system costs little more than a one-way system and will be far more useful in a small-craft harbor. If there is a question as to the need for individual service, telephone company recommendations should be obtained, especially regarding requirements for conduits installed initially in anticipation of future telephone service. One of the best communications systems that meets the requirements of most medium size marinas is a paging system, with speakers located within easy hearing distance of every slip and support facility, and a central telephone switchboard in the administration office. The harbor manager routes each incoming call to the phone extension nearest the party called and notifies that party by paging. Conversely, he handles all outgoing calls as instructed and charges a predetermined per-call fee for this service.

Conduits and waterlines should be extended out on each main walk in an easily accessible utility duct or a hanger system with supports at 8-foot intervals. A few berthing systems provide built-in utility ducts with easily removable covers just under the main walk decks (Fig. 109). However, this location tends to disrupt the structural integrity of a system. Most systems employ an alternative utility line location on the outside of the stringers, below deck level (Fig. 110) and, although not as convenient as a special utility duct, this is probably the best hidden site for the lines. The primary objective is to locate the lines for easy repair work. In some marinas the desire for convenience outweighs the desire for neat appearance, and conduits and waterlines are placed directly on the decks (Fig. 111). Duplicate lines on the sides of the main walk sometimes serve each row of slips separately. The common practice is to extend the waterline on one side and the electrical conduits on the other side, with underdock cross runs to serve outlets on opposite-side slips. Chaney (1961) should be consulted for a detailed treatment of water supply and electrical installations.

Sanitary sewer lines have been provided on some fixed-pier structures in marinas, usually where they serve restaurants, fish cleaning stations, or clubhouses built over the water. There has been little requirement for direct-connection sanitary systems to serve boats berthed in marinas. With the trend toward houseboats and live-aboard-berthed-craft practices, consideration is now being given to the use of shore-connected water supply and automatic pumpout facilities in many of the berthed craft. Craft with pumpout facilities may soon be in service, and the future need for installation of direct-connection sewerlines

Figure 109 Main deck utility duct with removable cover.

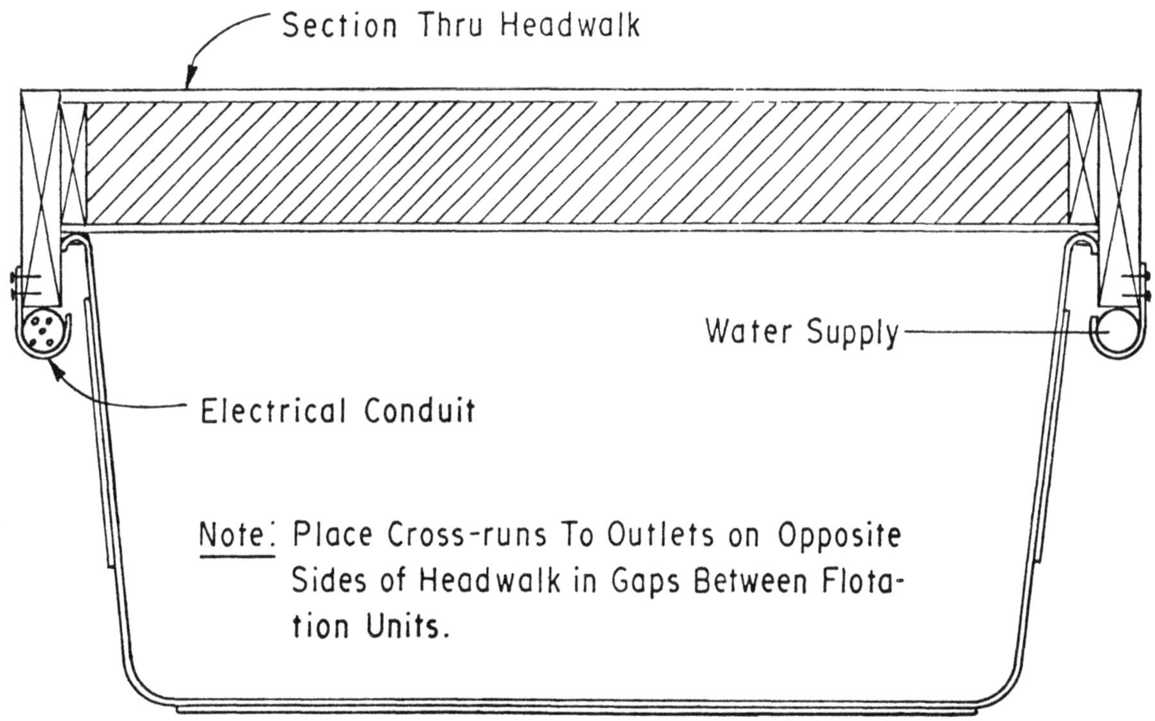

Figure 110. Utility lines hung under side stringers of a floating main walk.

Figure 111. Utility lines laid along sides of main walk. Note locker boxes with tread tops for boarding boats.

(on at least some piers of each new marina welcoming *live-aboards*) suitable for transferring the pumped out effluent into the landside sewer system may be necessary. Technical problems involved in such systems need to be worked out; in time, another utility line may be required.

Locker boxes are not only used by slip renters for gear storage, but often provide structural support for electrical fixtures or hose bibbs. The boxes may be located in the knees of floating systems (Fig. 112) or in some out of the way spot on decks of fixed systems (Fig. 113). Some are custom made of plywood or sheetmetal, but manufactured fiberglass boxes suitable for most sites are usually available for less cost and are more durable in a marine environment. Some marinas require tenants to furnish boxes, but for uniformity the modern trend is toward the initial installation of all locker boxes as an integral part of the system. This is imperative if locker boxes support utility outlets and light standards. Some are available with a conveniently located low-wattage light that not only illuminates the power outlets and locker interior but also keeps an ambient temperature, thus preventing condensation and resultant mildew or corrosion of stowed gear. Many boaters use the tops of locker boxes as workbenches; if the lid is not already suitable for this purpose, a special removable overlid cover can be devised for patrons that use them for workbenches.

Figure 112. Locker box with built-in outlet box and hose bibb. Note location on knee at junction of finger pier and main walk

Figure 113. Step along basin perimeter provides site for locker boxes

Provision of adequate firefighting equipment is a good precaution for any marina and a requirement of most controlling agencies. Although dock fires are extremely rare outside of fuel docks, they can occur. Water is a poor agent for fighting hydrocarbon and electrical short-out fires. For this reason, there is a trend toward elimination of fire hydrants and reliance on chemical firefighting equipment (Fig. 114). When used, chemical firefighting equipment should be placed at intervals of about 200 feet along each main walk unless closer spacing is required by ordinance. The cabinets housing the equipment should be painted red.

Figure 114. Cabinet for chemical fire extinguisher.

b. Fuel Docks and Pumpout Stations. The design and construction of a fuel dock differs from an ordinary loading dock in that the fuel dock must be more rugged, support the fuel pumps, and take aboard the lines from the buried tanks ashore. These requirements present no special problems in a fixed system, but a floating system also requires flexible fuel lines leading to the fixed lines installed in the dock. The simplest solution is to lease the entire fueling area to an oil company to develop and operate. Where the economics indicate an advantage in owner operation, a mechanical engineer specializing in piping systems should be engaged to work with the structural engineer in devising the system. An important consideration is a quick-disconnect provision at the high point in each fuel line that will prevent the siphoning of the storage tank contents into the basin in case of a line break below the stored-fuel level. Examples of a modern fueling station are shown in Figure 115.

Emphasis on water quality control has led to the requirement in many areas that all boats with heads be equipped with chemical holding tanks. Periodically, the tanks must be pumped out and fresh chemicals added. Several manufacturers now make pumpout facilities for this operation (Fig. 116). Facilities are often placed on or near the fuel dock so that fueling and pumpout can be accomplished successively at the same place. However, the pumpout station should not be too close to the fuel pumps and interfere with fueling operations. The recharge chemicals for the holding tanks are also sold by the station attendant. Every new marina should have a pumpout station unless one is already available nearby.

c. Cleats and Fenders. Every dock or slip requires a set of cleats for mooring lines and fenders to cushion the impact of moored or drifting craft against the dock. Manufactured metal cleats, either galvanized or of noncorrosive alloys, are available in several sizes. Cleats can also be made of hardwood (Fig. 117). Wooden cleats are preferred by some marina operators because when damaged or loosened after the nuts have rusted fast to the bolts that hold them, they can be split and chipped out, leaving the bolts exposed for easy removal with a hacksaw. A 10- or 12-inch length is preferred for boats up to 40 feet long; lengths of about 16 and 20 inches will serve cruisers up to 75 and 100 feet long, respectively. The cleats must be securely bolted to the framework with through bolts rather than lag bolts. Many lightly fastened cleats have pulled out under severe line stress.

The cleat-arrangement pattern may vary with different slip or docking systems, but generally one cleat near the knee and one at the end of a finger (on each side) will serve boats up to 35 feet long. For larger craft, one more cleat per side should be added for each additional 30 feet or fraction thereof. If cleats are used on loading docks or fuel docks designed for breast mooring, they should be sized for the largest craft anticipated and spaced at about 15-foot intervals. A better system is to provide a continuous curb rail supported on blocks spaced about 3 feet apart for tying mooring lines. In a double-boat slip system, two cleats about 3 feet apart should be secured to the edge of the headwalk and centered between the two fingers. The finger piers of many fixed systems are either not as substantial as the headwalk or the stringers are too far in from the edge to provide a secure cleat anchorage. The inboard or knee finger cleats must then be secured to the headwalk (Fig. 118). A tie pile will often substitute for the two cleats at the outer end of the finger (Fig. 119).

Many fendering systems of the early marinas were of makeshift quality, using old rubber tires, discarded firehose, or hemp hawsers for the bumper elements. Special synthetic extrusion or molded shapes are now being manufactured, which provide better appearing and more functional fendering for small-craft dockage in floating systems. The preferred

Figure 115. Typical fueling stations

Figure 116. A sanitary holding-tank pumpout facility (Courtesy of Kenton Equipment Company)

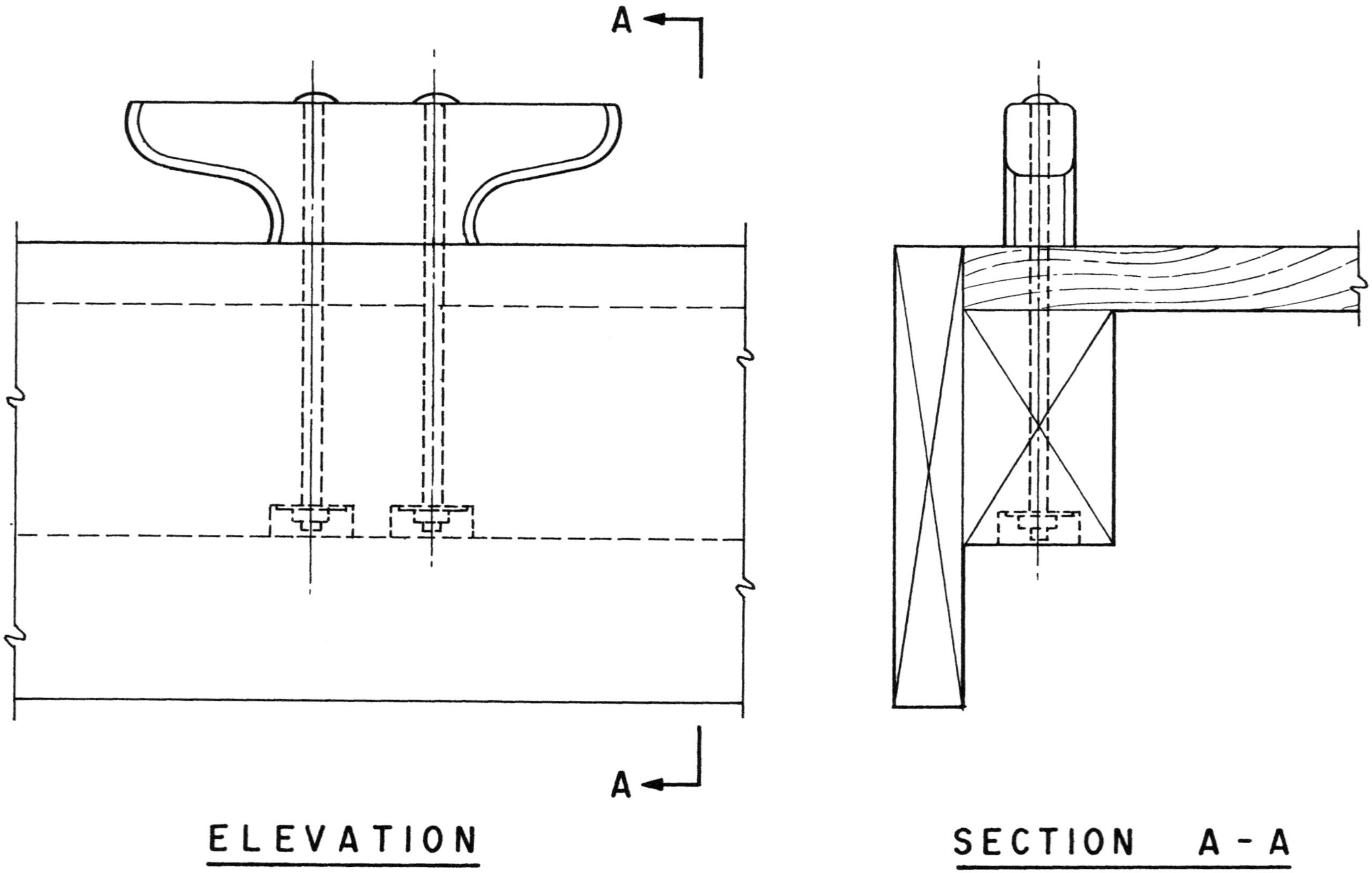

Figure 117. Hardwood cleat for small-craft mooring lines.

Figure 118. Cleats along edge of main walk for bowline ties

Figure 119. A row of tie piles beyond ends of finger piers.

type of bumper stripping for the edge of a finger or dock is a synthetic extrusion that runs along the top edge of the outside stringer with a lip extending over the top. The usual accident that destroys the continuity of such a fender strip is a sharp-edged bow or stern impact that bites into the synthetic material and tears it out at nailing points (Fig. 120). For heavy-duty use, an extrusion with a slight recess for metal fastening strips should be obtained to resist this tearing action (Fig. 121). Materials that have good weathering qualities and resilience retention are neoprene and butyl rubber. Other commercially available materials may also prove satisfactory, but many turn brittle or chalk off with age in an exposed environment.

Outside-corner fendering requires special consideration, because this is where most collisions occur; also, extrusions cannot bend around a sharp corner. Molded corner bumpers of the same material as the extrusion can often be obtained and will provide adequate protection for smaller craft. Large cruisers and sailboats can best be accommodated with corner wheels (Fig. 122), especially where the finger extends all the way out to an interpier fairway. Entry into the slip can then be made by easing the craft against the wheel before it is fully turned and then completing the turn like a large vessel uses a turning dolphin. Corner wheels are used primarily on floating systems, where they are always at the proper level with respect to the hull of the boat; seldom will they work properly on a fixed system at all water levels. Some floating fingers are manufactured with rounded ends, thus avoiding the sharp corner problem (Fig. 123).

Because of water level changes, the fendering of fixed piers usually runs vertically rather than horizontally. Projections beyond the hull of a boat such as a rub strake or overhanging gunwale coping may catch on any horizontal member of a fixed-fender system during a rising tide. The most commonly used fenders for fixed systems are vertical timbers spaced at 8- to 10-foot intervals along each side of a finger pier (Fig. 124). The size of the timber ranges from 3 by 4 inches to 8 by 8 inches depending on the size of the berthed craft. Lengths are determined by a need to extend upward above the highest part of a gunwale at extreme high water and to extend below the lowest rub strake at extreme low water. Pier attachments must be adequate to resist any moment that might be applied by cantilever bending loads up to the design moment of the fender section. Bolt heads or nut and bolt ends must be countersunk into the fender pieces to avoid scoring of boat hulls. This countersinking reduces the strength of the fenders and must be accounted for in the calculations.

A softer vertical fender is the plastic tube stretched between points of support or suspended from a top fastening with a heavy pendulum weight hanging below the lowest possible point of hull contact (Fig. 125). The tube works on the same principle as the vertical timber fender except that it cannot take cantilever loading. Its milder impact on hulls and long lasting quality makes the plastic tube worth the extra cost in many marinas. Special instructions for installation are supplied by the manufacturer.

Figure 120. Damaged bumper stripping along a floating dock.

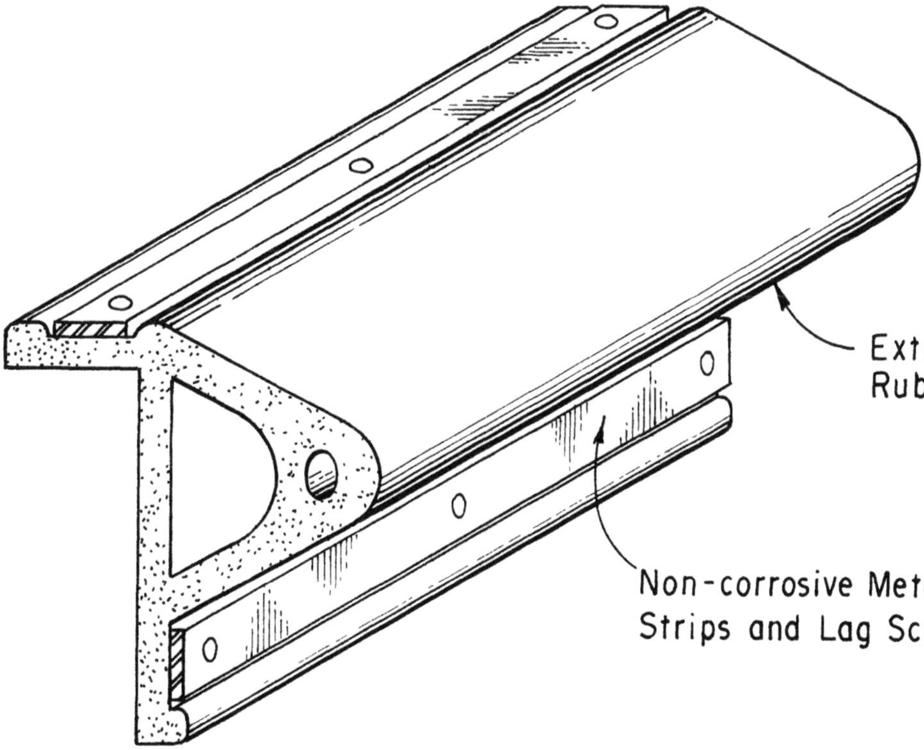

Figure 121. Bumper extrusion designed to resist tear-out.

Figure 122. Corner wheel on a floating finger end.

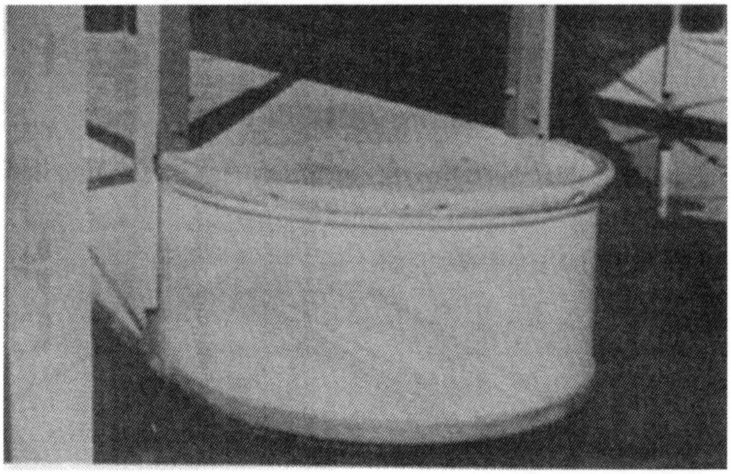

Figure 123. Rounded finger end facilitates entry to slip.

Figure 124. Vertical timber fenders for fixed concrete slips

Figure 125. Vertical plastic tube for soft fendering (Courtesy of Marina Products Manufacturing, Incorporated).

d. Launching Hoists, Elevators, and Ways. The most common types of equipment used in marinas for transferring boats between land and water are listed below:

Davits	Mobile cranes
Forklift trucks	Overhead-rail launchers
Jib-boom cranes	Stiff-leg derricks
Lift slips	Straddle-truck boat hoist
Marine railways	Vertical-lift platforms

Four of these types will not be discussed in this study for various reasons, and the reader may consult Chaney (1961) or some other text if detailed information is needed. Mobile cranes, either truck-mounted or crawler-mounted, are too expensive for boat launching alone, and can be justified only where used in conjunction with some ancillary operation that requires this type of equipment. Marine railways are rapidly becoming obsolete because of high cost, waste of space, and lack of versatility. Vertical-lift platforms are used mainly for large craft (25 tons and over) and are not normally a component of a small-craft harbor. Davits are for smaller craft and lack the versatility of other launching devices.

Of the six remaining types of boat-handling equipment, two are completely mobile and most versatile—the straddle-truck boat hoist and the forklift truck. The hoist is similar to the lumberyard truck, but is designed for boat handling service. It is limited to the boat size the hoist will handle by its load-carrying capacity and straddle clearance, and must be used in conjunction with a launching well or parallel piers (Fig. 126). Some models have one open end so that boats with masts or superstructures can be handled, as shown in Figure 126. All are proprietary vehicles, and the costs, dimensions, and capacities can be obtained from the manufacturer or distributor. Construction of the launching well must follow the principle outlines for construction of bulkhead walls, except that the well walls must extend down to the design depth of the basin and be capable of supporting the extra surcharge loading of the vehicle and the design boatload. The cost of constructing the well or parallel piers frequently exceeds the cost of the vehicle.

The forklift truck used for boat launching has a fork ladder with the capability of hyperextending down over a bulkhead wall to the depth required to place the forks under the hull of a floating boat. Because of the eccentric loading, forklifts are only available in capacities for lifting smaller craft (up to about a 30-foot length). All outboards and some inboard cruisers have most of their weight near the stern and can be lifted stern-to, with the keel parallel to the forks. This greatly facilitates placing the boats in dry storage racks (Fig. 127). Because of a limited downward reach, forklifts are best suited for use in marinas with little water level fluctuation. In marinas with large tidal ranges, they may be used for repair and maintenance work on the boats if not inconvenient to await a high tide. However, this would be impractical for daily operational use. Although no boat well is required for a forklift, the bulkhead wall for launchings and retrievals must be designed for the additional

Figure 126. Straddle hoists. Top photo shows hoist at a boat well; bottom at parallel piers. Note storage rack in background of bottom photo.

Figure 127. Hyperextendable forklift truck for boat launching

surcharge loading of this operation. Also, a specially contoured curb must be located on the exact line that will positively prevent any further forward motion of the vehicle and yet will allow the fork ladder to clear the top of the wall when lowered into the basin. Such details must be obtained from the manufacturer.

Overhead-rail launchers, either monorail or duorail, function the same as a straddle vehicle except that they are not mobile (Fig. 128). For daily trailer-boat launching where versatility is not required, this system provides a rapid operation at less cost per cycle than can be obtained with mobile equipment. The overhead rail is pushbutton-operated, hence requires less operational skill, and is easier to maintain. However, an electric power source is required, and because of the extensive structural framework, the installation cost may equal or exceed that of a straddle-vehicle system. Unless the framework is high, the overhead rail cannot be used for masted or high-superstructured craft. Some proprietary rail-hoist systems include the structural framework, and others have only mobile and hoisting components, so that the design of the supporting framework can be tailored to the requirements of the site.

Because only one rail hoist can normally be afforded at any given marina or launching site, and the cost increases rapidly with design lifting capacity, the selection of size is critical. A forecast of probable use will indicate the largest craft to be handled in sufficient numbers to make a rail-hoist operation feasible. A rail hoist of this capacity should then be installed; larger craft must go elsewhere or be handled by a stiff-leg derrick or other large-capacity equipment that may also be installed at the same marina for such oversize use. A graph for selecting the proper capacity hoist for various lengths of power boats and sailboats is shown in Figure 129.

The counterpart of the rail hoist for sailboat launching is the jib-boom crane (Fig. 130). The crane is generally less costly than a rail hoist of the same capacity and can handle power boats and sailboats. Two main drawbacks of the crane are the relatively short reach of the boom and the slower operation because of the need to rotate the craft as the boom swings. The rotation is done by hand to avoid any contact by mast or rigging with the boom and to lower or raise the craft without scraping the hull on the bulkhead wall. These hazards are not in a rail-hoist system. The excessive cantilever bending moment at the base of the crane, caused by the eccentric load, requires a rugged foundation that must be close to or made a part of the bulkhead wall without endangering stability. Unless the foundation is designed initially, the later addition of a jib crane could require costly modification of the bulkhead at the crane site.

The stiff-leg derrick is usually less costly than other launching equipment with equal lifting capacity (Fig. 131). The derrick has the rotational problem of the jib crane and also requires greater skill by the operator and usually a tag-line crew to control the craft while in the air. Derricks are usually installed in larger capacities for handling larger craft that cannot be handled by the faster operating but lighter equipment. In very large capacities, cost is less than the other types, and with a long boom can operate over a fairly large area, thus

Figure 128. Duorail hoist boat launcher

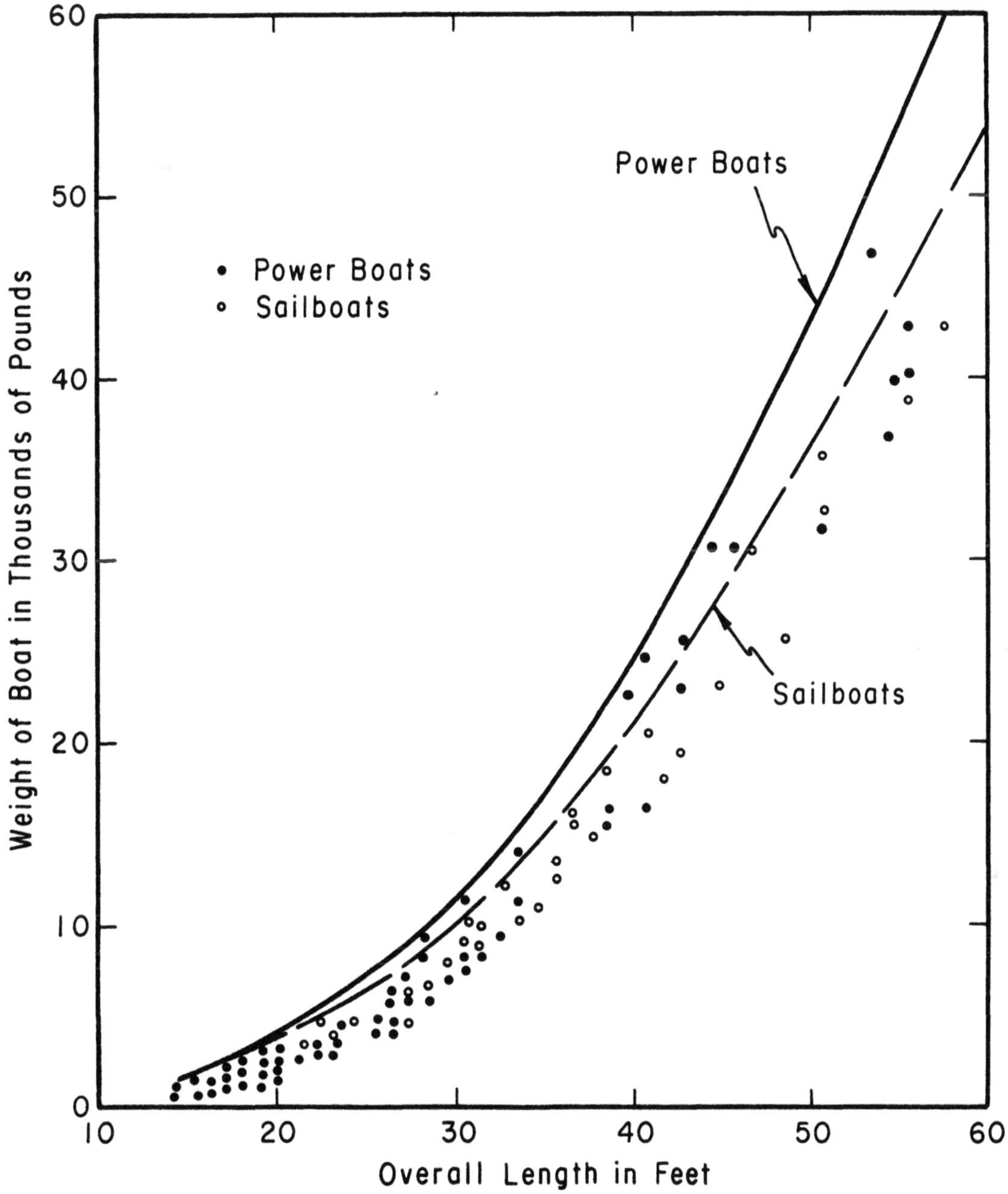

Figure 129. Weight versus length of recreational craft.

Figure 130. Jib-boom crane for sailboat launching

Figure 131. Stiff-leg derrick hoist.

increasing versatility. Stiff-leg derricks of various types and sizes can be purchased complete, or designed and built with readily available structural members and mechanical components to fit any given situation.

Any type of lifting equipment requires a spreader bar to avoid crushing a hull or springing the structural framework of the craft. Heavy-duty straps of woven nylon or other strong synthetic fabrics are available for any size boat normally handled by such equipment. Straps must be examined periodically and replaced if signs of severe wear or dangerous cuts indicate incipient failure. Some sailboats and small motorboats have lifting eyes that are integral with the structural framework of the hull and can be lifted with hooks rather than straps. Many yacht clubs and a few marinas require that all craft have lifting eyes to standardize and speed up the launching operation.

Three types of lift slips used in some marinas do not transfer boats to dry land. One is the elevating work slip that merely raises the boat out of the water so that the hull may be washed or minor damages to hulls, rudders, or screws repaired (Fig. 132). The boat remains on a cradle or in a sling in raised position above the water only until the operation is completed. In the second type of lift slip, a pair of hoists that remain as part of the berthing facility hold the craft above the water while not in use (Fig. 133). The third type is used in some berthing sites along the shores of large lakes or bays that cannot be protected at reasonable cost against the moderate wave climate of the area. This is merely a fixed-framework hoist that will allow the waves to pass through the supporting legs without overturning. The berthed craft is hoisted high enough to clear the worst "sea" condition anticipated (Fig. 134). Another reason for using the second and third types of lift slips is to prevent hull damage due to barnacle incrustations, corrosion, or ice formation.

e. Launching Ramps. Hoist-launching operations involve the use of mechanical equipment and usually trained operators. Many small-boat owners resent paying high fees for such service; therefore, some marinas provide only a launching ramp to get any boat that can be transported by trailer into or out of the water. Some boating facilities consist only of a launching ramp and a car-trailer parking lot, with no provision for berthing small craft in the water, except temporarily for boarding purposes. Regardless of the purpose for which they were installed, all fixed launching ramps have essentially the same design criteria.

The slope of the ramp ranges between 12 and 15 percent. Few trailered boats can be launched with a ramp slope flatter than 12 percent without submerging wheel hubs of the pulling vehicle. Slopes steeper than 15 percent are dangerous for all but the most skilled drivers. Many trailered sailboats cannot be launched without hub submergence, even on a 15 percent slope. The best alternatives are to use a trailer-tongue extension or to launch the craft with a hoist.

The surface of the ramp should be paved down to an elevation of about 5 feet below extreme low water level; the top should be rounded over on a 20-foot vertical curve until it becomes nearly level at about 2 feet above extreme high water. The bottom of the ramp should end in a level shelf of loose gravel so that a vehicle losing brakes or traction would be stopped before sliding deeper into the water.

Figure 132. Large and small floating boat lifts

Figure 133. Hoist-out dry storage for a covered slip.

Figure 134. Hoist-out storage rack for overwater mooring

The wetted part should be paved with portland cement-concrete, as asphaltic or bituminous paving does not hold up well from traffic in submerged areas. Unpaved ramps will soon deteriorate under even moderate use. A single-lane ramp should not normally be narrower than 15 feet, and a multiple-lane ramp should not have raised divider strips. Lane-marking is not necessary and may lead to less than optimum use during peak hours.

Ample maneuver room should be provided beyond the top of the ramp, usually 40 to 60 feet, on a gentle rampward slope (about 1 on 50) for proper surface drainage. About 50 pullthrough parking spaces should be provided for each ramp lane, with a clearly marked traffic-circulation pattern between the parking area and ramp.

Boarding docks should be provided, preferably on each side of the ramp and extending out into or along the sides of the basin, with a total boarding length of at least 50 feet for each ramp. The ramp should adjoin fairly quiet water, although not necessarily as quiet as that needed for a berthing site. Ample protected *holding area* in the water just off the ramp and boarding dock location should also be available for boats awaiting their retrieval turn during peak hours.

Ramps leading into saltwater or polluted waters should have a conveniently located washdown facility, just outside the maneuver area if possible. This area should be large enough to accommodate one car and trailered-boat per ramp. A waiting area of about the same size for boats just retrieved should also be provided. Freshwater piped from the local supply main and washdown hoses of ample length should be available at this site, together with an adequate drainage system. Boaters will desire to wash saltwater and pollutants from the boat hull and from the wheels and hubs of the trailer and car. In washing, some oil, grease, mud, or other debris will be hosed off the undercarriages of the vehicles. For this reason most controlling agencies require that the washdown area drain into a debris-trap sump and then into the sanitary sewer system of the area rather than back into the basin. The sewer intake must then be provided with a rain valve to avoid flooding the sewer system with rainwater. Whether or not the use of detergents is permitted with this operation is largely a decision to be made by the controlling agency. Ramp users often beach their craft temporarily before retrieving them. Accordingly, users will appreciate having a small sandy beach reserved for this purpose near the launching area. A typical launching-ramp facility is shown in Figure 135.

Where the ramp area can be unwatered down to the lowest elevation, a poured concrete pavement is normally the easiest and least costly to construct. The surface should be finished with deep, square-shouldered grooves molded into the surface parallel to the contours (Fig. 136). Raked, rough-broomed, or other coarse-grain finishes without the deep grooves will soon lose traction. Although the grooves may fill with mud, pneumatic tires will force enough mud out to bring full bearing on the squared concrete shoulders and thus prevent skidding. The grooves are usually made with 1- by 1-inch oiled sticks located about 3 inches apart on a frame that is pressed into the freshly poured surface and then carefully

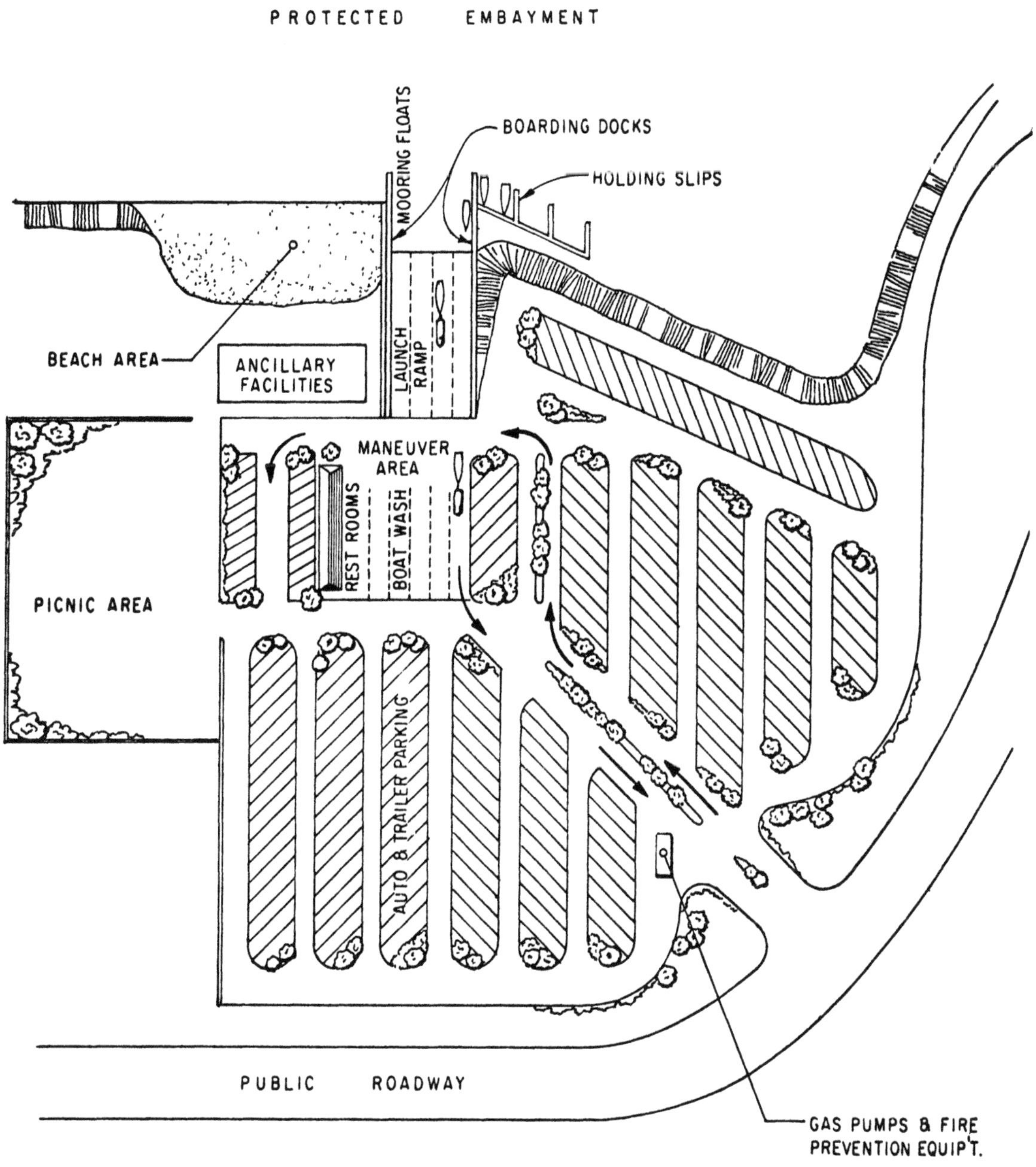

Figure 135. Layout of a typical launching ramp facility.

Figure 136. Nonskid concrete ramp surface.

pulled out just before the concrete has initially set. The sides of each stick should first be planed to about a 0.062-inch taper before being nailed to the framing boards to facilitate frame removal. A tap on the end of each stick before lifting will help to break any bond that might form.

Any part of the ramp that must be placed underwater should be precast in sections of a size that can be conveniently slid or lowered into place onto a carefully prepared gravel bed about 6 inches thick. One method that has been used successfully is to precast 6- by 12-inch slabs that are a lane-width long, space them 3 inches apart perpendicular to the slope, and fill the gaps with coarse gravel. Some submerged ramps have been tremie-poured, but often with poor results and at a cost greater than that of a precast-slab ramp. Large concrete bricks and building blocks have been used, but they tend to become dislodged easily if the subgrade is soft. (See Portland Cement Association, 1965, for more detailed information on ramp design.)

A partial listing of manufacturers offering products related to small-craft harbors or ancillary facilities is presented in Appendix E. All the manufacturers listed aided in the data preparation by contributing solicited information about products, samples, or photos of their proprietary items or systems. However, this is not intended to be a comprehensive list of all marine product manufacturers, nor a special endorsement for those listed.

f. Support Buildings and Ancillary Structures. The most important support feature of a small-craft harbor is the administration building. Criteria for location have been presented under "Layout Planning," in Section V, 3 but the internal allocation of space also requires careful planning. Although only two or three rooms are required for a small installation, many rooms are needed for the headquarters of a large marina. The functions to be handled and features to be included are generally as follows:

(1) A clerical reception office with a counter for transacting business with clients, such as giving advice to visitors, making assignments of slips, and collecting rental fees. Free literature concerning the harbor and other informative matter should be displayed prominently in racks on or near the counter.

(2) A large record board showing a graphical layout of the harbor, the number or letter description of each slip and docking area, and the occupancy status.

(3) Record files containing slip-rental records, employee records, facility-maintenance records, and utility-service records. Important records or documents should be kept in a fireproof vault.

(4) An office where the manager can work, shielded from disturbances of the clerical office, or can discuss in private with public officials, salesmen, employees, and special visitors.

(5) A communications center for relaying incoming and outgoing calls and operating the paging system.

(6) Restroom facilities for the staff, and in large installations, for the visiting public as well.

(7) In large installations, such supplemental facilities as a boardroom, a coffee-break room, an engineering room for planning and designing and for filing record drawings, and a storage room for items not currently used.

Some examples of administration building exteriors are shown in Figure 137; Figure 138 is a typical floor plan for a large administration building.

The harbormaster's office, if not located in the administration building, should be nearby for convenience or have a duplicate occupancy-status chart. In a small marina the manager may function also as harbormaster, but in a large installation, the harbormaster's duties are so demanding as to require one individual's complete attention and are usually handled by another person. The harbormaster is responsible for all navigational aspects of harbor operations, including compliance of all boaters with regulations for security of the entire complex as well as the berthed craft, for dissemination of information to harbor occupants regarding navigation and berthing matters, and often for supervision of regattas and other boating activities. For these reasons, he should have a patrol boat berthed close to his office and a vehicle for land travel so that he can quickly reach any part of the harbor in emergencies and take charge of the situation. He supervises the work of all harbor patrolmen, and he or some of his subordinates are often uniformed and deputized as peace

Cape Cod

West Palm Beach

Seattle

Figure 137. Examples of administration building exteriors.

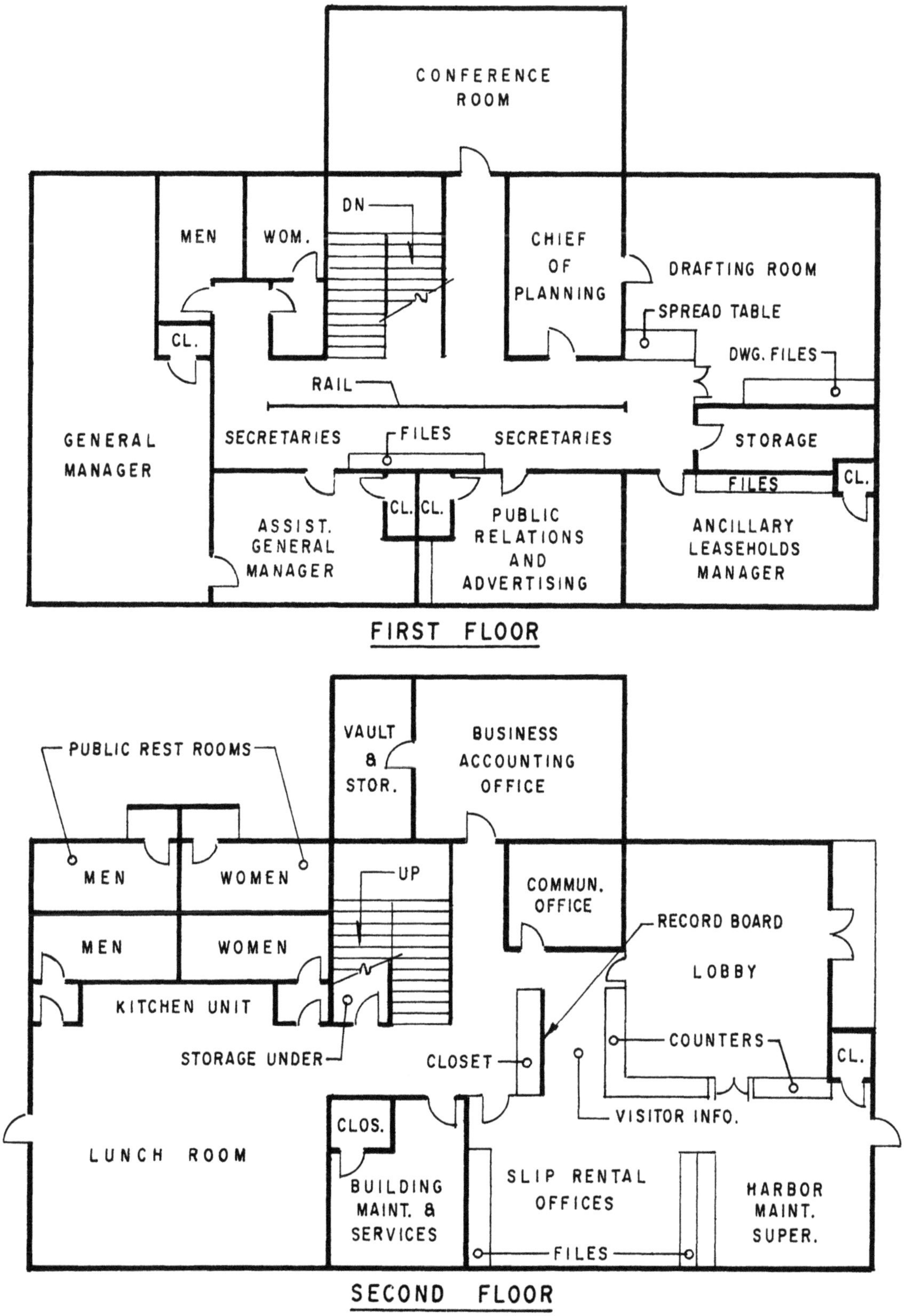

Figure 138. Typical floor plan for large administration building.

officers. The harbormaster's office must therefore include all facilities needed to carry out these functions, such as dressing rooms with lockers, a place to train and brief subordinates, stowage space for gear and equipment, and record files for traffic counts, accident reports, vandalism records, and regulations and ordinances established by the harbor management and higher authority.

Every small-craft harbor requires a maintenance building and yard for storage of vehicles, equipment, and spare components and parts used to keep the installation in good condition. If the marina operates a boat repair facility, the harbor maintenance functions may be combined with that facility, and one building and yard will satisfy both requirements. If the boat repair and maintenance function is leased for private operation, a separate maintenance facility is required. The facility should include work benches, equipment, tools, and supplies needed for grounds-keeping, for vehicle maintenance, and for repair of docks and berths. Replacement items for the harbor utility system, the firefighting equipment, and the berthing system should also be stocked.

Replacement items for the berthing system should include fendering components, mooring cleats, and for floating systems, extra flotation components, pile-guide rollers, and hinge pins. Most harbor maintenance buildings include a paint and sign shop. A small office will be needed for keeping maintenance records on all items requiring periodic attention, for preparing requisitions for needed parts and supplies, and for transacting business with salesmen and delivery personnel. Some open storage areas will be needed for large items that do not require cover, including a paved entry road leading to an unloading dock for receipt of supplies. In a large marina, a small machine shop may be needed for repair and maintenance of a fleet of mobile land equipment and vehicles, together with a washdown apron for keeping the equipment clean and operable. A typical maintenance building and yard layout is shown in Figure 139.

A boat repair installation is a necessary adjunct in every large marina and a desirable ancillary facility in a small-craft harbor. The facility is usually a part of, and immediately adjoining a hoist or elevator. Various methods of transporting the boats from the retrievel area to the working area have been used. In small installations where only lighter craft are handled, four-wheel, castered dollys are usually the most satisfactory (Fig. 140). However, large-capacity elevators are usually accompanied by a rail system with a transfer pit for shunting the boats on fixed-axle dollys from one track to another (Fig. 141). The repair installation normally has a covered work area with a gantry crane for removing engines and other major components, as well as tools, work benches, and other equipment needed for engine and hull repair (Fig. 142).

A marine hardware supply store is a welcome ancillary facility at a marina; it serves a basic need for harbor patrons, and may also be a supplemental source of revenue for the overall project if handled properly. In a small installation the store is often run by the management, but frequently handled on a franchise-lease basis in a large harbor complex.

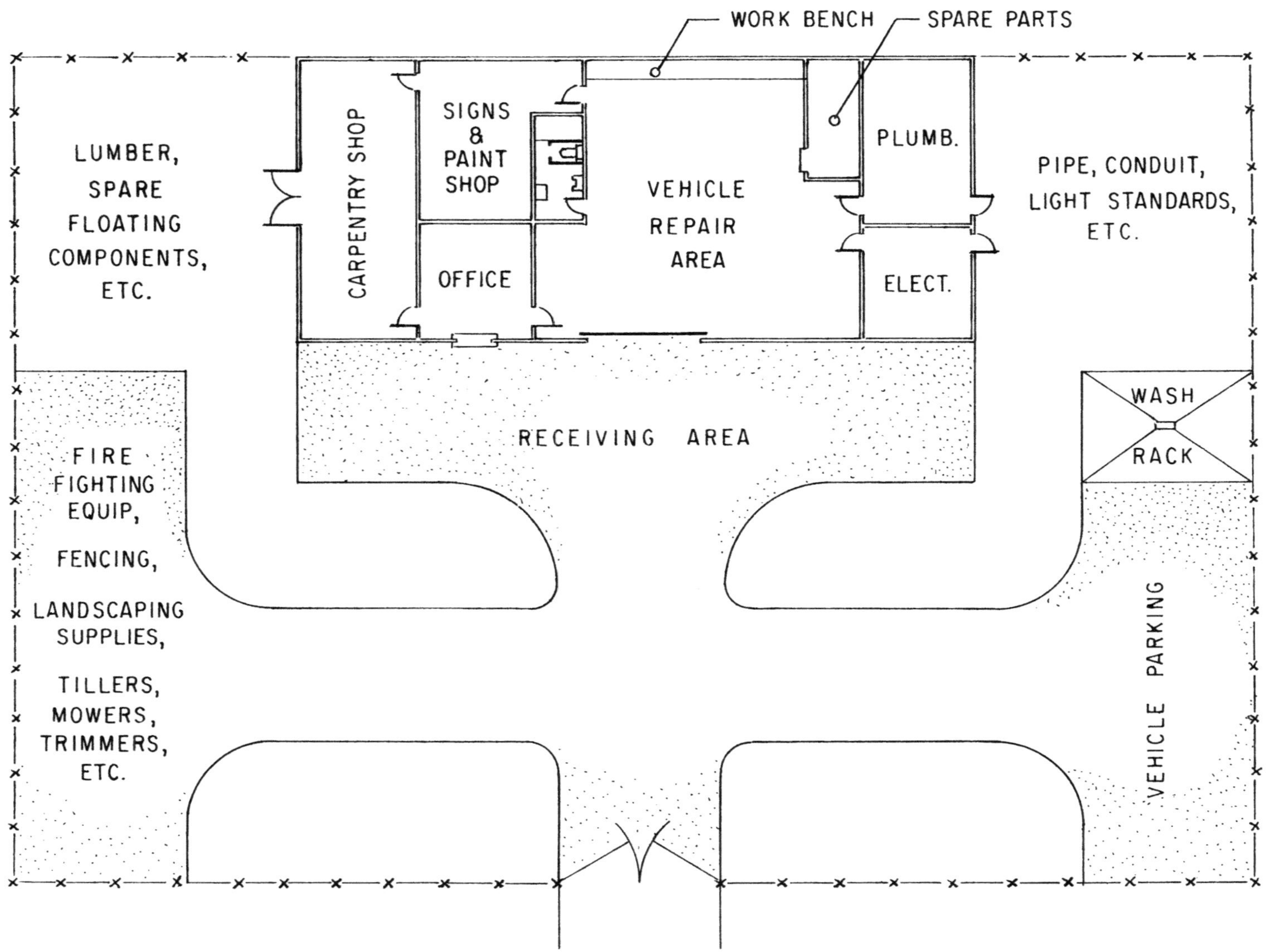

Figure 139. Typical maintenance building and yard layout.

Figure 140. Castered boat dolly for use in boat repair yard.

Figure 141. Rail transfer pit serving boat lift.

Figure 142. Large covered boat repair facility.

The building that houses the store should either be built by the harbor management and leased, or the site should be leased on a long-term basis and the lessee required to secure management approval of the design. No special criteria are offered, but the architecture should be in keeping with the decor of other facilities at the installation and the interior functionally suited for the intended purpose. Boat sales or yacht brokerage enterprises may either be a part of the boat service installation or headquartered in a separate building.

Public restrooms must be provided as part of the harbor complex on the basis of about 1 toilet for every 15 berths (with equal numbers of men's and women's facilities) unless local authority specifies a different ratio. Restrooms of such capacity should be located no farther than about 1,000 feet from any slip. In areas prone to vandalism, it may be necessary to keep all the restrooms locked and provide passkeys for the slip tenants. Where this practice is prohibited by ordinance, all fixtures should be of a type that cannot be twisted off or otherwise easily damaged. Pushbutton or spring-closing valves should be used in lieu of ordinary faucets, and washbasin drains should have no plugs. In some areas it has been necessary to install warm air hand driers in place of paper towels that can be used to form plugs to stop up drains. Some marinas provide showers in restrooms, but always requires a passkey for entry. The architecture of the restrooms should agree with that of the other facilities of the complex.

Trash containers must be provided, and should be located near slips with easy access for collection and, if vandalism is a problem, be located behind a locked gate or otherwise made secure against removal, overturning, or entry. The containers are usually at or near headwalk landings, often alongside a small shed or enclosure used for storage of pier maintenance equipment such as brooms, pails and hose. In many marinas, patrons have poured crankcase oil into trash containers; this can be avoided by providing special oil containers.

Shopping centers, transient housing facilities, restaurants, snack bars, and recreational facilities within the marina complex should be planned along with the more basic facilities. The architecture and floor plans should be checked by the marina management or an architectural review board, and adequate access routes, parking, lighting, and public restrooms provided.

The entire land area of the marina must be served by a road and walkway system (with adequate lighting) that provides good traffic circulation and pedestrian access to all the facilities in the complex. The need for adequate parking has been mentioned, and parking-lot design will be discussed later. Because many nonboaters go to marinas just to observe the activity, observation points, some in the form of kiosks, are often provided at convenient points.

In most residential or commercial projects, the road and walkway system usually provides the routing pattern for utility lines, storm drains, and sanitary sewers in the land area of the complex. The design of such a system is beyond the scope of this manual, but the planner and engineer can obtain the necessary criteria from several texts on the subject.

Esthetic considerations normally require that the electric power and communications system of a marina be underground. Therefore, advance planning of underground vaults and conduits is essential. Small-craft harbors in saltwater environments often have chemically or galvanically active soils that require care in routing direct burial lines and conduits.

One small item, not usually considered before marina construction, is that many boaters have pets. A conveniently located dog run will help preserve the cleanliness of the complex. Some unused part of the land area that is generally hidden from view is best and should perhaps be left to develop natural foliage without special landscaping effort.

Certain land areas of the complex may require fencing. At some marinas, a complete perimeter fence with entrance gates kept locked at night may be required for security against theft and vandalism. The amount and nature of the fencing will vary with the proximity of the harbor to public parks and beaches, depressed or high-density residential property, or busy commercial and industrial areas. In planning the fencing and security of the complex it is advisable to consult with local police authorities. In some public marinas where fencing was not installed to allow complete freedom of access for good public relations, theft and vandalism have been reduced by permitting or increasing the number of live-aboards in the slips.

g. Dry Storage. The need for dry storage along with wet storage at a small-craft harbor arises from either a lack of sufficient space in the protected waters or a desire by some boatowners to keep craft dry when not in use. Because of limitations on size and weight that can be handled safely by ramp or hoisting equipment, dry storage during the operating season is normally limited to boats not exceeding 2 tons. Most sailboats under 16 feet are unmasted and stored by hand, keel-up in racks (Fig. 143). Powerboats under 24 feet are often stored with forklift equipment, right side-up in racks (Fig. 127). Larger powerboats and sailboats are usually stored on special trailers built for transport over public roads (Fig. 144); or on adjustable cradle dollys furnished by the marina, and moved either by hand or small tractors.

If dry storage is on trailers, the facility may be only a designated area where the trailers may be parked. The boatowner is then required to move the trailered craft to and from the hoist or ramp with his own car. Where the dry storage is in racks or on dollys, the facility operator must launch and return the craft to storage at the owner's request, usually for a predetermined fee. The primary advantages of this system are that less space is used for storage, and the owner can phone an advance request to have his craft launched and ready for boarding at a certain time. This procedure saves waiting time for the owner and allows the operator to spread his work more evenly over the time available for a series of launchings. The owner can leave his boat at the boarding dock for the operator to return to storage whenever time permits. A hoist-launching and dry-storage layout designed for such a system is shown in Figure 145.

Figure 143. Small sailboats unmasted and stored keel-up in racks

Figure 144. Trailered sailboats dry-stored at the Washington, D.C Sailboat Marina.

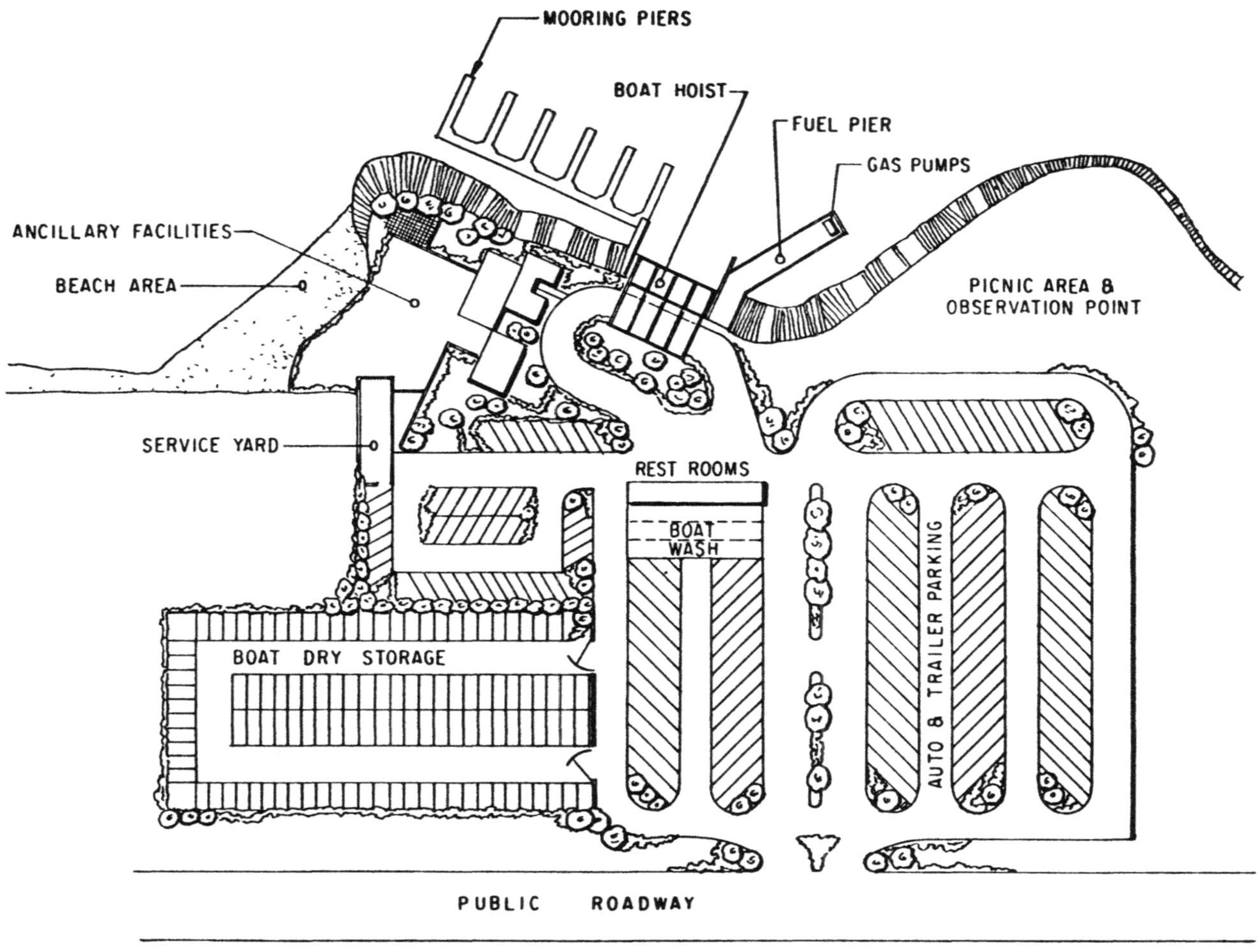

Figure 145. Hoist-launching facility with dry storage yard.

Many dry-storage systems have had a poor experience record because of a relatively long launching cycle. A system using certain proprietary equipment will increase the storage capacity and at the same time speed up the launching cycle (Fig. 146). The system requires a storage building extending about 40 feet into the basin on piling and lined with boat racks up to 8 tiers high. A swiveling stacker crane, suspended from a traveling gantry frame, moves between the tiers. The operator, riding with the stacker, guides the forks under a boat on either side of the building, lifts the boat from a rack, and launches directly into the water. An associate can then move the boat to an unused part of the boarding dock or to a holding slip. This process is reversed when the owner returns the boat.

In saltwater areas the hulls of dry-stored boats should be hosed with freshwater before storage; these provisions should be incorporated in the system. A special method can be used with the system described in the preceding paragraph. The retrieved boat is placed on a dolly, attached to a moving endless chain that pulls the boat through a washdown area outside the building and then back into the building. At a certain point, the dolly is automatically disengaged from the chain and remains on the floor of the building until the stacker has an opportunity to pick up the boat and place it in a rack.

Many yacht clubs use dry storage extensively to accommodate increases in membership. Figure 147 shows a typical sailboat stored on a trailer on the concrete apron adjoining a pair of jib hoists (Fig. 148). Trailered boats are moved to and from the hoist area by a special powered trailer-puller (Fig. 149). Where no handling equipment or hardstand storage space is available, boats may be rolled up on a beach for dry storage on a sand dolly (Fig. 150).

A novel but very practical system of launching small boats from dry storage is the hinged floating ramp (Fig. 151). The boat is placed on a two-wheeled dolly (Fig. 152), and then moved down the ramp until the wheels rest against the bottom stop curb. The boat is then pushed off into the water. This device can be used for dinghys, small sailboats, and small outboards.

Many small craft larger than 25 feet carry or tow dinghys when cruising, so they can go ashore in areas or at destinations that have no landing docks. In home port these little craft must be dry-stored. The owners usually prefer racks on the headwalks near the slips, but the use of space over the water for this purpose is frequently unjustified. Dry-storage racks for dinghys, similar to the sailboat racks shown in Figure 143 but smaller, should then be provided on shore near the slips they serve. The number of racks required will vary from one marina to another and from region to region. The requirement is so variable that no universally applicable formula can be given for predetermining the number needed in a marina. It is recommended that the designer of a new facility check the experience records of several nearby harbors before establishing the number of dinghy racks to be installed.

h. Parking Lots. After the required number of parking spaces has been determined and the sites of the parking lots delineated, they must be designed for optimum functional use and pleasing appearance. Rectangular lots are the easiest to design, but other shapes are often imposed by terrain limitations and space availability. Some of the criteria normally

Figure 146. Swiveling stacker crane for launching and storing boats (Courtesy of Marina Associates)

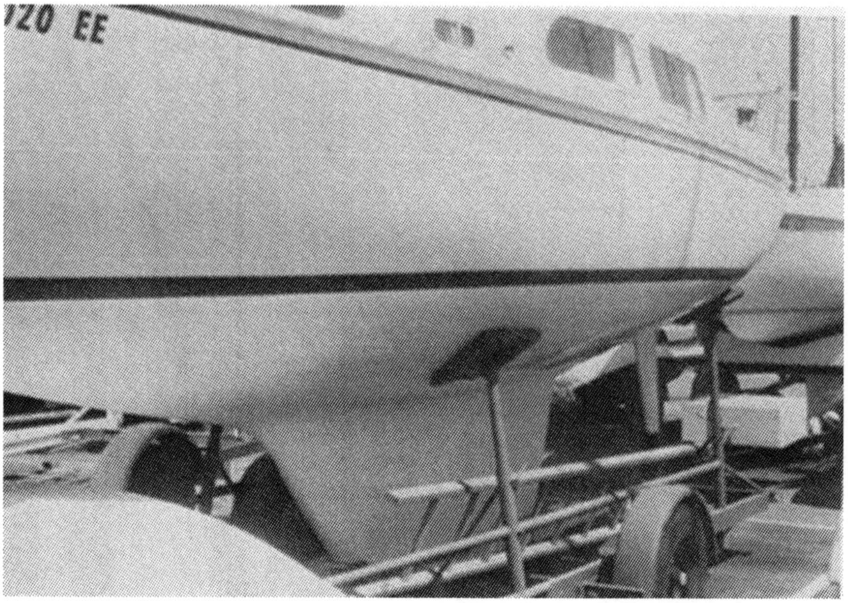

Figure 147. Deep-keel sailboat stored on trailer at yacht club.

Figure 148. Jib-boom hoists and large boarding dock at yacht club

Figure 149. Powered trailer-puller used to move trailered sailboats to and from hoists.

Figure 150. A well used sand dolly.

Figure 151. Hinged, floating sailboat-launching ramp.

Figure 152. Two-wheel dolly used to launch small sailboats

used for parking lot layouts are shown in Figure 153. Detailed criteria can be found in specifications issued by some controlling agencies that have strict regulations for such layouts. Other agencies leave the design to the engineer or architect. Generous use of curbs and planters will improve the appearance of the lots. Maintenance of clearly visible lane striping and provision of bumper-stops will help control the neatness of the parking lots and ensure optimum use of available space. If necessary, painted direction arrows will help keep traffic flowing in the intended patterns. The basic pavement in nearly all parking lots is about 3 inches of asphaltic concrete placed on a carefully prepared subgrade of compacted earth, or if the ground is too soft, on a 6-inch bed of rolled gravel.

Waterlines and powerlines under the parking area pavement must be carefully planned to supply water to the sprinklers in the planters and power to the light standards. If not otherwise specified, lot lighting should be the intensity of not less than one foot-candle. This should be increased to 5 foot-candles in areas of high vandalism. Lights on high standards should be shielded and not cause glare outside the lot and especially in water areas of the complex. The surface slope of any parking space should not exceed 4 percent, and the entire lot should be graded to drain properly.

It is not customary to charge a slip renter a parking fee at a marina. Slip-rental rates should be set high enough to cover the cost of providing adequate vehicle parking space for the boatowner and his guest. Some marina operators have tried reserving parking spaces by slip number, but that usually leaves too many spaces unused. The most common practice is to post signs in the parking lot designating the spaces reserved for boatowners and those available to visitors. The number of reserved parking spaces is usually about half the number of berths in the marina, which is adequate for normal use. On peak days, the overflow boatowner parking requirements are accommodated in *visitor* spaces or nearby parking areas designed for ancillary facility patrons. A large restaurant parking area, for example, may sometimes be located to handle overflow parking needed for boating patrons during peak hours.

Many launching ramps require a use-fee, which may be more easily collected by charging the user for his car-trailer parking space. Such spaces can either be paid for by a coin-operated gate or a ticket-dispensing gate opener with a pay-as-you-leave checkout booth. Coin-actuated parking meters are seldom used at a marina or launching ramp because of the difficulty of monitoring meters, and most patrons do not know how soon they will return. In most regions both patrons and visitors at a boating installation resent having to pay for parking, and the general trend is not to charge for parking privileges.

i. Sign and Bulletin Boards. The generous use of neatly painted signs in a small-craft facility will save many hours of explanation to visitors and new patrons. Signs usually found at a marina are: (a) a welcoming sign at the entrance giving the name of the facility and owner, (b) direction signs where appropriate, (c) signs designating buildings, and (d) signs concerning parking regulations, slip-rental rates and schedules, pier and slip designations, ramp-and hoist-use regulations, and sanitation and antipollution regulations. A little

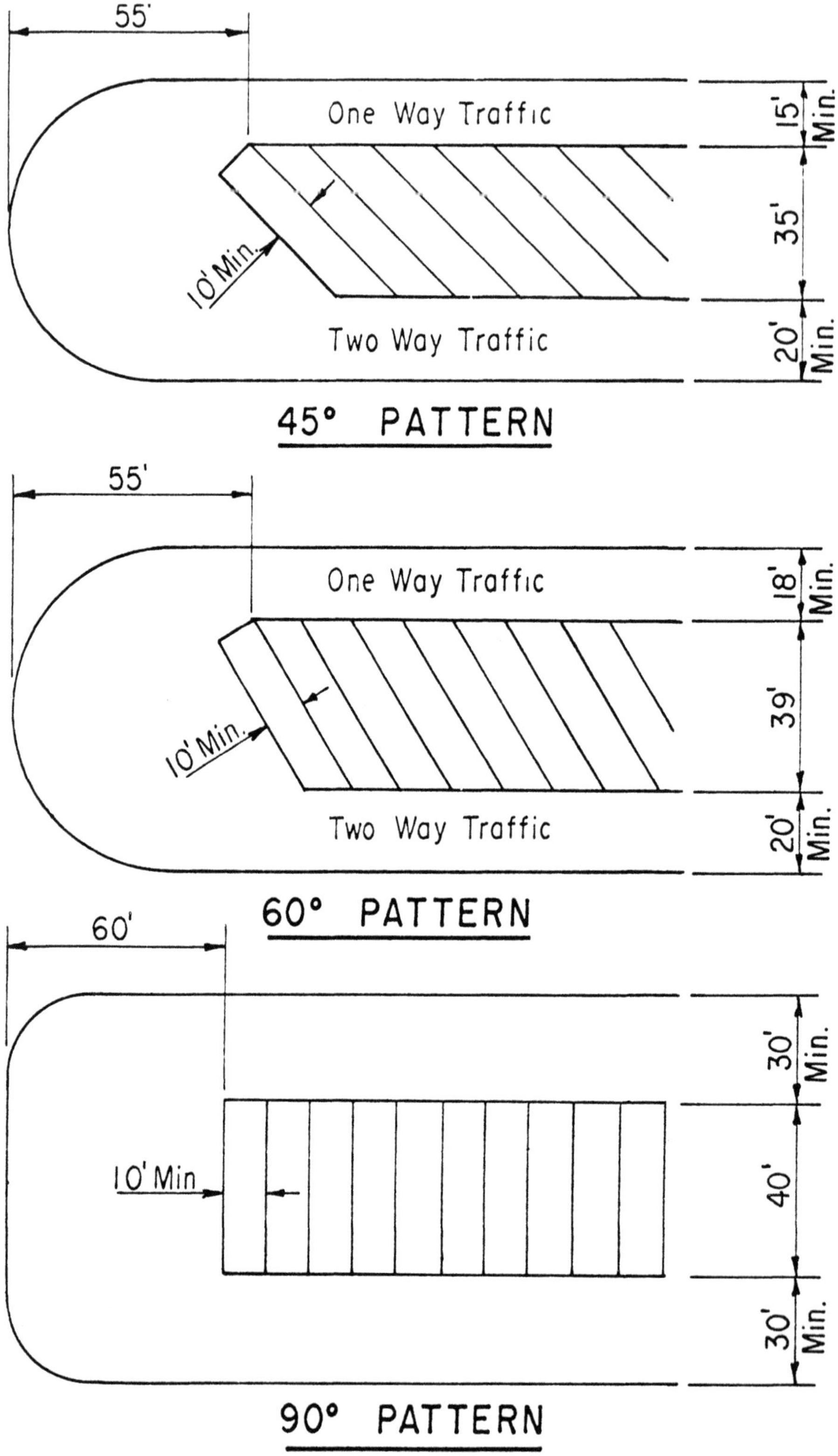

Figure 153. Standard parking lot layout criteria.

planning will develop a functional array of signs for the initial opening, but certain repetitious questions at the information counter or evidence of misuse of any of the facilities or grounds may soon indicate a need to supplement this initial effort. Some examples of permanent sign posting are shown in Figure 154.

A large bulletin board, located where harbor patrons frequently congregate or where most normally pass to reach their boats, can disseminate a large amount of information. Notices of regattas or other water-sports events, classes in boat handling and boating safety, meetings of boating clubs, new places to go by boat, reciprocation privileges shared with other marinas, and numerous other items can be posted on the bulletin board. It is important that the management keep the board current, regulate its use to best serve the need of the marina patrons, and ensure that all are made aware of events and items of interest. In marinas having a large segment of permanent patrons, a printed newsletter mailed periodically may be needed to reach owners who do not use boats regularly. A copy of the latest newsletter should be posted on the bulletin board.

6. Environmental Protection.

a. Snow. In the colder climates snow loads may contribute to the problems of a small-craft harbor. Floating systems are seriously affected by excessive snow loads, and only the covered floating berths are apt to be submerged by a heavy snowfall. Most roofs of floating sheds cannot be made steep enough to ensure that the snow will slide off. Snow normally weighs about 0.10 as much as water, volume for volume. Hence, snow may be assumed to weigh about 6 pounds per cubic foot. The area of the roof multiplied by 6 times the anticipated highest snow load in feet will give the extra weight in pounds that is carried by the flotation. If this exceeds the design live load, the amount of flotation should be increased enough to support the difference without submerging the flotation units. For example, if the roof area is four times the deck area and the design live load is 20 pounds per square foot, the snow load will equal the live load at about 10 inches of snow fall. Beyond that limit about 1 cubic foot of foam must be added to the flotation for every 10 square feet of roof area covered by an additional foot of snow.

Boats berthed in the open during the snow season in cold regions should be covered with tarpaulins, fitted canvas or reinforced plastic covers. This is the boatowner's responsibility, but the marina management should notify slip tenants when coverage is needed or desirable. When the snow load on the covered berths reaches dangerous limits with regard to flotation or structural safety, the load should be lightened by removing some of the snow. Adequately designed fixed-pier systems are not likely to fail under heavy snow loads, but floating systems may be submerged. When this appears imminent, the snow load should be reduced either by hand shoveling or with small mechanical equipment. In most cold regions, boats are removed from the water to dry storage before the winter season sets in. A common practice is to winter-store the larger craft on cradles in the vehicle parking lot and cover with tarpaulins. Smaller boats are often stored keel-up for better shedding of rain and snow.

Figure 154. Example of permanent sign posting

b. Ice. In regions where temperature drops are not excessive and natural freezing does not cause a thick ice sheet, ice formation can be prevented near piles and floating slips by forced convection currents. A few typical systems that have had some success are illustrated in Figure 155. These systems work on the principle that water reaches its greatest density at about 39°F. and tends to stratify in layers, with the heaviest on the bottom. Forced convection of this warmer but denser water from the bottom to the surface when the surface temperature approaches 32°F. will prevent or at least postpone freezing. Unfortunately, the internal circulation of water in a berthing basin may prevent the layer stratification that makes a bubbler system work. Although success is achieved in some areas, the system is ineffective in others, and a careful study of thermal interchange in a basin should be made by an expert before installing such a system.

If a bubbler system is not used, sheet ice damage can be reduced or prevented by proper design. Piles can be driven deeply enough in some materials to develop sufficient withdrawal resistance to prevent lifting by ice, especially if the piles are cladded with metal sheaths in the ice-formation zone so that the ice slides on the smooth surface as it rises. The flotation components of floating systems can be designed with rounded or tapered bottoms so that the pinching effect of the ice merely squeezes them upward (Fig. 9).

Damage to perimeter walls and revetments due to ice thrust can be prevented or greatly reduced by selection of a perimeter treatment that is less susceptible to such damage and by correct construction methods. A smooth concrete pavement on a perimeter slope will fare much better than a riprapped slope. Thorough compaction of the backfill behind a vertical wall will help to resist ice thrust. Careful design and construction will eliminate many crevices, a major source of frost-expansion damage. On rivers and lakes where ice floes occur, protective breakwaters may provide sufficient ice-impact protection. If not, deflecting booms made of logs or heavy timbers can often be used and sited to protect the berthing area from drifting ice. Chaney (1961) presents details for constructing two types of ice breakers that may be used in rivers to help break floes into smaller pieces and offers other suggestions for combating ice problems.

c. Hurricanes. In regions with a history of hurricane winds, some additional design considerations are suggested beyond the increased windload criteria used in structural design. Guide piles and cable or chain-anchorage systems for floating docks should be capable of accommodating the design fluctuations of the water level and resisting the lateral forces of the wind. Special attention should be given to safety provisions for the fuel dock to prevent fuel and oil spills. As the electric power supply is subject to failure during the storm, the installation of standby power-generating equipment should be considered for facilities or floodlights that might be needed at that time. All major and minor improvements of the harbor such as decking, roofs, and outside trash containers, must be anchored down or fastened together to prevent movement and possible collision with berthed craft or harbor structures. Low profile, streamlined designs for structures should be adopted, hurricane-type shutters should be installed, and structural projections that increase wind stress should be avoided.

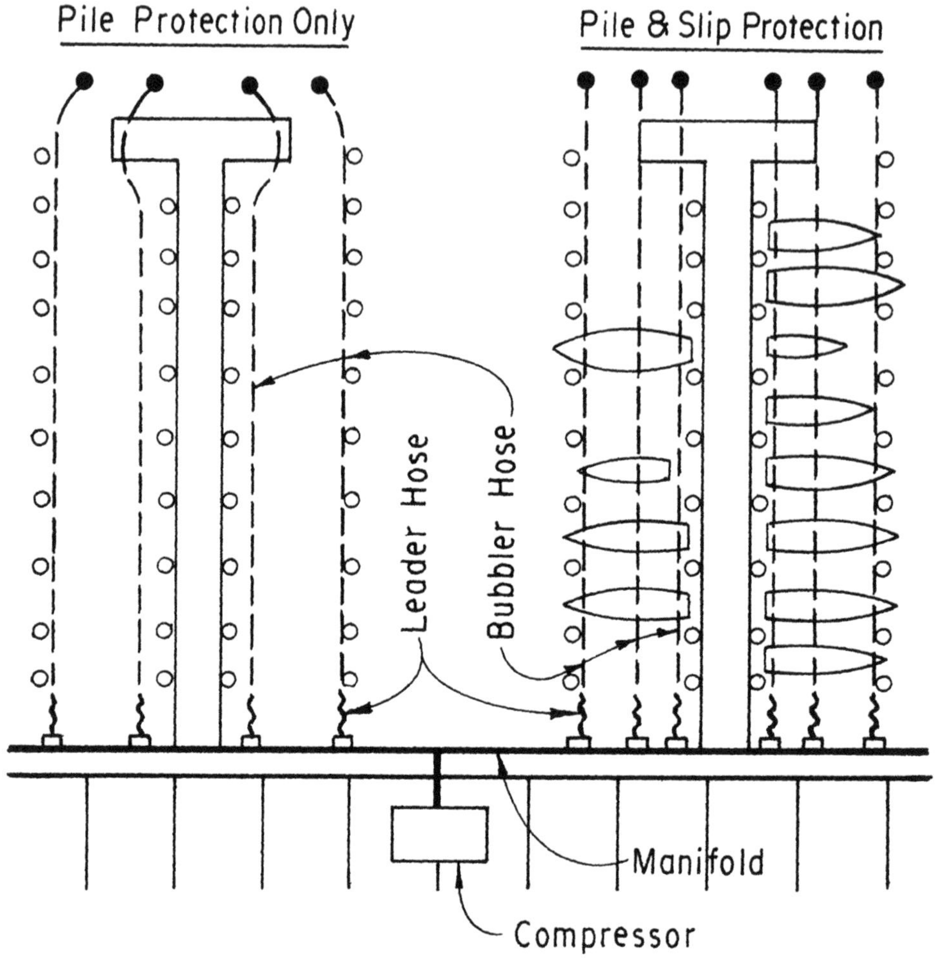

Figure 155. Examples of ice-prevention systems (Courtesy of Schramm, Incorporated).

Other precautions to avert or reduce the severity of hurricane damage and to prevent injury include: (a) removing all loose or fragile items to a protected area, (b) opening a window or door on the lee side of each building to balance pressures, (c) preparing all emergency equipment and vehicles for immediate use, (d) tightening or reinforcing the mooring lines of all berthed craft, (e) lashing racked dinghys or other small dry-stored boats that cannot be moved indoors, (f) devising a system of lifelines for harbor personnel who must check the installation during the storm, and (g) disconnecting all electrical appliances not needed during the emergency.

d. Fire. The use of noncombustible materials in as many components of a harbor as possible is the best precaution against fire damage, but economic considerations often prevent doing this. The need for adequate firefighting facilities has already been mentioned. It is important, however, that everyone concerned know how to use them, especially the harbor security and maintenance staff. Certain regulations should be posted and enforced such as prohibiting use of charcoal burners on wooden decks and smoking at fuel docks. Extension cords should be checked periodically for signs of insulation failure. During the harbor planning phase, the local fire chief should be consulted about fire safety that could be incorporated in the design. An occasional visit by a fire department inspector after the harbor is operating may uncover fire hazards.

e. Oil Spills. Large oil spills within a small-craft harbor are unusual, but outside oil spills that move into the berthing area can create a major problem. If the harbor has a narrow entrance, a simple method of keeping the oil out is to release a continuous curtain of air bubbles from the bottom all the way across the entrance. The bubbles set up a local forced convection current system that moves the surface water away from the curtain in both directions and returns it at the bottom. Because oil remains on the surface and cannot pass the screen against the current, boats can pass through without upsetting the barrier action.

Once spilled oil enters the harbor, however, removal must be done mechanically or hydraulically. Floating booms have been designed with which large areas of oil slick can be encircled or guided into an unused part of the harbor. Oil skimmers then suck up most of the oil into containers. The remainder is removed by the straw-absorption process. The U.S. Coast Guard has been working on this problem for several years and has developed various means of removing spilled oil that are suitable for different situations, and should be contacted whenever an oil slick threatens to penetrate the harbor.

f. Floods. The danger of flood damage is present only at harbors on or near a river. Harbors on the river usually have floating docks, designed so that the anchorages are secure at any elevation of the water surface. Because steep-gradient streams are poor cruising courses, most riverside marinas are on the sluggish segments of rivers as they pass through flat valleys. For this reason, flood damage is usually in the form of accumulations of debris and collisions of floating objects with structures. Therefore, berthing areas must be sheltered from the main current so that floating debris during floods will stay clear of the entrance. In flood stage, the river will usually carry a heavy silt load which remains in suspension only in areas where the current speed is relatively high. This material tends to deposit in the more quiet areas of the watercourse, such as sheltered berthing areas. The best protection against both the debris and shoaling hazards is a training wall. Orientation of the wall will cause floating debris to be deflected away from the entrance and silt deposits will occur in an area that can easily be dredged out after the flood.

Off-river basins may have minor shoaling and floating debris problems, but will be concentrated at or near the basin entrance. A greater danger is the possibility of bank scour or overflow of the main river into the basin during a major flood. This has happened a few times, with devasting results (Fig. 156). The best precaution against such an occurrence is to protect the riverbank at and upstream from the basin so that it cannot fail by scour or overtopping in that area.

g. Vandalism. The looting of boats and malicious damaging of boats and marina property by vandals is considered both an environmental and a sociological problem. The environment of the harbor site is an important factor to be considered in planning the security measures required. Several suggestions have been made for tightening the security of various harbor components and for fencing the entire installation. Since these measures increase the cost and restrict freedom of action without improving the functional efficiency of the harbor, they should be held to a minimum, and consistent with the nature of the environment and the behavioral pattern of the more antisocial elements of the local society. No set of criteria can be offered to apply in all situations, and every site must be analyzed separately. The best advice is to discuss the problem with local police officials during the harbor planning stage.

VI. PUBLIC AGENCY PARTICIPATION

1. Federal Government.

a. General. The Federal Government exercises control over navigable waters of the United States and structures built in those waters, and maintains certain navigable waterways by periodic removal of shoals and obstructions. The Federal Government also has a number of programs under which the States, their political subdivisions, individuals, groups, and associations may qualify for assistance in developing boating facilities. This assistance involves various degrees of credit, cost-sharing, technical aid, educational services, and research. The purpose of this section is to inform interested parties of the basic provisions of pertinent Federal assistance programs. Details concerning any specific program may be secured from the administrating agency. The headquarters addresses for these agencies are listed in Appendix E.

b. Primary Agencies Involved, Authorization, and Programs.

(1) U.S. Army, Corps of Engineers. This branch of the U.S. Army is responsible, among other things, for the development and maintenance of navigable waterways within the United States. Navigation improvements are authorized by Congress to assist in the

Figure 156. River flood damage, Ventura Marina, California

development and conduct of waterborne commerce. Coastal harbor improvements include channels and anchorages for both deep-draft and shallow-draft shipping, harbors to provide refuge for small craft, and breakwaters and jetties to provide protection against wave action. Shallow-draft navigation includes commercial fishing, recreational boating, and barge traffic. Improvements of inland waterways consist of deepening, widening and canalization where locks and dams are required. In addition, the Corps has the responsibility for administering the Federal laws relating to the protection and preservation of the navigable waters of the United States. These responsibilities include granting permits for structures, not including bridges, in navigable waters; establishing regulations for use of navigable waterways; removal of wrecks and other obstructions to navigation, and preventing pollution. Procedures and guidance on permit application can be obtained from the nearest U.S. Army, Corps of Engineers District Office.

The Corps is also authorized to study, and if economically justified, to aid in the construction of public small-boat harbors and other navigation improvements. This construction is limited to the waterway system of the harbor, namely breakwaters, jetties, general navigation channels, turning basins, and anchorage areas. Under present law, such Federal participation in a small-craft harbor project may be obtained in three different ways, depending on the estimated cost to the Federal Government: (a) if the Federal share of the project is less than $1 million, the project can be approved and funds allocated by the Secretary of the Army acting through the Chief of Engineers (Section 107, 1960 River and Harbors Act, as amended in the 1970 Act), (b) Section 201 of Public Law 89-298, 1965, authorizes the Secretary of the Army, acting through the Chief of Engineers, to construct projects with a Federal share of less than $10 million without specific Congressional authorization, if there is little or no controversy over the project, and if approved by resolutions of the Public Works Committees of the Senate and House of Representatives, on the basis of reports submitted to Congress, and (c) projects in which the Federal share exceeds $10 million must be approved individually by Congress, and then after review and approval by the Appropriation Committees of the Senate and House Committees, be individually funded. The Federal financial share for construction of these general-navigation features, according to present Congressional policy, varies from 100 percent for features serving commercial navigation to 50 percent for purely recreational features. Construction costs for combined harbors are usually prorated according to relative economic benefits.

Because of background knowledge developed through numerous small-craft harbor studies, the U.S. Army, Corps of Engineers may also be of assistance to any public agency in assessing the need for a new harbor at any given site and in providing advice on permit procedures, even though the project may not qualify for Federal participation.

(2) U.S. Coast Guard. This Federal agency is responsible for the regulation of boating and the control of boating safety in all Federal waters. The Federal Boat Safety Act of 1971 increased the U.S. Coast Guard's responsibility and authority in this field. Initially, the U.S. Coast Guard bore the entire burden of regulation in coastal waters, but the Federal Boating

Act of 1958 allowed relinquishment of this duty to the individual States in cases where State legislation was in conformity with the provisions of the Federal Act. Subsequently, most States passed conforming legislation, established a State Boating Law and appointed an administrator to carry out the law's provisions. As a result, the U.S. Coast Guard has developed the role of coordinator in this field and has broadened the scope of safety and regulatory activities into a comprehensive program that starts with small-craft manufacturers and continues through dealerships, owners, and marina operations.

In the development of small-craft harbors, U.S. Coast Guard assistance is limited primarily to information programs. Program aims are to promote public safety on navigable waters, to promote efficiency in the operation of powered craft and sailboats, and to facilitate U.S. Coast Guard operations in other areas. Free public courses are provided which emphasize safety, small-craft handling, and basic seamanship. A courtesy motorboat examination provides the boatman with a free check of all required and recommended equipment. Marina clientel can also avail themselves of films and lectures, provided to groups by U.S. Coast Guard personnel upon request.

(3) Bureau of Outdoor Recreation. This Bureau established in 1962 within the Department of the Interior, serves as a focal point for coordination between State and local governments, other Federal agencies, private organizations, and individuals with the aim of providing increased opportunities for outdoor recreation. The Bureau supplies financial assistance to State and local governments for acquisition and development of boating and other public outdoor recreation facilities through the Land and Water Conservation Fund Program. Projects to qualify for assistance, must be in accordance with the State Compresensive Outdoor Recreation Plan (SCORP) and approved by the Secretary of the Interior. Funding to public agencies is a 50-50 matching basis, as established by the SCORP, with project priorities determined by State liaison officers who are appointed by the Governors of the respective States. The Bureau also provides limited technical assistance in the form of referrals and information from a technical assistance clearing house.

c. Programs of Other Agencies. Several other Federal agencies offer programs related to development of small-craft facilities that meet certain unique criteria and other specific requirements. The Farmers Home Administration of the U.S. Department of Agriculture will provide low-interest loans to qualified farmowners who wish to develop income-producing recreational enterprises on farmlands. The loans may be used to develop land and water resources, repair and construct buildings, buy land and equipment, and pay operating expenses. Almost any type of outdoor recreational enterprise will qualify, including developments for golf courses, campgrounds, nature trails, and small-craft facilities.

Another program that could include marina development is the Department of Housing and Urban Development's program for *Model Neighborhoods in Model Cities.* This program provides grants and technical assistance to plan, develop, and carry out comprehensive programs for rebuilding or restoring slum and blighted areas through coordinated use of all

Federal programs, State, private, and local resources. Grants may include up to 80 percent of the project cost. To be eligible for grants and technical assistance, plans must cover not only housing, jobs, health, and education, but also associated problems such as social and recreational services. A marina development that is planned as the nucleus for a total recreation complex might qualify for assistance under the master plan for that complex.

The Economic Development Administration (EDA) under the Department of Commerce is a third unique source for possible assistance in small-craft harbor development. EDA is authorized to provide financial assistance to areas of high unemployment and low median family income to enable them to stabilize and diversify their economies and create new and permanent job opportunities. Such assistance includes: (a) grants and loans for public works and facilities development projects, (b) loans on attractive terms for business development projects, and (c) technical assistance. Table 2 is a partial listing of EDA-funded projects involving marinas or boating facilities.

Public Works grant and loan assistance by the Secretary of Commerce may be extended to tourism or recreational projects if it is shown that the project directly contributes to the fundamental growth of tourism in an area where tourism was previously unimportant, or that there is a lack of sufficient private capital to overcome a shortage of accommodations in an area already dependent on tourism. It must also be shown that the proposed facilities will provide services or have special attributes that tend to induce a net increase in tourist visits to the area, or that it is part of a larger area-wide program.

A number of other Federal agencies sponsor programs which under certain circumstances may provide aid and assistance in developing small-craft facilities. A comprehensive list of all Federal agencies that relate in any way to small-craft harbors is presented in Appendix F.

2. State Governments.

a. Agencies Involved and Control Exercised. State agencies that exercise a primary measure of control or authority over small-craft harbors or activities of small craft are listed in Appendix G. The scope of authority and degree of control exercised by these individual agencies vary considerably from State to State. A summary of the more significant services, programs, and controls offered or executed within each State is given in Appendix H.

Several State agencies have developed programs to the extent that they possess universal jurisdiction over all marine matters in both inland and coastal waters. Some States have agencies that confine their control to public waters only, or waters that are confined within or adjoining State-owned land; others exercise authority only over non-Federal waters, leaving all tidal waters to the control of the U.S. Coast Guard.

A few States have no special programs or controls of marine matters, or have but a few laws or statues that apply uniquely to development of small-craft facilities or to activities of small craft. Vermont, for instance, upholds a statute concerning land under public waters, and Maine has no laws that directly affect marinas.

Methods of administering marine affairs vary between the individual States. Several States have attempted to centralize control under one agency, such as the California Department of Navigation and Ocean Development, which is responsible for the California

Table 2. Economic Development Administration (EDA) Projects

Location	Description
Alabama	
Scottsboro	Outdoor Recreation Complex
Arizona	
Maricopa County	Recreation Facility/Community Building
San Carlos	Recreation Facility (Soda Canyon)
Window Rock	Reservoir for Tourism
California	
Dobbins	Feasibility Study Recreation
Fort Bragg	Mooring Basin/Boats
Illinois	
Shawneetown	Port Facility-Marina
Maine	
Jonesport	Public Landing and Boat Facility
Michigan	
Mackinaw City	Recreation Boating Facilities
Manistee	Marina
Marquette	Marina
Naubinaway Village	Marina
Northport	Marina Facilities
Ontonagon	Marina
Petoskey	Marina
Port Huron	Public Marina
Presque Isle	New Marine Facility
Minnesota	
Cass Lake	Marina/Trailer Park Exposition
Grand Marais	Improve Marina
Knife River	Construct Harbor/Marina
Onigum	Recreation/Marina/Camp
Onigum Point	Tourism/Marina Complex
Tower	Tourist Complex/Marina
Montana	
Browning	Lodge/Restaurant/Marina
Crow Agency	Marina and Facilities
Nevada	
Pioche	Recreational Complex
New York	
Ogdensburg	Feasibility/Marina Complex
North Dakota	
Fort Berthold Reservation	Recreation Complex/Motel/Pool
Utah	
Fort Dushesne	Tourism/Recreation Complex
Washington	
Ilwaco	Mooring Basin
Wisconsin	
Bayfield	Harbor-Marina Development
Bayfield	Refuge Harbor-Marina Development
Saxon	Improve Marina

Boating Law. Other States divide the responsibility and authority between several agencies, e.g., the Department of Fish and Wildlife Resources of Kentucky has jurisdiction over 24 lakes that are owned and managed by the Department, while the Department of Public Safety exercises control over most other waters.

The majority of States have either adopted or are in the process of adopting State Boating Acts. These Acts officially delineate the laws and codes concerning small craft and navigation and safety standards that will be enforced in the State. Subjects normally covered by State Boating Acts under the headings of laws, codes, or navigation and safety standards are listed in Table 3.

Many State general laws, codes, and statutes also bear directly on small-craft activities or facilities, such as zoning laws. Many States are now zoning land areas adjacent to water, including the water areas, with restrictions favoring commercial fishing, sport fishing, water recreation, water conservation, and commercial development. Comprehensive land planning, development, and management programs have recently become favorably inclined toward establishing marinas as hubs for recreation complexes or parks.

Most States rely on standard design and construction codes to cover the development of small-craft marinas or launching facilities, although many States also adopt specific standards to cover the special problems of these facilities. Hawaii, for instance, publishes 70 pages of regulations specifically for small-boat harbors.

Ecological factors have become a major concern of most State-regulating agencies. Stringent requirements concerning the ecological features of environmental impact statements that are prerequisite to any construction project are not uncommon. State agencies are also concerned with water quality, pollution prevention, and waste and sewage disposal problems in marinas and other areas frequented by small craft. Even emissions from outboard engines have come under critical scrutiny.

In Florida, the Division of Marine Resources has as one of its primary goals the minimizing of the possible adverse impact that marina and harbor development may have on marine biological resources. Therefore, in planning for development, consideration is given to protecting marine grassflats, mangrove fringes along the water's edge, and the maintenance of good water quality in marina areas. In design, the division favors riprap or sloping revetments over vertical bulkheads because they reduce wave reflection and provide more area for attachment of marine organisms.

Similarly, the Nebraska Game and Parks Commission concedes that regulations covering the construction of small-craft harbors and marinas are mainly directed toward control over ecological factors, and has adopted very stringent water quality and pollution standards. The State of New York requires a review of all marine engineering projects, investigates their impact on the environment and on navigation, and requires provisions for sanitary facilities and pumpout stations at all marinas. New York is also among the few States that require an economic feasibility study showing a favorable benefit-cost ratio for all proposed projects.

b. Assistance and Development Programs. The range in scope and degree of participation varies widely between the States in assistance and development programs for

Table 3. Subjects Covered by State Boating Regulations

Age of operators
Application of State law; local regulations
Boat and water funds
Capacity of watercraft
Device for arresting backfire
Duties of sheriffs, conservation officers, county boards
Enforcement for violations
Fire extinguisher; ventilation
Hazards to navigation; removal of buoys or structures
Horn or whistle; siren
Lights
Liquor, drugs, physical or mental disability
Marine toilets
Muffler
Navigation of watercraft on waters of State
Obstruction of navigation; advertising; buoys
Operation generally
Personal property taxes
Race or other competition or exhibition
Reimbursement of county sheriffs for search and rescue operations
Rental of watercraft
Riding on gunwales or decking
Safety equipment
Stopping at scene of accident; reports, liability
Swimming or bathing areas
Temporary structures and buoys
Water skiing; scuba diving
Watercraft licenses
Waterway markers

small-craft facilities. Most of the programs are partially or totally financed through revenues derived from taxes on marine fuels, registration fees, and various tariffs.

State-supported grant programs are the most common forms of assistance. The Ohio Department of Natural Resources provides grants for civic marina and civic launching-ramp development to a maximum of two-thirds of the total development costs. The Oregon State Marine Board and the Washington State and New York State Parks and Recreation Commissions furnish grants covering 100 percent of the same qualifying developments. A common practice for States that offer grant programs is to reserve rights to review engineering designs and proposals and to enforce construction and design standards.

The same reservation of rights is evident in State-sponsored loan programs especially where loans are not restricted to civic marina developments. The following excerpts from the California Boating Law (1971), shows the control exercised by many States concerned with marina planning or construction loans:

"(1) Feasibility Study: A report containing sufficient information and detail to demonstrate that the project is engineeringly and financially feasible, and economically justified. The report shall include, but not limited to:

(a) A project plan and map which establishes the project area and location.

(b) Preliminary layout and designs of project features in sufficient detail to develop accurate cost estimates.

(c) A plan for operational and fiscal management of the project throughout the loan period.

(d) The proposed method and means of retiring the loan, meeting other financial obligations of the project, and, if the project is to be undertaken with funds in addition to the construction loan applied for, a funding plan indicating the source of such additional construction funds.

(e) A report on the effect the project would have on the environment. The department will provide the sponsor with a copy of the latest law on this subject.

(2) Feasibility Review: The applicant's formal application will be judged as to the following principal considerations:

(a) Engineering feasibility, including a determination as to whether or not the project can be developed within the total amount of funds to be made available.

(b) Economic justification, including an evaluation of the benefit-cost ratio in accordance with the department's criteria.

(c) Financial feasibility, including an analysis of the availability of capital to finance construction to completion, user's willingness and ability to pay anticipated berthing and other charges used in estimating revenues, and evaluation of the sufficiency of revenues to cover annual cost on a year-by-year basis, including the amortization of the applied for loan."

About 10 percent of the States fund and administer development programs have *no strings attached.* Alaska probably has the most comprehensive program of this type. Other than a few private marina facilities in Ketchikan, Juneau, and Sitka, nearly all small-boat harbors have been provided by the State. The Division of Water and Harbors is the primary agency responsible for the development of small-boat harbors and other shallow-draft vessel facilities in the State. The division provides public docks, floats, grids, launching ramps, and other associated harbor improvements. Small dredging and filling projects are included if sufficiently justified and are not covered under any Federal program.

The Alaska harbor program is a traditional carryover from the former territorial government and is geared to tax income from marine fuel sales. In keeping with a general State policy of delegating management responsibility to local interests, the facilities are leased to local government entities wherever possible. As a matter of policy the lessee is responsible for the installation of electrical utilities, sanitary facilities, parking areas, and other supporting shoreside facilities and minor maintenance. Of the 60 separate facilities in Alaska, 42 have been leased to the local community for the nominal fee of $1 per year and the remainder are under management control of the State. Responsibility for major repairs at all facilities rest with the division. To maintain some degree of uniformity throughout the State, the standard lease agreement specifies minimum moorage rates, together with regulations regarding other basic operational procedures. In general, however, the community has broad authority to operate the harbor as it may deem appropriate.

The goals of the Alaska Division of Water and Harbors are to bring all existing harbors to optimum capacity and to provide new harbors where needed. The objectives of the program are to: (a) reduce or prevent the incidence of property damage or loss of life by providing protected small-craft moorage facilities, (b) enhance and promote the development of fishery resources, and (c) develop opportunities for the enjoyment of the recreational potential of the coastal and inland waters of the State.

Boating facilities in the State of Hawaii are generally designed, constructed, operated, and maintained by the State Department of Transportation, Harbors Division. Much of the design and construction of main channels and protective structures is accomplished in cooperation with the U.S. Army, Corps of Engineers. The State encourages private enterprise to enter this field, but to date the private sector has been discouraged by several factors, including the shortage and high costs of land and protective structures.

In the Commonwealth of Pennsylvania, the Fish Commission administers a comparable program. The commission enters into agreement with local governments under which the State takes a long-term lease on a harbor site to construct a small-craft facility. The completed facility is then returned to the local government for operation and maintenance. The State funds each project in entirety as long as the facility is designed to serve only the needs of the fishermen and boatmen (from whom the funds are derived). If the local government desires supplemental facilities beyond the exclusive needs of fishermen and boatmen, local funds may be provided.

Many States do not have programs as extensive as the three described, but do provide significant assistance. The Illinois Department of Conservation collects marine fuel taxes in excess of $2 million per year, which are used for developing boating facilities. The department does not construct small-craft harbors, but their program is oriented toward construction of access facilities to serve the needs of boaters. Statistics show that 90 percent of the boaters of Illinois trailer their boats and use the State-provided launch ramps, parking areas, comfort stations, drinking water, lighting, and small picnic areas. Facilities are located primarily on navigable public rivers of the State and Federal-State impoundments, and the interior lakes of the State. New Mexico, North Carolina, and Minnesota have programs similar to that of Illinois. Other States offer little assistance or none at all.

c. Information Programs. Many States provide significant assistance in the form of information centers and aids to planning small-craft harbors and other boating facilities. These States usually have compiled, or can gain access to data and reports that can help justify either local-public or private participation in small-boat harbor development. Reports on traffic patterns, annual income figures for existing facilities, boating and fishing seasons, and fish populations are particularly useful in the economic evaluation of any given site.

Most States publish data on availability of small-craft facilities and on boating activities within those States. Many States sponsor special boating activities such as sea festivals and regattas, and some develop marine parks. Florida has an underwater State park on one of its Keys. These State activities and sources of information, together with such ancillary services as navigation and boating safety courses offered by agencies of many States, complement the Federal programs and local development efforts in furthering the advantageous use of navigable waters throughout the Nation.

3. Local Governments.

a. General. Most marinas, with the possible exception of those being constructed by the Federal and State government on their own lands, will require some degree of interface with the local government. The relationship of a marina to the local community is based on mutual needs and cooperative goals of the parties involved, including the community. Communities have developed rules and regulations that must be respected, and many also have master plans for economic growth and for land development and management. These plans must also be respected. A new waterfront development, including a small-craft harbor or marina, must be compatible with the established plans of the community or it will not receive the support of civic leaders or administrative officials.

Many municipal and county planning boards have already provided for waterfront development, and a small-craft facility may be the logical improvement of an undeveloped waterfront. A city or county can derive as much benefit from a marina as from a park, playground, golf course, swimming pool, or any other recreational facility. Many marinas, often combined with one or more of these other facilities, have proved to be assets to the local community.

From a strictly fiscal standpoint, many communities have experienced increases in land values around marinas that have measurably broadened their tax bases. In several cases the expansion of waterfront boating facilities has been directly responsible for a new and significant prosperity in the surrounding area. The many locally sponsored information and development programs testify to the value placed on marinas. The time and effort devoted to the drafting and adoption of regulations controlling the operation of small-craft facilities also demonstrate this favorable concern on behalf of local governments.

In some instances, a small-craft harbor is located to serve not one but two or more cities; in others, the harbor serves only the coastal area of a large city or county, and people living inland have little or no interest in boating and will not support community efforts to establish boating facilities. For either situation, the formation of a harbor district or some other improvement district may be the best solution. Most States have laws under which special districts can be formed, and when established by a vote of the people living within the designated district boundaries, they are vested with such taxing privileges and development and regulating authority as to enable them to provide the locally desired harbor facilities. These districts have all the capabilities of a city or county government in the area of boating-facilities development and come under the same general category of local government insofar as small-craft harbors are concerned.

Some special districts and authorities that have been established for other purposes or for recreational facilities development in general also are endowed by law with small-craft facilities development capabilities. They include port districts, recreation districts, water-resources districts, and port authorities; each may be considered an agency of local government in later discussions.

b. Regulations and Permits. The regulations and permit requirements originating at local government level are generally intended to complement the State and Federal regulations applicable to a given area. The local regulations are usually more broad in scope and more detailed in technical coverage because they must take into account those characteristics of the local environment that may have a special impact on local construction or development. The design, construction, and maintenance of a marina complex must comply with the regulations established by the local public entity governing the area in which it is built. The minimum standards and requirements for marina construction will be set forth in ordinances, codes, rules, and regulations of the public entity, whose administrators will usually measure compliance therewith against a set of documents such as: (a) Uniform Building Code, (b) Uniform Plumbing Code, (c) Uniform Wiring Code, (d) National Electric Code, (e) Water and Sewerage Standards, (f) Standard Land Development Specifications, (g) Standard Design Specifications, and (h) Uniform Fire Code. Local site characteristics may dictate additional regulations intended to reduce damage from hurricanes, tornados, earthquakes, and extreme weather conditions. Some local codes apply specifically to marinas, such as berthing facility and fire protection standards for boatyards and marinas.

Many local governments that have joint public-private marina development programs publish standards with titles such as *Design Criteria for Construction by Lessees*, which deal solely with marina design and construction. These standards include topics about site work, buildings, signs, and mooring facilities. Site work includes controls over items like those listed in Table 4. The regulations in the *Buildings* section are usually directed toward achievement of harmony between structural design and space utilization, and use of architectural themes to enhance the harbor complex. This type of control imposes no restriction on architectural design; its purpose is to achieve a proper blend of esthetics and functional adequacy. Design requirements could include establishment of public viewpoints overlooking the entrance and harbor areas and use of building materials that will create weather-resistant, low-maintenance structures. They could also cover screening from public view such items as utility and mechanical equipment, trash-collection centers, and service areas, and provide criteria for setback of structures, and use of color on exteriors.

The section on *Signs* should include topics on sign posting, sign design, restraints on advertising signs, required permanent signs, and controls over use of temporary signs. A typical design that may be required for temporary signs at a marina is shown in Figure 154.

The section on *Berthing Facilities* will normally establish all controls and regulations pertaining to the design and construction of the interior harbor facilities previously discussed in Section V.

In addition to preparing the technical regulations, the local entity must designate the proper procedures for submitting designs, acquiring permits, requesting plan reviews, obtaining appeals, and executing resubmittals. Data formats and required time schedules must also be provided for the guidance of prospective developers.

If several small-craft harbors are to be constructed or already exist, the local city or county government may elect to regulate their operations by ordinance. A model for such an ordinance, as drafted by the California Marine Parks and Harbors Association, is presented in Appendix I.

c. *Development Programs.* The municipally sponsored marina is often financed by bond issues. Because public funds are used, it is under mandate to provide the best possible service to the public. Thus, municipal authorities must ensure that the resources of each municipal installation are used to the fullest potential, preferably on a self-sustaining basis, either as a civic enterprise or as a leased operation.

Many municipalities provide the inner basins, perimeter protection, and entrance to a harbor, but allow private enterprise to provide the berthing and ancillary facilities under lease agreement. For ancillary facilities, only the site is provided by the local government and the lessee develops it completely except for utility and sewer systems.

Leases to the private sector are usually granted to provide services not normally performed as government functions, but which will enhance public use and enjoyment of the harbor. When the harbor administrative authority determines that a service should be

Table 4. Site Work Items Requiring Design Control

Curbs, Gutters and Walks	**Perimeter Screening and Barriers**
Configuration	Falls
Material	Fences
Mixing, forms and finish	Other barriers
Drainage	**Grading**
Catch basins, storm drains, sinks	Required slopes
Outfalls	Tie-in to universal datum
Curbs and gutters	Fill compaction
Hydrology study	Fill materials
Landscaping	**Area Lighting**
General plan	Luminaries and standards
Lawns	Circuitry and switching
Small shrubs and plants	Height
Trees	Density
Sprinkler system	Shielding
	Safety
Signs	
Utilities	**Parking**
General	Boat slips and launching areas
Sewers	Ancillary facilities
Electrical	Employees
Water	Surfacing
Telephone	Curbs, planters, tire bumpers
Rubbish collection	Surface markings
Public address system	Traffic patterns

provided through a lease agreement, definitive specifications for the scope of the lease are prepared. Lessees are then selected either through direct negotiation or from a list of qualified parties who have submitted proposals that comply with the lease scope specifications. It is not always necessary or even advantageous for the lessee to be selected through competitive bidding procedures. Applicants are usually evaluated on the basis of financial status, experience, and competence as well as on the plans submitted and the proposed development schedules. The design and construction must meet all requirements specified in the local agency's *Design Criteria* document, and the lessee's development will be subject to review and controls of the local authorities.

Lessees usually are required to pay for general maintenance and repair of landscaping, property and equipment, and any alteration or improvement of the leased area must be in accordance with an approved master plan. All leases are subject to periodic review to ensure that reasonable maintenance, modernization, and operational practices are followed.

Leases are granted for as short a term as is consistent with sound economic policy. The object of a lease in the eyes of the governing authority is to provide the best possible service to the general public at the least possible cost commensurate with quality merchandising, operations, and service. The lessee must be allowed to realize a reasonable return on his investment while paying a fair rental to the municipality. Leasehold fees may either be assessed at a flat rate or computed as a percentage of gross incomes with a yearly minimum. The initial fees are often set low to allow the lessee adequate time to develop the project into a profit-making enterprise. After it is operating with a profit margin, the rental fee may be increased sufficiently to allow the city to realize a fair return on its overall investment in the harbor, which will probably include many nonprofit features.

Normally, long-term leases are executed under an agreement whereby all improvements become the property of the city on termination of the lease period. In the event of cancellation before the termination date, the improvements may be bought by the municipality in accordance with a predetermined amortization schedule.

Another method of local government sponsorship is the *lease-back* arrangement whereby the facility is built with public funds for some special or revolving account and then immediately purchased under prior agreement by a large financing firm. The purchase price is returned to the special account where available for other purposes, and the financing firm leases the facility back to the public agency for operation and maintenance on a rental basis. The primary objective of this financing system is to allow the city to develop the harbor and control its operation without tying up a large expenditure of public funds over a long period of time.

Various standard provisions often included in local agency-private sector lease agreements are: (a) audit of records by agency, (b) quiet possession by lessee, (c) posting of performance bond by lessee, (d) entry and inspection privileges of agency, (e) reservation of easement rights by agency, (f) no subletting by lessee, (g) compliance with all laws, (h) adequate insurance, (i) breach of contract terms, and (j) responsibility to receive notices.

d. Information Program. Local governments may be able to achieve some small-craft facility development goals simply by disseminating information concerning opportune sites for such development and then offering further advice and planning assistance wherever private enterprise is favorably inclined. The prominent display at *City Hall* or other public buildings of a master plan showing the areas where such development is desired, and occasional feature articles in the local press covering success stories on similar projects elsewhere, may plant the seeds of future implementation in the right places. Some of this work may be sponsored by the local chamber of commerce and other civic-minded groups in cooperation with municipal officials.

The planning department of the local government may assemble certain statistical data that will help a private developer to evaluate the potential of the site. The road department should be prepared to discuss access routes to new facilities and possible public road extensions to accommodate the new development. The local Department of Building and Safety may establish a special section to handle marine affairs and prepare a special set of minimum standards for berthing and marine ancillary facilities that will assist the developer in estimating his construction costs. This section should also be prepared to answer questions concerning disposal of surplus dredged or excavated materials and the special requirements of the cognizant water quality control and environment-impact agencies for waterfront construction. A cooperative attitude on the part of civic officials may pave the way to securing the desired waterfront development with a small expenditure of public funds.

Private organizations offering information in small-craft facilities are listed in Appendix J.

VII. ECONOMICS IMPLEMENTATION AND OPERATION

1. Feasibility Studies.

a. Purpose and Procedure. The decision to investigate the feasibility of constructing a new small-craft harbor is usually reached through consideration of more favorable aspects. Unless the need was fairly obvious or the economic potential attractive, the project would never have reached the first planning stage. The purpose of the preliminary feasibility study is to determine whether the initial indicators of desirability will remain valid in the light of all the many factors that must be considered, some of which may prove to be unfavorable. The scope of the study should not normally cover more than is required to establish beyond reasonable doubt that the project is either worthwhile or should not be undertaken. In some instances, the scope must be extended to determine the optimum size of the new facility or, if the size initially considered proves infeasible, whether a larger or smaller size will show feasibility.

The first step in the study is to search out any factors that may disqualify the project on grounds other than economics. It could be found, for example, that expansion of a nearby existing harbor will accomplish the same purpose at less cost. An initial rough evaluation of the environmental impact of the project may show that it cannot qualify for a permit under established local regulations. An initial evaluation of the physical characteristics of the site may reveal construction or operational problems too great for further consideration of the project. If an initial analysis fails to reveal any potentially disqualifying factors, each of the factors discussed below should be investigated, generally in the order listed. The report should then summarize the findings, list the conclusions, and present recommendations for further studies needed in project planning.

b. Selection of Site. If more than one potentially feasible site is available, the first efforts of the study should be directed toward eliminating those that are less desirable and reaching a decision as to the best site as soon as possible. Each site should first be subjected to the elimination tests suggested earlier; the surviving sites should then be analyzed only to the extent necessary to establish which is the best, all factors considered. If two or more sites appear to be about equally well qualified, each must be evaluated sufficiently to enable the owner or the sponsoring public agency to reach a decision.

c. Availability of Public Assistance. If the sponsoring agency is a public entity, such as a city or a harbor district, some form of contributory assistance may be available from one of the higher-level government programs described in Section VI. In some instances, the feasibility of a project may depend on whether it qualifies for assistance, and in any event, a feasibility report is usually required as a feature of the application for assistance. It is important to ascertain at the beginning of the study the items of information that must accompany the application and the leadtime required to obtain the assistance offered. If it appears questionable that the project actually qualifies, or if the indicated leadtime is so great that it is doubtful the assistance will arrive in time to be helpful, the feasibility study should present an analysis of the economics both with and without assistance.

Direct public assistance is seldom available to a private developer. However, many projects that have been initiated by private enterprise include public features, the desirability of which was first pointed out in a developer's preliminary feasibility report. For example, a large privately owned tract of land adjacent to a good harbor site may be worth developing only if the harbor is also developed. A city (or other local public entity) may desire to develop the harbor, but cannot qualify for State or Federal assistance until it owns some of the perimeter land and can demonstrate full-use potential. The developer's feasibility study is then prepared on the premise that if he deeds a part of his land to the city and develops the remainder into a residential-commercial complex, the full-use potential of the harbor can be realized and the city can qualify for assistance. Other possibilities of private-public cooperation can often be found that will benefit both parties. Such possibilities must be sought out and analyzed in the early stages of the feasibility study, as they may set the course for the remainder of the study.

d. Socioecological Aspects. An environmental impact statement must be submitted on nearly every small-craft facility project in most States. It is important to learn very early in the feasibility investigations whether a study is actually needed or is likely to cause the project to be rejected. This information can often be obtained simply by asking the proper agencies. The local water quality control agency may know immediately whether the project will be acceptable from a water quality standpoint. If not, it will outline the information that must be submitted to obtain a ruling. A ruling will probably have to be obtained if the project requires dredging, placing fill material in the water, dry excavation to be flooded later, or disposal of waste materials in nearby waters.

The probable effects of harbor construction on the ecology in the project area must be reported in the impact statement if required. Local building and safety permit authorities can usually provide information on other local or State agencies that exercise jurisdictional control over local environmental matters and whether projects of the type under consideration have a good or poor acceptance experience. If the experience is generally good, the impact statement may be deferred to a later stage of planning. If experience has been poor, an early submittal may result in project rejection before much time and effort have been spent on planning. If no information can be obtained at the local level, a letter request to the Environmental Protection Agency (Federal) should produce the necessary guidelines for covering this field in the feasibility studies.

Another aspect of an environmental statement is the possible effects of harbor operations and boating on noise levels (e.g., racing engines), water pollution (due to oil spills, surface drainage into harbor, and littering), and vehicle traffic. The existence of a harbor in lieu of an undisturbed natural shore-area environment has been protested as a downgrading of the esthetic qualities of an area. These are all matters that may influence a controlling agency in judging the value of the project to the community. On the other hand, existing conditions or ongoing activities in the surrounding area may have a detrimental impact on harbor operations which, when pointed out in the feasibility report, could discourage the developer from further consideration of the chosen site. Examples of such impact would be the close proximity of large concentrations of disruptive elements of society that would tend to sabotage or plunder the completed project, or nearby factories that pollute the air or water to a degree that would be harmful to exterior finish of small craft or to persons living or working at the harbor. These detrimental factors are usually discovered before site selection, but, if not, they must be revealed and evaluated in the very early stages of the feasibility study.

e. Area and Access Availability. The first evaluation of the site should define the exact area available, both on land and in the water. By applying general planning principles and rules outlined in Section V, a good estimate can be quickly obtained for the number of boats that can be accommodated. This application may place a physical limitation on the capacity of the project which, in areas of great need for harbor facilities, may make it necessary to evaluate the market potential. However, if ample area is available, the market potential must be evaluated to determine the size of the facility.

A most important part of the area study is the evaluation of access routes, both by land and by water. Good vehicle access routes will be required, including some offsite road construction. Whether this must be done as part of the project cost or will be a cooperative effort by the local public road department, must be determined at this stage of planning. Similarly, if an entrance channel is needed from the main water body to the harbor site, the study must reveal whether it will be provided by others or will be a project obligation.

Ownership and acquisition problems must also be evaluated at this time. If the developer does not already hold all the land and water area in full ownership, the probable cost of aquisition must be determined or the terms of a possible lease agreement negotiated with the current owner. If the water area is owned by the State or Federal Government, some arrangement for its proposed use as a small-craft berthing or launching-basin facility must be agreed upon during the feasibility studies.

f. Geophysical Characteristics. After the proposed boundaries of the site have been established, certain field investigations will usually be needed to support a construction cost analysis. The extent of exploratory work done at this time will vary with the geophysical characteristics of the site and prior knowledge obtained from construction in adjacent or similar land formations. The purpose is to reveal information needed to determine the relative difficulty of dredging or excavating wherever the work is to be done, and to evaluate bearing capacity, slope stability, and other soil mechanics of the various formations to be encountered. Some exploration and soil testing may be deferred to a later stage of planning, but on a small project or one that appears feasible, it may be more economical to obtain the field data while mobilized for such work.

Other geophysical data to be aquired at this time should include wind, weather, tide, and wave statistics or data on river flows or lake levels needed for design of protective structures, berth construction, and anchorages. Temperature trends must be analyzed to determine what kind of seasonal changes in harbor usage to expect and whether snow, ice, or frost will be troublesome. Extremes of heat in summer often place covered berthing systems in high demand.

g. Market Potential. A study of boating habits and experience records of other small-craft facilities in the area will often provide the information needed to determine the probable requirements for berthing facilities, ramps, and hoists. In an area newly opened to boating, such as a reservoir on a lake or bay to which vehicle access has recently been provided, the numbers, types, and sizes of craft that will be attracted to the facility must be estimated. This will require a study of distances to population centers, boating recreation opportunities offered at the body of water served, and experience records of similar installations at other newly opened areas under similar circumstances.

In urban areas where other facilities are available within easily traveled distances, a study must be made of spheres of influence of such facilities in relation to the one under consideration. Because boating is still an expanding form of recreation in most areas, the harm that will be done to existing facilities by a new installation competing for patronage is usually short lived. Enough new craft will usually be purchased by families living in the market area to make up for lost patrons at existing harbors within a short time. Nevertheless, a statistical survey of boats registered within the market area should be made to determine whether the ratio of boats to population is nearing the saturation level by a comparison with other strong boating-oriented communities in a similar environment.

In any area or situation where a real question exists concerning the types and numbers of craft that will be attracted to a new installation, the market analysis becomes the most important feature of a feasibility study. If the proposed facility is fairly large, an economic consultant specializing in market analysis should be engaged to assess the market potential. An economist will frequently have better sources of information than the developer, and because of his training and experience can easily obtain the necessary data to better judge the true meaning of the final assemblage of statistics. The analysis will be interpreted in terms of probable numbers of different types of craft coming to the new facility shortly after opening, and the annual increase in patronage to be expected. This information is vital to determine the proper phasing of the various components of the facility. It will ensure that no components are built in the initial effort only to be idle for several years and deteriorate before actually needed.

Another objective of the market analysis is to determine the ancillary facilities, both to satisfy the needs of the boating patrons and to provide additional revenue for the developer. In many areas the going scale of slip-rental rates is so low that the marina management must rely on ancillary facilities for an adequate net return on the overall investment. Here again, the economic consultant is best prepared to evaluate the ancillary-facilities potential of the installation. A little imaginative forethought along these lines coupled with an objective analysis of available statistics will often result in a flourishing complex of facilities and activities in an area that might otherwise become devoted to a purely boating enterprise that must struggle for economic existence. A perusal of the case histories in Section VIII will readily demonstrate this point.

h. Predesign Planning. As soon as basin dimensions, channel requirements, and protective structure needs are determined by analysis of the geophysical characteristics of the site and consideration of the numbers and types of craft to be accommodated, the actual planning of the installation can begin. In the preliminary feasibility stage, this planning is carried only to the point necessary to prepare a valid construction cost estimate with schematic layout drawings and sketches to convey to cognizant permit-approving agencies and prospective financing institutions a fairly clear idea of what is intended. It must also show the numbers of boats to be accommodated and locations of all proposed support and ancillary facilities.

It is customary at this stage to show all reasonably valid alternative layout schemes and to present a brief analysis of each, leading to the selection of the best plan. Consideration of alternatives is not essential with a private development, but serves a very useful purpose in a public project. Concerned citizens will often point out what they think are errors in planning if only a single concept is presented. Alternatives will not only prevent criticism, but will give the general public a sense of confidence in what its officials are doing. If by chance a better plan was overlooked in the preliminary studies, the presentation of alternatives may stimulate general thinking to the point that someone will submit this better plan before project planning has reached the stage where a change would be costly.

i. Cost Estimate. The initial construction cost estimate must be prepared in sufficient detail to ensure that no major item is overlooked and that the total is adequate to cover the actual cost. The time required for construction of each major component should be estimated for any large project, and a rough construction-phasing scheme prepared showing how much funding will be needed at any given time. Because of a time lapse before actual construction begins and a possible change in plans, this preliminary estimate is usually based on major-item quantity estimates–square feet of piers, docks, buildings, and parking lots; lineal feet of perimeter walls, streets, and walks; cubic yards of excavations, fills, and dredging; lengths of pipelines or conduit runs for utilities, water supply and sewers; and various other shortcuts for estimating additional items of lighting, landscaping, and sign-posting. Since this method may fail to include certain minor items that will be necessary later, and conditions at the site may result in higher costs than elsewhere, a contingency item of 10 to 20 percent should be added to ensure that the total is adequate for budgeting purposes. Preparation data for the estimate must be given so that any escalation in the construction-cost index for the region can be taken into account if the project is delayed for any length of time.

j. Method of Financing. A major indirect cost is the interest that must be paid on the capital required to fund any project. The best method of financing must be made in the feasibility study to determine the approximate amount of the carrying charges on the investment. If the project is to be funded with capital already on hand, a determination must be made of the returns which that capital would bring if invested elsewhere, and these potential returns must be charged against the project. If funds are not readily available, several possible sources should be investigated. Private developers can often obtain financing through a major marina construction firm by agreeing to use that firm's system to the exclusion of all others. Public agencies, however, are required to obtain competitive bids for any project construction and must have an established source of funding before such bids can be solicited. The usual sources of construction funds are loans from government agencies and bond sales.

Although the Federal Government does not loan funds for local agency construction projects, some of the grant programs discussed previously for the Federal and State Governments may reduce the financing load of the local public agency. If the project can qualify for a State loan, this source will usually carry a lower interest rate than bonds or institutional loans. General obligation bonds, guaranteed by the taxing power of the local agency, will bring the lowest interest rate of any bond-sale method of financing. However, a two-thirds vote of the electorate is usually required to authorize bond sales, and the allowable bonded indebtedness is limited to a percentage of the total tax base of the agency as established by its organizing charter.

A simple majority vote of the electorate will authorize a revenue-bond sale, which does not obligate the taxpayer and is guaranteed only to the extent of the revenues that the project will bring. Because of the greater risk, these bonds are normally bid at a rate considerably higher than the general obligation bond rate for a similar project. The actual amount of this rate will be determined to a large extent by the findings reported in the feasibility study. Because of this, it is not unusual for the local agency to have the preliminary feasibility study upgraded to the status of a prospectus for revenue-bond sales by the addition of such further marketing analysis and cost studies as may be considered necessary to give the prospective bond buyer a better insight into the project's economic potential.

k. Cash-Flow Analysis. An important device for evaluating economic potential is the cash-flow table that sums up the estimated expenditures and receipts for each year of the project from inception to full amortization of indebtedness. The table must take into account the probable debt-servicing charges, advertising and insurance costs, and all operational and maintenance expenses (including the costs of replacing any short-life components) as compared to the estimated annual income from berth rentals, launching fees, and ancillary-facility leaseholds. If phased construction is anticipated, the dates on which additional capital input will be needed must be shown, together with the resultant increases in debt-servicing charges during the ensuing years. If a financing institution disagrees with the debt-servicing estimates, but accepts the other elements of a cash-flow analysis, it can simply substitute a figure for debt servicing and reach an independent conclusion as to the economic potential of the project. To gain a better understanding of the many problems to be faced in managing a small-craft facility, and to make a more realistic appraisal of the economic factors involved, the prospective developer should carefully review the following principles, and the case histories of actual marina projects described in Section VIII.

2. Project Implementation.

a. Timing of Design and Construction. If it appears certain during the feasibility study that the project will be implemented, or as soon as funding is guaranteed, preparation of final plans and specifications for the facility should begin. Any site investigation work left for this stage of project planning must be completed to provide the necessary design criteria. If the findings reveal a need to modify any features of design assumed in the feasibility study, such as the protective structures or the dimensions of channels or basins, these facts must be evaluated at once to determine their effects on economic feasibility. Detailed planning can then be pursued, but the order in which the various components are designed and constructed may be important in several respects.

Sometimes protective features such as breakwaters and jetties must be constructed before work can be done in the interior water areas. Then, it is customary to complete the design of these protective features first, so that they can be constructed while planning continues on the interior features. Approach roads and some interior road work may be necessary to provide access to the site for the working equipment. These roads should be planned and constructed early in the project. If the site is remote from developed areas, water supply and power transmission lines may have to be designed and built separately as another early feature of the project. Certain components may require a longer time to construct than others, e.g., long entrance channels and large excavated basins. Design and construction must be started in time for completion along with the rest of the project.

On small projects, the best procedure may be to design the entire facility before construction begins so that it can be accomplished under one contract. On larger projects, planning and construction time estimates become important in setting up work schedules that defer all expenditures until they are actually needed to prevent delaying project completion. The Program Evaluation and Review Technique (PERT) is often employed to relate all the various events in design and construction to the time required to accomplish them. A similar technique is the Critical Path Method (CPM) by which all activities in project design and construction are analyzed graphically to determine which ones become critical in accomplishing all the work in the least possible time. The scope of this text does not permit a detailed explanation of these procedures, but they are covered in various engineering publications and may be found useful in some large harbor construction projects.

All the major components with the exception of certain proprietary systems require complete design detailing. With the wealth of specialized knowledge in the fields of floating-dock systems and boat-handling equipment, considerable design effort can be saved by merely designating the areas available for such systems and the approximate dimensions of docks, piers, and slips, or the required capacities for the hoisting equipment. Proposals are then solicited from the various equipment and systems manufacturers, and the most desirable selected early in project planning. These systems will have certain requirements for anchorages, gangway landings, utility line connections, local perimeter treatment, and foundations that must be considered in the design of basins and perimeter walls. If a bubbler system is installed in a cold climate marina, it may have additional requirements for line connections and the compressor site. All of these requirements must be obtained from the manufacturers or system installers and satisfied in the final design of any other components that may be affected.

b. Contracting Procedures. The degree of detailing in the construction drawings and specifications will vary with the contracting procedure used, but a minimum amount is required regardless of the contractual arrangement. For this reason it is advantageous to both the developer and the contractor to provide a minimum of detailing to ensure the proper quality of materials and workmanship and to avoid misunderstandings about what is expected. The basic rule is to inform the contractor exactly the ultimate product desired, but allow him maximum freedom of choice as to how this goal will be reached. Certain critical items must be specified in great detail; other items may be roughly dimensioned and specified with instructions that the contractor submit shop drawings for approval before proceeding with construction. This is especially important for features where modular units must be fitted together or where multiple choices are available that will give approximately equal results.

Most public agencies and large private developers have standard contract forms to protect them against mistakes, accidents, and inabilities of the contractor. If the agency or developer has had little experience in contract administration, the following is a checklist of protective items normally covered in the bidding and construction-contract documents.

(1) Items Covered in the Invitation to Bid.

(a) A deposit for each set of plans and specifications taken out covering their reproduction costs. The deposit is reimbursable when the items are returned with the contractor's bid. Extra copies may be turned over to the successful bidder, as several sets will be needed during the construction period.

(b) Bid bond in the amount of about 10 percent of the bid price to ensure that, if successful, the bidder will sign the contract.

(c) Prevailing minimum wage rates for labor (usually required by a public agency's charter, but not necessary for a private developer).

(d) A listing of the bidder's equipment to be used on the project.

(e) Reservation of the right to reject any bids, to waive informalities in bids, and to make awards as the interest of agency or developer may require.

(2) Items Covered in the Contract Form.

(a) Time for commencement and completion of contract with specification of a *Liquidated Damages* assessment for each day of overrun in the estimated amount that such delays in completion will cost the owner in terms of administrative expense and (in some cases) loss of revenues. Certain provisions for time extension must be included to cover delays that are not the fault of the contractor.

(b) Acknowledgment of prior site investigation by the contractor and ensurance of his complete acquaintance with conditions that may affect his work.

(c) Declaration of full responsibility of the contractor for all the work specified including that done by subcontractors and their employees.

(d) Requirement for the contractor to comply with all applicable laws and to secure all necessary permits.

(e) Provision for holding the owner harmless against law suits or other actions that may be brought against the contractor as a result of his performance of the contract.

(f) Provisions for making changes in the work and for increasing or decreasing quantities as shown in estimates.

(g) Provisions for inspecting the work and final acceptance, including correction of all defects and resolution of disputes.

(h) Statement of conditions for partial progress payments and final payment.

(i) Provisions for termination and default in case the contractor fails to fulfill contract provisions.

(j) Requirement for posting a performance bond in the contract amount to ensure fulfillment of contract.

(k) Provisions for auditing of contractor's records.

(l) Requirement that the contractor be properly insured with regard to workman's compensation and general liability.

c. *Advertising and Berth Reservations.* While the small-craft harbor is under construction, future availability must be advertised to the boating community well in advance of the opening date. This can be done by encouraging feature articles concerning the new facility, both in the local press and in boating magazines. Full use should be made of any allocations budgeted for advertising in project funding. The harbor administration office should be opened as soon as the advertising program begins and a large layout map of the berthing basin prominently displayed. Reservations for berths should be accepted on a reasonable downpayment basis and posted on the layout map. Initially, a small staff will be required to handle reservations and answer questions. If ancillary facilities are developed by leasehold, they should also be advertised and the staff instructed as to the proper disposition of inquiries concerning them. The sooner leases can be negotiated, the greater will be the ensurance of a successful enterprise and the estimates will become more accurate for proper timing of successive stages of project development.

3. Operating Principles.

a. *Staffing.* The initial staff of a small-craft harbor should only be large enough to: (a) handle the problems and administration requirements of the slip renters at the time of opening, (b) exercise control over activities within the harbor boundaries, (c) operate launching equipment, and (d) maintain the premises in good condition. The usual staff consists of a manager, an administration clerk, a harbormaster, a maintenance supervisor, and enough additional help to operate equipment and other manual work required. In a small installation, some of the functions can be combined and handled by one person. In a large installation, the manager and each department head may require one or more assistants to perform the supervisory work.

The patrolling duties of the harbormaster and his deputies will become more demanding as occupancy of the harbor increases, but the work required of the administrative staff and maintenance personnel may be heavy from the start. The reason for the early administrative load is the need to set up the record files and handle an influx of new harbor patrons during the early growth period. When the harbor approaches occupancy capacity, the administrative work becomes more routine and although the number of patrons has increased, the files have been established, billing procedures have been systematized, and fewer questions are being asked at the reception desk. The early demands on maintenance personnel are usually of a troubleshooting nature—new equipment is being broken in, minor flaws develop in the berthing or utility systems and must be corrected, and surface drainage problems arise and must be solved. Most of the landscaping is omitted from the initial construction stage to reduce costs and is left to the maintenance staff, which is not expected to be too heavily burdened until the newness of the facilities begins to wear off.

b. Administration and Operating Procedures. An important part of the early administrative work is the systematizing of recordkeeping, including preparation of standard forms to be used for this purpose. A separate file should be set up for each slip renter by name, and crossfiles established containing the more important records by boat or slip classifications for statistical and slip-rental billing purposes. A cost-accounting system must be established to keep a complete record of receipts and expenditures for auditing purposes. A personnel and payroll section must be established to keep a permanent record of each employee, prepare paychecks, maintain withholding accounts, and perform all other functions normally assigned to such a section. A system of purchase orders must be established for procuring maintenance items and replacement parts. These are the major administrative considerations, but others may develop and all will require careful scrutiny until they are being handled smoothly and routinely. National Association of Engine and Boat Manufacturers (1967) contains some excellent suggestions for handling administrative details and operating a small-craft harbor; typical examples of forms used in cost accounting and recordkeeping are also shown.

c. Liability and Insurance. The manager of a small-craft harbor must be aware of all the laws and regulations covering his own operations and the activities of his patrons. If any part of the water area is under State or Federal jurisdiction, the cognizant agency must be consulted on requirements for aids to navigation and other steps to be taken to ensure compliance with applicable laws and regulations. In the navigable waters of the United States, the U.S. Coast Guard is responsible for upholding the established rules of navigation and the U.S. Army, Corps of Engineers for maintaining channels and removing hazards to navigation. In inland lakes and on intrastate canals and rivers, the State or a Federal agency other than the U.S. Army, Corps of Engineers may exercise control over waters for navigation, fishing, skiing, and other water sports. The relationship between such agencies and the harbor manager or owner should be one of mutual cooperation rather than law enforcement and compliance. The common objective is to increase the enjoyment of boating (or in some instances the benefits of commercial small-craft navigation) with the least hazard to life and property and without damaging the environment or the quality of the water.

Activities in the land areas are also regulated, but usually by the laws and ordinances of the local governing authority. The manager may find that the U.S. Coast Guard or the State boating agency regulates activities in the water, whereas the city or county police and fire department enforce the laws and protective ordinances on land. In both land and water areas the manager can best protect himself and the owner against property damage, injury, and law suits by making patrons clearly aware of the prevailing rules and regulations and by insisting that the harbormaster enforce them courteously but firmly. The judicious use of sign posting and dissemination of written information will be helpful, but personal contacts through public meetings will often be the best way to educate the patrons. Classes in navigation and safety should be promoted with the assistance of the U.S. Coast Guard, the U.S. Power Squadron, or the local yacht club. National Association of Engine and Boat Manufacturers (1967) contains a sample set of rules and regulations, and, if modified to reflect local needs, could be adopted by the management of a small-craft facility.

Regardless of the efforts toward patron education and elimination of physical hazards, the owner of a small-craft facility may still be sued for damages and injuries due to accident or other causes. The most common types are: (a) damage to berthed craft while unattended, caused by disasters, acts of other patrons, or vandalism, (b) dropped craft at hoists, (c) vehicles skidding down into the water at an alleged improperly designed or maintained ramp, (d) any kind of fire damage to berthed or dry-stored craft in the owner's absence, (e) damage due to wave or surge battering of berthed craft (claiming inadequate protection), (f) hull or propeller damage due to collision with submerged obstructions, especially at low water level, (g) falling injuries due to steep gangways, gaps in handrails, tripping obstacles, and other possible causes, (h) injuries due to accidents in using equipment provided for harbor or launching operations, and (i) accidental drownings in the harbor area.

The marina owner should have insurance against liable suits for any of the above-listed or other accidents to users of the harbor, and also against accidental damage to or loss of the facilities and equipment and injuries to employees. The common types of insurance are fire, automobile liability, public liability, workmen's compensation, and marina operator's liability. National Association of Engine and Boat Manufacturers (1967) contains a detailed discussion of insurance policies and the operator's liability in all areas of small-craft facility operations. The insurance coverage of the harbor should be reviewed periodically to ensure that it is adequate. Insurance companies change their insuring policies from time to time and may offer endorsements to cover special needs as they arise. Some insurance agents are not completely familiar with small-craft harbor operation and may fail to realize all of the potential hazards. In the final analysis, the manager is responsible for seeing that insurance policies held by the owner provide all the protection needed.

d. Maintenance. The limited scope of this report does not permit inclusion of a comprehensive discussion of all the problems commonly encountered in maintaining a small-craft harbor. Maintenance of buildings, landscaping, roads, walks, parking lots, and utilities in the land areas of marinas is similar to a land-based enterprise. Maintenance of basins, channels, piers, docks, perimeter structures, and protective works is different from landside maintenance and warrants the following remarks:

(1) Breakwaters and jetties are frequently maintained by the U.S. Army, Corps of Engineers or the State. If maintenance is a local responsibility, detection of any early signs of deterioration is important. Excessive toe scour of a structure in a high-wave energy environment should be corrected by filling the scour pockets with stone of adequate size before the integrity of the structure is jeopardized. Steel structures usually deteriorate as a result of corrosion unless their coatings are maintained or their cathodic-protection system are kept in working order. Abrasion near the sand line is also a possibility, but prevention is difficult. Cladding with concrete is a method by which steel abrasion has been successfully prevented. Concrete structures usually deteriorate as a result of progressive cracking and spalling. Major damage can often be prevented by pressure-grouting the cracks with an epoxy cement. This process requires the equipment and skills of a concrete repair specialist. Timber structures usually deteriorate rapidly if connectors loosen and begin to *work* and *flex* under wave agitation. Periodic tightening of nuts may prolong the life of a timber structure. Leaching of pressure-injected preservatives will expose the timber to marine-borer attack or dry rot. Cladding with concrete has prolonged the life of some timber structures. Various synthetic membrane wrappings have been used for protection of timber piles, often with considerable success.

(2) Armored perimeter slopes usually fail either by plastic flow or sliding of the embankment after a rapid water level drop or by having the fines pumped through voids in the armor as a result of an inadequate filter. Several types of corrective action may be taken depending on the type and severity of the failure. Trenching and installation of drains behind the slope may solve a sloughing tendency in clayey soil. Grout-sealing a leaking riprap may prevent further pumpout of fines. If a riprap deteriorates under prevailing wave and current action even though the filter remains effective, it may either be grouted with concrete or stabilized by the replacement of an additional layer of heavier stone.

(3) Vertical bulkhead walls frequently fail because of inadequate depth of penetration of sheet piling or improper filtering of soils under and behind a poured-in-place concrete wall. Saturation of the retained fill coupled with pulsations in the hydrostatic head differential due to wave action results in the fill being pumped out from under the wall. This action can sometimes be stopped by filling the sinkholes behind the wall with a mixture of straw and gravel or quarry waste, which eventually clogs the leak and effects a filtering action. If this corrective action fails, it may be necessary to excavate a wide trench behind the wall to the depth of the mud line at the face of the wall and place a layer of filter stone in the bottom of the trench before backfilling.

If the leakage of fines is through joints between concrete or timber sheet piles, sealing of the joints may be possible. One method is to drive about a 3-inch pipe between the wall and the fill at each joint using a special sacrificial driving tip that forces the tip to follow the joint. Grout is then forced through the pipe as the pipe is withdrawn. If the gap is so wide the grout escapes through it, a special rubber or plastic tube must first be inserted in the pipe to act as a form to retain the grout after the pipe is withdrawn.

(4) Shoaling of channels and basins may present a difficult maintenance-dredging problem. Ordinary excavating equipment such as draglines, clamshell-dredges and the usual hydraulic dredges cannot safely work near armored slopes, vertical walls, fixed piers, and docks or under piers and docks whether they are fixed or floating. A new type of dredging suction head will now solve this problem. It uses high-pressure water jets to loosen the bottom material and compresses air to operate pneumatic slurry pumps in the suction head which force the material in high concentration mixture with water through the discharge line. The suction head has no moving parts and is operated entirely through flexible tubing leading down to the suction head from a floating or landside water pump and air compressor. Thus, it can be more easily guided into areas that would normally be difficult to reach without endangering either the suction head or nearby structures. It is only necessary to guide the suction head at constant depth to all parts of the shoaled area to restore the desired uniform depth of the basin or channel.

(5) Fixed piers and other structures built on piling need the same maintenance as landside structures except that in a moist air environment they may require more frequent painting and rust-prevention treatment. Moreover, the supporting piles and bracing members must be examined periodically for evidence of deterioration. Damaged or badly deteriorated piles may have to be removed and replaced occasionally, and bracing members should be replaced when broken or otherwise cease to be functional.

(6) Floating piers and docks may deteriorate in several ways: (a) loss of flotation, (b) *working* of bolts against wooden side stringers or connections of finger to headwalks, (c) loosening of hinged connections, (d) fatigue failure of structural members or their connectors due to continuous flexing under wave action, and (e) abrasion of deck surfacing.

The most troublesome problem is usually flotation loss, which may be partial for foam floats but sometimes total in hollow-shell floats. A periodic check of the freeboard in various components of the floating system will indicate where flotation loss has occurred. Nonleaking hollow floats sometimes take on water by internal condensation of airborne moisture or possibly by infiltration of surface water through breather caps and only need to be pumped out occasionally. Some foams are partially water-absorbent and may stabilize at a satisfactory level after becoming saturated over a period of time due to capillary intrusion.

In seawater, an incrustation of marine growth may cause some loss of flotation, although most plant and animal life have the same density as its water environment. The shells of crustaceans cause most of the settlement due to marine growth, and have to be scraped periodically to hold the desired freeboard. In some marinas, scuba divers may be hired to clean the hulls of boats and bottoms of flotation units supporting floating piers and docks with water jets and hydraulic vacuum cleaners. The bottom surfaces of some foam floats have intentionally been left uncoated to ensure automatic self-cleaning in seawater. After a few years of incrustation the growth becomes heavy, tears loose and drops to the bottom. When this occurs, a small amount of foam will be removed, but not enough to cause appreciable loss of freeboard over the normal life of a floating structure. A greater danger in some areas is an infestation of screw worms or other burrowing biota that reduce the effective buoyant volume of the foam. In turn, birds and small animals cause further loss of foam by searching out these worms for food. In areas known to be infested with such burrowing biota, all foam surfaces should be coated.

Loosening of bolted connections is a major cause of deterioration of floating systems in harbors with more than minimal wave action, especially if the joined components are heavy. Nuts often corrode fast to bolts and cannot be tightened. The resultant differential movements of floats and stringers may then quickly enlarge the boltholes to the extent that both the connecting hardware and the stringers must be replaced. Hinge-connected floating components often stress the hinges severely under wave action. Worn hinges not only become noisy but the resultant peening of metal against metal as stresses are reversed by the wave cycles may accelerate the wear so that the hinges break long before the components become unserviceable. Such peening action may be prevented by inserting block rubber expanders between the ends of the hinged components to keep the hinge continuously in tension. This also eliminates noise, provided the hinge is kept well greased.

(7) Pile guides in a pile-anchored floating system deteriorate more rapidly than other parts of the system and if not repaired may damage the piles and threaten the integrity of the anchorage system against strong wind or current-induced stresses. When this occurs, the damaged guides should be replaced with heavy-duty roller guides such as the one shown in Figure 86. Two or three guides of this type may satisfy the needs of the system and prevent wear in the other guides that are forced into contact with the piles less frequently because of the peculiarity of the guide-pile geometry system.

Cable anchorage for floating-slip systems is so variable that no single aspect of such a system can be singled out as being universally prone to early maintenance trouble. Unlike the customary multiple anchorages, cable systems usually rely on not more than three or four cable ties per floating system for maintenance of position. Hence, it is important that all winches, lines, and anchors be inspected regularly, as one lost anchor or line may allow the entire system to drift and collide with some other object. The results of such a collision could be disastrous.

(8) Utility lines and outlets on fixed piers differ slightly in maintenance from similar installations on land. On floating piers, however, utility conduits and pipes are usually much closer to the water surface and hence exposed to more wave splash and moisture-laden air. Loose connecting stringers may allow enough flexing along a headwalk under wave action to break the conduit or pipes if they are attached firmly rather than loosely supported. In

basins subjected to large water level fluctuations, flexible hose connections at gangways are often early points of failure in the utility-distribution system. Conversion to the hanging U-configuration shown in Figure 102 will prolong the life of such hose connections.

Valves and washers in hose bibbs require periodic replacement, and an adequate stock of spare parts should be kept on hand. The trend toward greater power consumption frequently overloads the electrical distribution system of a marina. Wherever break-tripping becomes a frequent occurrence, larger capacity wires, outlets, and circuit breakers should be installed. With a large amperage installation, special attention should be given to electrical safety precautions. Slip-renters must be cautioned against the use of undersized or defective extension cords, and a sharp surveillance of all electrical equipment should be maintained to prevent overheating, shorting, or sparking.

In some saltwater marinas, a high rate of pitting of propeller blades and water-seal bearings has occurred. This may be due to stray currents of low potential but high amperage escaping into the water from the electrical distribution system. An electrical engineer should examine the system to determine and eliminate all probable causes of this electrolysis.

VIII. MARINA SURVEY AND CASE STUDIES

1. General. In developing information and data for this report, it was not practical nor even desirable to include every unique design feature encountered in the course of the investigation. Critical response from marina owners and operators, however, can be extremely helpful in evaluating the validity of present design concepts, and can also help to pinpoint some of the significant and recurrent problems relating to marina design, development, or management. In an effort to obtain such response and to reveal problems that need greater resolution, a survey was taken of 394 marina owners and operators throughout the 50 States, Washington, D.C., the Virgin Islands, and Puerto Rico. This survey was based on the questionnaire presented in Appendix K. A summary of the survey-sample characteristics and the case studies developed from the data obtained, are presented in this section. The last part of the marina survey questionnaire solicited statements concerning any design feature that had worked better or worse than expected, or any problems that in the operators evaluation could have been avoided by better planning. These statements are also summarized in this section. Several of the concepts mentioned have already been covered to some extent in this report, but are repeated because of their high incidence of occurrence in this survey.

2. Survey Sample and Data Resolution. About 60 percent of the marinas contacted during the survey were chosen at random from a commercially prepared mailing list. The remaining 40 percent were recommended by State agencies as being typical or representative of their respective State, or were selected from various publications such as boating almanacs, boatman's guides, and boating periodicals. Fifty-seven marinas across the United States were visited, with onsite interviews conducted during the visits.

The size of the facilities covered by the survey ranged from a six-slip installation for a birdwatching club to a 6,000-boat complex. The average marina size was 255 slips. The marinas were sited on a wide variety of water bodies, with about 36 percent located on bays, 22 percent on rivers, 16 percent on lakes, 14 percent on the ocean, and the remainder on watercourses and basins such as the Intracoastal Waterways along the east coast, reservoirs, manmade canals, and locked basins.

Boat-mix trends within the survey sample can best be described as following an intuitive pattern. Sailboats were more prevalent on larger bodies of water and in warmer climates. Powerboats outnumbered sail craft in more restrictive waters such as rivers, in colder climates, and in areas reputed to have good sport-fishing waters. Commercial tour boats, fishing boats, and rental craft were encountered generally throughout the survey sample. Boat-mix trends were difficult to discern and any results concerning boat-mix must be considered inconclusive.

In berthing areas, the average depth was near 12 feet, but ranged from a low of 2.5 feet to a high of 250 feet. The water surface fluctuations due to tides, drawdowns, and seasonal variances ranged in extremes from 0.5 foot to 20 feet daily and from 1 foot to 40 feet yearly. The most extreme range of water surface fluctuation on record for any body of water by a marina in the survey was 65 feet.

These basin characteristics had a predictable effect on the respective berthing facilities. Fixed-pier systems were generally used in locations where the basin depth did not exceed 20 feet and the water level fluctuation was never more than 4 feet. Sixty-five percent of the sampled facilities used floating-dock systems, primarily self-manufactured. About two-thirds of the floating system were secured by guide piles, the others fixed by cable or chain systems, or by hinged struts, usually due to greater basin depths.

A freshwater supply to the docks was almost universal and in all cases free. Four out of five marinas sampled provided 120-volt power to the berths, and one-fourth also provided 240-volt service. A majority of the marinas had public address systems, dock lights above waist level, and phone service only in the office or at public pay stations.

All of the marinas in the sample provided one or more ancillary facilities. Major ancillary facilities listed in order from most common to least frequently provided and percent of sample marinas providing these facilities were as follows:

Ancillary Facility	Percent
Fuel station	60
Boat sales	55
Major boat repairs	42
Snack bar	34
Minor boat repairs	32
Restaurant	26
Boat rentals	24
Bait and tackle	14
Motels or accommodations	10

Other services offered included marine supplies or ship's chandlery, engine repairs, showers, cocktail lounge or bar, laundry, grocery, ice, licenses, scuba equipment and tank refills, electronics shop, and trailer parks. A summary of average marina characteristics derived from the survey data follows:

1. Marina Location: Bay
2. Basin Depth: 12 feet
3. Capacity: 255 slips
4. Boats per Slip: 2
5. Average Daily Water Level Fluctuation: 4 + feet
6. Type of Docks: Floating
7. Anchored by: Guide piles
8. Utilities:
   a. 120-volt power supply to docks, included with slip-rental
   b. Water provided free
   c. Phone available in office
   d. Lighting system, located above waist level
   e. Public address system
   f. No sanitary pumpout facilities available
9. Waste Collections per Week: 2+
10. Car Parking Spaces per Slip: 1–
11. Ancillary Facilities Available:
    a. Fueling station
    b. Snack bar
    c. Boat sales and repairs
12. Most Probable Operational Problems:
    a. General maintenance
    b. Parking
    c. Vandalism (minor)

3. Design Problem Summary.

a. The Master Plan. Many of the problems encountered by the marinas surveyed could be attributed to inadequate planning either because no master plan existed or because the existing plan was not broad enough or not followed.

By far, the most frequent problem discussed was the lack of slips. The demand for slips at the surveyed marinas averaged 140 percent of the marina capacities. One marina complex of 150 slips had a waiting list of 130 patrons. Another had a projected waiting period of 5 years. This situation has forced some patrons desiring slips at this marina into *paper partnerships* in which they purchase a craft from an owner who already has a slip under contract. Although the boat is purchased outright, the previous owner's name is kept on the title as a 50-percent owner and will not be required to give up the slip presently occupied by that craft. It was evident that the demand for boating facilities greatly exceeded their availability and that new slips were being filled as fast as they were constructed. The Bureau of Outdoor Recreation predicts that this problem will worsen. In 1965, boating was the 10th most popular outdoor recreation activity according to the Bureau. The Bureau's publication, "Outdoor Recreation Trends," projected an increase of 215 percent in boating participation from 1965 to 2000. The Bureau's statistics show that the number of pleasure boats in use in the United States almost tripled between 1950 and 1971 from about 3 million to 9 million, exceeding the 215-percent projection growth rate. This trend should be taken into account in the master plan, with the number of slips being maximized within any project and provisions for expansion given closer consideration.

Many of the marinas experiencing this problem of being too small also related difficulties in correcting the situation. A common complaint was that the piecemeal development, afterthought design, and scab-on expansion projects tended to dislocate the operational routine of a previously smooth-functioning operation. Maintenance procedures were often complicated when added facilities consisted of proprietary systems that were not similar to those in the original installation. In several instances, owners stated with regret that no provisions had been made for future expansion and the now obvious growth potential cannot be realized. However, detailed planning of projected future facilities long before they are scheduled to be built is also unwise. Changing concepts of marina development indicate that it is best to plan specific facilities before development. This allows the planner to take advantage of the most current state-of-the-art.

Other problems revealed by the survey were not easily predicted; they were based on trends less obvious and more difficult to forecast. Nevertheless, they now plague many of the surveyed marinas. The trend to wider-hulled craft and craft with high prows has caused problems with fendering systems and berth spacings. The emphasis placed on pollution problems is being felt and the majority of large installations cited problems in keeping the marinas clean. Sixty-two percent of the marinas surveyed specifically mentioned the growing demand for sanitary pumpout facilities, whereas only nine marinas indicated that they had such facilities.

Surprisingly, much discussion concerned the planning trade-offs made between functional design and esthetic design. Several marina operators stated that practical planning had been preempted by esthetic consideration with resultant inconvenience to patrons. They felt that appearance should not be the major consideration in locating fueling docks, or administration and supply buildings. The location of utility runs in inaccessible places was another point of contention, particularly with maintenance personnel. The operators believed that most marina patrons were little concerned with the appearance of utility systems as long as they functioned properly.

b. Parking and Traffic Circulation. The most universal, although less vital problem area revealed by the survey was that of vehicle parking. The average number of parking spaces per boat slip was slightly less than one. From the number of parking problems cited, the adequacy of this ratio must be questioned. Most of the parking conflicts seem to stem from patrons of major ancillary facilities on the marina premises competing with slip tenants for convenient parking spaces. A few marinas indicated that this problem was alleviated with a more extensive sign system. Others have resorted to decal-controlled or card-key lots. Another major problem was the need to protect vehicles left for extended periods of time by boatowners on long cruises. Many marina operators state that they are increasingly pressed to provide camper parking and hookups for recreation vehicles.

With the overcrowded conditions experienced at marinas, especially during peak-usage times, a few operators have cited traffic circulation problems. Wider roads and left-turn stacking lanes were recommended solutions by several marina managements. At some larger installations, the trend toward bicycle riding has also caused problems, and the management of the marinas involved suggested that consideration be given to bicycle lanes in marina layout planning.

c. Ice. The largest regional problem was ice formation in the cold climates. Consideration was frequently given to ice buildup in harbor planning, but that ice floes could be windblown or moved by currents into the berthing areas was overlooked. Many of the marinas surveyed claimed that free-floating cake ice was the greatest cause of ice damage. Two distinct problems were involved. In a fixed-pier system supported on tapered timber piles, it is customary to drive each pile tip-down. Sheet ice and rising water can readily extract piles. Where ice formation is anticipated, the piles should be driven butt-down if practicable, so that as the water level rises, the ice sheet will slide off the pile taper rather than wedge to it and pull it out. Also, when the water level drops, the ice will tend to hang up on the piles before breaking, and keep the piles well seated.

In a floating-pier system the greatest damage was reported to be caused by ice crushing the floats. Therefore, many marina operators removed their floating systems from the water during the winter season; most who did, commented unfavorably on the work involved and time consumed in the seasonal removal, storage, and reinstallation effort. Because of these problems, many operators are now using forced-convection or bubble systems to combat ice formation. Most of the marinas involved report that these methods of ice prevention are quite effective both for fixed and floating systems. One marina in the Chicago area reports that the system was able to clear ice layers up to 4 feet thick. The systems, however, are an added expense, requiring 35 compressors to keep 400 slips ice-free. The system took 1 month to install.

d. Miscellaneous Common Problems. The most numerous complaints about system components at marinas concerned utility services, particularly electric power. The problem usually resulted from the severely corrosive environment, especially seawater, which has an insatiable appetite for metals. Many of the surveyed marinas that have low level lighting systems reported failures due to sea spray in high wind conditions. Unprotected connections or outlets are reported to incur high maintenance costs. One operator who switched to an environmentally protected electrical system complained of the lack of standardization in marine electrical equipment and the many interface problems that developed with what he had assumed were salvageable parts of his original system.

Difficulties with water supply was also reported. Because of the frequency with which marina water valves corrode shut, T-type faucet handles were recommended over the circular type. It was found that women patrons could open stock T-handled valves with a sharp blow or kick, whereas most could not budge the circular handles. One operator also reported several incidences in which a circular-cast handle broke or failed in the patron's grip, resulting in severe hand injury. It was also found that with a copper-tubing system, the vulnerability of the tube risers caused many sweated joints to be broken, either by accident or inadvertently during repair work. Rejoining of the components could then be accomplished only after the entire tubing system had been drained and dried out. The maintenance superintendent at the reporting marina recommended that all hose bibbs and risers have threaded connections with the mains or cross runs.

Many comments were directed at the use of untreated wood for decking or structural members of fixed piers. In most cases the maintenance and replacement costs required for untreated wood eventually exceeded what would have been the initial cost and subsequent maintenance cost of treated wood components. Generally, the use of inferior construction materials will lead to future economic problems that must be faced while the marina is in operation. Replacement of inferior construction will prove costly, and income will be lost while the defective facility is repaired.

Chemonite and cellon treatments of wood components received favorable comments. Nonbattened creosoted timber, however, was castigated for ruining clothing or staining personal property of the patrons and for polluting the surrounding water areas.

Floating-pier systems evoked two particular complaints. The first was the difficulty of adjusting chain or cable anchorages during water surface fluctuation cycles and the tendency of these systems to drift beyond desired limits. A related problem was maintaining shore access when a pier drifted or its anchorage was adjusted. The second complaint was the time and difficulty involved in the seasonal removal and storage of floating systems at locations requiring winter security.

Some marina operators questioned the reliability of fuel line connections to floating-pier stations. Fixed-pier stations, however, experienced an operational problem in fueling boats during extreme low water levels. Two marinas claimed that it was advantageous to locate their fueling facilities on land behind a bulkhead rather than on a dock or floating pier. This location also permitted gasoline sales to motorists, with the same pumps and service crew handling both marine and landside sales. Several other marinas, having adopted this fuel-service setup, now regret their choice, contending that it was more difficult to service the boats from pumps located behind a bulkhead. Careful consideration should be given to the many alternatives for locating support facilities to determine which is best for the general layout and development plan.

Finally, two very common problems, vandalism and boat-wake waves, although not directly related to design, should be given consideration in the early planning stages. Vandalism was a major problem in 30 percent of the marinas surveyed, and about one-half of the managers attributed the problem to improper siting. These marina locations were variously described as "undesirable areas, heavy industry regions, tough waterfront locations, and slum sections." Most of the 70 percent who found that vandalism was not a major problem at their facility attributed this to the marina policy allowing live-aboards, or the remote location of the marina, away from densely populated areas.

Fifty-three percent of the survey respondents complained of boat-wake problems. The primary complaint was that existing speed regulations for craft operated near berthing areas were not being enforced.

4. Unique Design Features. Most of the favorable design features mentioned by the marina owners and operators were of a general nature and have already been covered. Several unique features deserve mention, even though they cannot be universally applied.

The choice between floating and fixed docks is usually influenced by local customs and other nontechnical criteria. Floating systems are usually more expensive than fixed systems where the water depth does not exceed 20 feet, but each location must be evaluated in light of several other factors that would qualify this generality. The primary factor is water level fluctuation. As previously stated in the discussion of design, with all other factors favoring a fixed-dock system, the maximum acceptable fluctuation is about 5 feet. Other factors include bottom configuration, basin side slopes, soil characteristics, wind and wave conditions at the site, and storm history.

Occasionally, conditions at the site are borderline indicators between fixed and floating construction. One enterprising engineer combined the advantages of both systems in a 96-slip installation featuring an adjustable fixed-dock system (Fig. 157). The system consists of fixed timber walkways and fingers supported by steel stringers. The stringers are attached to steel pipe piling by large U-bolts so that the entire system can be adjusted to lake level. The system is part of the Lakeside Marina (freshwater) complex at South Lake Tahoe, California. Lake Tahoe has a yearly water level fluctuation cycle of 2 to 3 feet; however, over a longer cycle of 6 to 10 years, and dependent on the annual rainfall records, the fluctuation, though very gradual, can have a much greater range. The prevailing winds at this location are from the west and generate fairly consistent waves of 2 to 3 feet high. These factors and the shallow basin depth (10 feet maximum) suggested this unique solution.

Figure 157. Adjustable "fixed-dock," Lakeside Marina, South Lake Tahoe, California

Ice problems in cold climates may present rare opportunities for unique solutions. Burns Harbor Marina in Indiana takes advantage of the hot water return from a conveniently located steel plant to keep the berthing area ice-free during the winter months. Many large industries or powerplants require large volumes of water for heat exchange systems and are normally located near large bodies of water. With proper planning, the Burns Harbor solution could be repeated in many areas, and the result would be a more efficient use of a common resource.

Another innovation that has proven useful and effective is shown in Figure 158. The major function of these conically shaped plastic pile caps is to keep birds off the pilings and the marina neat and clean. Before pile cap, sharpened spikes were used to keep birds off pilings, but a few unwary creatures were impaled on the spikes. The change to these ecologically acceptable items solved the problem and improved the esthetics of the marinas where they have been installed. In large marinas, cones with distinctive colors have been used in different berthing areas to aid patrons and guests in finding their slips.

Figure 158. Conically shaped, plastic pile caps.

Finally, Figure 159 shows a schematic of a gangway counter-balance system encountered at the Royal Vancouver Yacht Club, British Columbia. This system is designed to support most of the deadweight at the lower end of the gangway and thus reduce the amount of flotation that would otherwise be needed for a landing on the floating system.

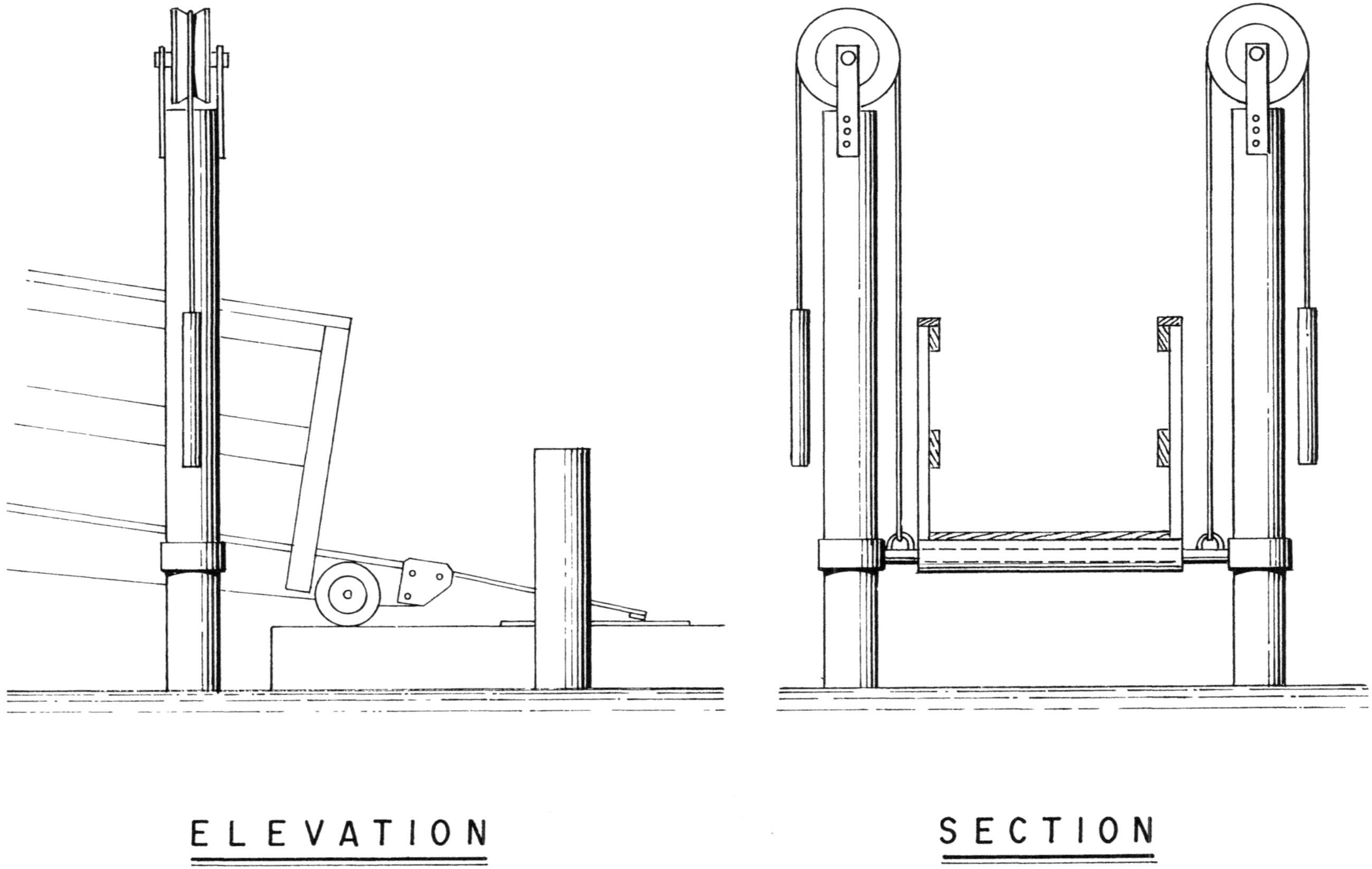

Figure 159. Gangway counterbalance system.

5. Case Studies.

a. Planning.

(1) Hawaii-Kai. Hawaii-Kai is a private, planned community development on the southeastern tip of the island of Oahu, Hawaii, about 15 miles from Honolulu. Presently, about 20,000 people live in this 6,000-acre Kaiser-Aetna development.

Initially, the site was undeveloped except for a few scattered farms, houses, and a small crossroads market. Thus, the planners had an opportunity to completely master-plan an entire new community from start to finish. The developers followed the principle that the feasibility of a master-planned project depends on the ultimate completion of the whole plan, and if parts of a plan are arbitrarily eliminated, the entire plan may be compromised as a result. This principle ensured that community services that were to be subsidized by the developers would not be omitted during later development stages.

The evolution of Kuapa Pond into the private Hawaii-Kai Marina has been a part of the Hawaii-Kai master plan since its inception in the early 1960's. A large swamp in the center of the project was dredged, and the surrounding land converted into residential waterfront, a harbor, and a 200-foot-wide access channel to the open ocean. This water-oriented community is similar to those at Huntington Harbor, California, Coral Gables Estates, Florida, or Nantucket Island, Massachusetts. The marina berths a total of 1,600 small craft, including those berthed at the Koko Marina docks and at individual berths of waterfront residents.

Over 1,400 marina dwelling units have already been constructed in Hawaii-Kai. Before completion, it is anticipated that about 2,500 additional marina units will be built along the waterways. Waterfront property owners pay a marina maintenance fee of $6 per month, and provide their own docks, which must be approved and not extend beyond the pierhead line. Owner-residents pay only $1 per year for boat registration. Residents who live off the water do not pay the monthly maintenance fee, but must pay $73 per year for boat registration and dockage at the Koko Marina.

The average depth of the water basins is 6 feet; the geographic configuration of the water system is shown in Figure 160. The developed area is surrounded by hills, providing a wide variety of terrain and environs. Residences range from marina waterfront homes to sheltered valley homes, and from large estates to small hillside view homes. Although Hawaii-Kai comprises 6,000 acres, only about one-half of this acreage is developable. The remainder includes conservation areas along the mountainsides and the 258-acre marina.

The objective of the Hawaii-Kai master plan was to develop a water- and recreation-oriented community for residents to enjoy Hawaiian living to the fullest. This project is a good example of how a marina can become the nucleus of a residential community in a previously underdeveloped area. It also exemplifies the implementation of a master plan that included comprehensive land planning, development, and management techniques.

(2) Marina Del Rey. The first feasibility study for construction of a small-craft harbor at the Marina Del Rey site in Los Angeles County, California, was submitted by the

Figure 160. General layout of Hawaii-Kai Marina Development, Oahu, Hawaii

U.S. Army, Corps of Engineers in 1949. This report proposed an 8,000-slip complex at a total estimated cost of $23,600,000. The report was taken under consideration and in 1953 the Los Angeles County Board of Supervisors sponsored State legislation that eventually granted the county a $2 million loan from State tidelands oil revenues to assist in the purchase of the present harbor site (Fig. 161).

The following year, Public Law 780 was signed, making the Marina Del Rey harbor an authorized Federal project, which committed the Federal government to a 50-50 sharing with Los Angeles County for costs of all the main navigational features of the development. Construction began in 1957 and the entrance channel and jetties were completed in 1958. The county's share of the costs and the remainder of the project was financed through revenue bond issues.

Construction delays plagued the early development and when the harbor finally opened for operation it suffered storm damage so severe that an emergency program was initiated to provide greatly increased protection from wave action. Fortunately, based on early indications of excessive vulnerability of the harbor to wave action, a model study was already underway at the U.S. Army, Corps of Engineers Waterways Experiment Station at Vicksburg, Mississippi, when the first physical damage occurred in the winter of 1962-63. With cooperation of the Federal Government the study program was expedited on a *crash* basis and the model was used for a feasible interim solution. The county proceeded immediately to construct temporary protective sheet-pile baffles in the turning basin of the main channel to give vital protection for continued development and operation pending the completion of permanent protective work by the Corps of Engineers.

Meanwhile, the results of the model study indicated a requirement for an offshore breakwater, and construction began in 1963 and was completed in 1965 at a cost of $4.2 million. Although 4 years behind the original development schedule, Marina Del Rey successfully surmounted major development problems and formal dedication was held on 10 April 1965. As indicated in Table 5, over 5,500 boats are presently berthed in the various marinas at the complex; hundreds of additional boats in dry storage also claim Marina Del Rey as their home port. The launching facilities have also made the complex a harbor of opportunity for many of the 100,000 trailered boats located in the Southern California area.

Occupancy of the existing apartment units is at 97 percent of capacity (1973) in a new community having a projected permanent population over 10,000 and a seasonal peak day population of 30,000. Construction of additional ancillary facilities is following a sound growth pattern. Private investment in facilities now exceeds $107 million, and long-range plans indicate a conservative total of another $65 million.

Accrued project income to the county from lease revenues is over $3.5 million annually (1973) and steadily increasing to a projected $4 million level. The tax base of the county has substantially expanded in an area where previously the tax return was insufficient to cover the costs of a required mosquito-abatement program. Total investment of public funds in the basic project is shown in Table 5.

Figure 161. Marina Del Rey, Los Angeles County, California

Table 5. Marina Del Rey Fact Sheet

Public Recreational Facilities	Residential and Transient Facilities	Boating Facilities
Public Beach and Picnic Area Spectator Events: Regattas, Crew Races, Boat Parades, Sailing Races Sportfishing, Harbor Cruisers Boat Rental and Sailing Instruction Fisherman's Village Theater Shopping Centers Direct Freeway Access via Marina Freeway	Hotels and Motels 16 Apartment Complexes 22 Restaurants	5,600 Boat Slips Beach launching area for hand-carried boats Launching facilities for trailer-borne boats (both by ramp and high-speed hoist) Dry storage of trailer-boats Yacht Clubs Repair Yards, Fuel Docks and Pumpout Station, Live Bait

Statistics:

780-acre site (405 acres of water, 375 acres of land).

2,340 feet of offshore breakwater.

2 miles of main channel, 1,000 feet wide and 10 feet deep.

3 miles of side basin, 600 feet wide and 10 feet deep.

7.5 miles of concrete bulkhead.

6 miles of landscaped boulevard roadways.

All utilities underground

Future Development	Private Development of Facilities		Public Development of Site	
	Facility	Cost	Site	Cost
Hotels, Motels, Office Buildings, Library, Public Park, Fire Station	Slips	$ 8,250,000	Federal Government: 50 percent of main navigation features	$ 4,600,000
Construction activity estimated at $42 million for 1972	Apartment Units	86,900,000	State Government: Site acquisition	2,000,000
	Hotels and Motels	4,350,000	County Government: Land acquisition, 50 percent of main navigation features	15,875,000
	Restaurants	4,000,000	Proceeds from sale of revenue bonds: Site preparation	13,000,000
	Commercial	7,700,000	Motor Vehicle Fund: Perimeter roads	775,000
	Yacht Clubs	1,800,000		
Total Cost		$113,000,000		$36,250,000

Economic Impact on Area:

Taxes paid by lessees provide significant new tax revenue for county, city schools, etc.

New community with projected permanent population of 10,000 and seasonal peak-day population of 30,000.

Over 200 individual businesses, providing 6,000 to 8,000 new jobs.

Leads the way in the area planning for full development potential of Santa Monica Bay.

Income:

Revenue derived from leases (estimated over $3 million annually) will service bond debt, pay operating expenses and repay State and County loans.

A unique feature of Marina Del Rey's development has been the joint public-private investment partnership in which capital from both sectors was used to develop a facility that benefited the entire community. The lessees are given a set of guidelines to follow and then are allowed to develop the berthing, support, and ancillary facilities under a permit-approval procedure. Lessees are also responsible for maintenance of all lands and facilities installed in their respective leasehold areas. This has effectively reduced the county's overhead and allowed the marina management personnel to function in a purely administrative capacity. The County Fire and Sheriff Departments maintain facilities and personnel at the marina, which greatly simplifies the work of the harbormaster. This is an extremely practical compromise in a harbor complex of this size, and has resulted in one of the most successful ventures ever undertaken by the County of Los Angeles.

(3) Salmon Harbor. Salmon Harbor, located off the mouth of the Umpqua River in Winchester Bay, Oregon, is unique in two respects: original development evolved primarily as a result of the single activity of salmon fishing, and the harbor is in the midst of an extensive expansion program (Fig. 162).

In 1951, Douglas County and the Port of Umpqua initiated planning for the project, directed toward: (a) protecting the waterfront of the town of Winchester Bay from future erosion, (b) providing a base for the local commercial fishing fleet, and (c) providing a safe, adequately equipped pleasure-craft harbor and marine recreation area. Construction began in 1952, with various dredging and filling operations completed in 1955, 1957, and 1963. Work was financed jointly by Douglas County and the Port of Umpqua.

The two authorities agreed, in 1952, to appoint a management committee to supervise the operation of Salmon Harbor. One member was appointed by the Douglas County Park Board, another by the Port, and a third by mutual agreement of both parties. The committee has functioned for the past 17 years, without pay, directing the operation of the harbor, and handling such varied problems as moorage rates, land leases, harbor maintenance, and *Fleet Days* activities.

The present basin, shown in the lower part of Figure 163, has a 650-slip capacity. The basin experiences little surge and only minor maintenance has been required. The entrance channel has been dredged twice in 20 years.

Spruce logs were used for flotation in the original docking system, but are gradually being replaced by polystyrene units with concrete decks. Creosoted timber piles are used for guides primarily because of their low cost in this region. The slip rental rate is $1 per foot per month, with slips ranging from 18 to 50 feet. Water and power are included in this rate.

Few clients live aboard their craft, and most remove them from the water for home storage during the winter months. The present facility has a launching ramp, two cranes, and a boatyard, and has had enough work to support it during the 5 years of operation. According to the manager of the installation, the only real problem is parking the many campers that use the facilities.

Most revenue from the Salmon Harbor operation has been from charter boats, slip rentals, boat launching, car parking, and real estate leases. Gross receipts for a 17-year

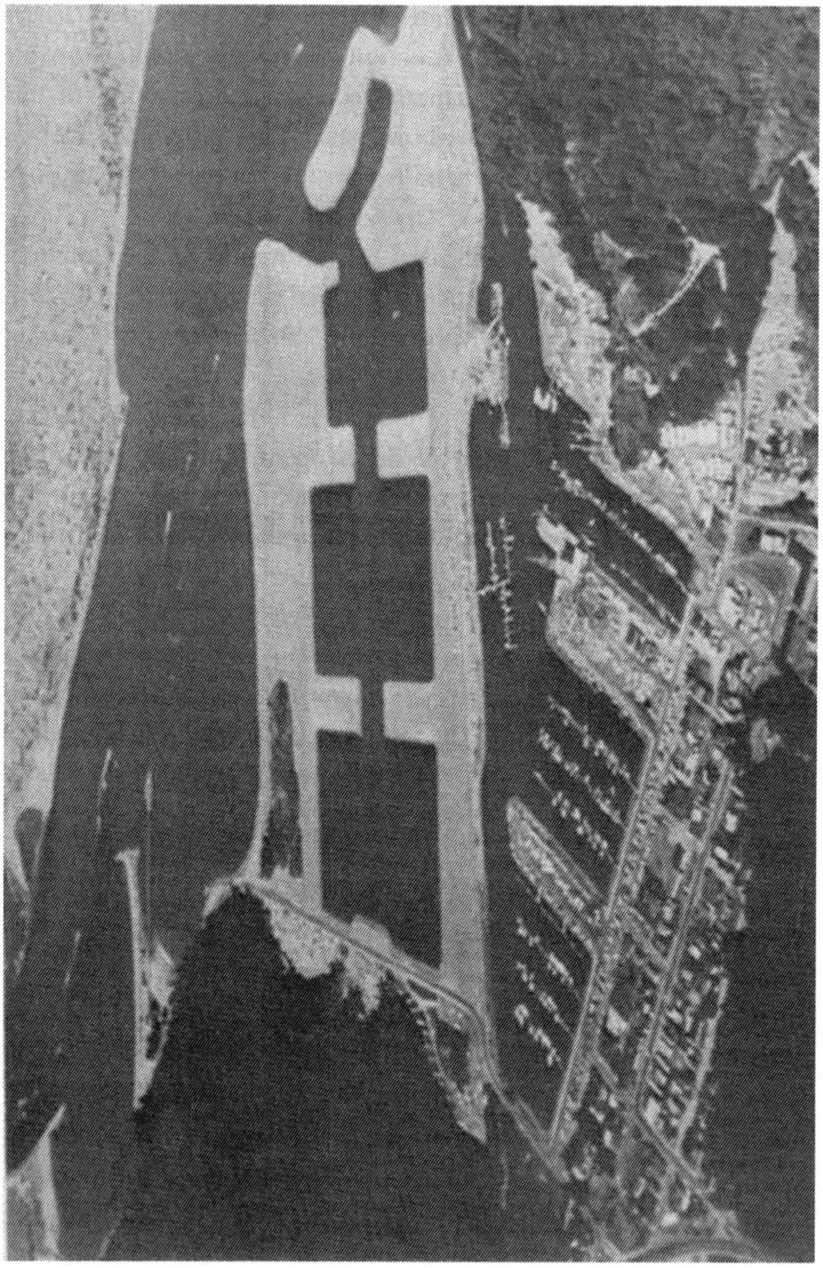

Figure 162. Salmon Harbor, Winchester Bay, Oregon

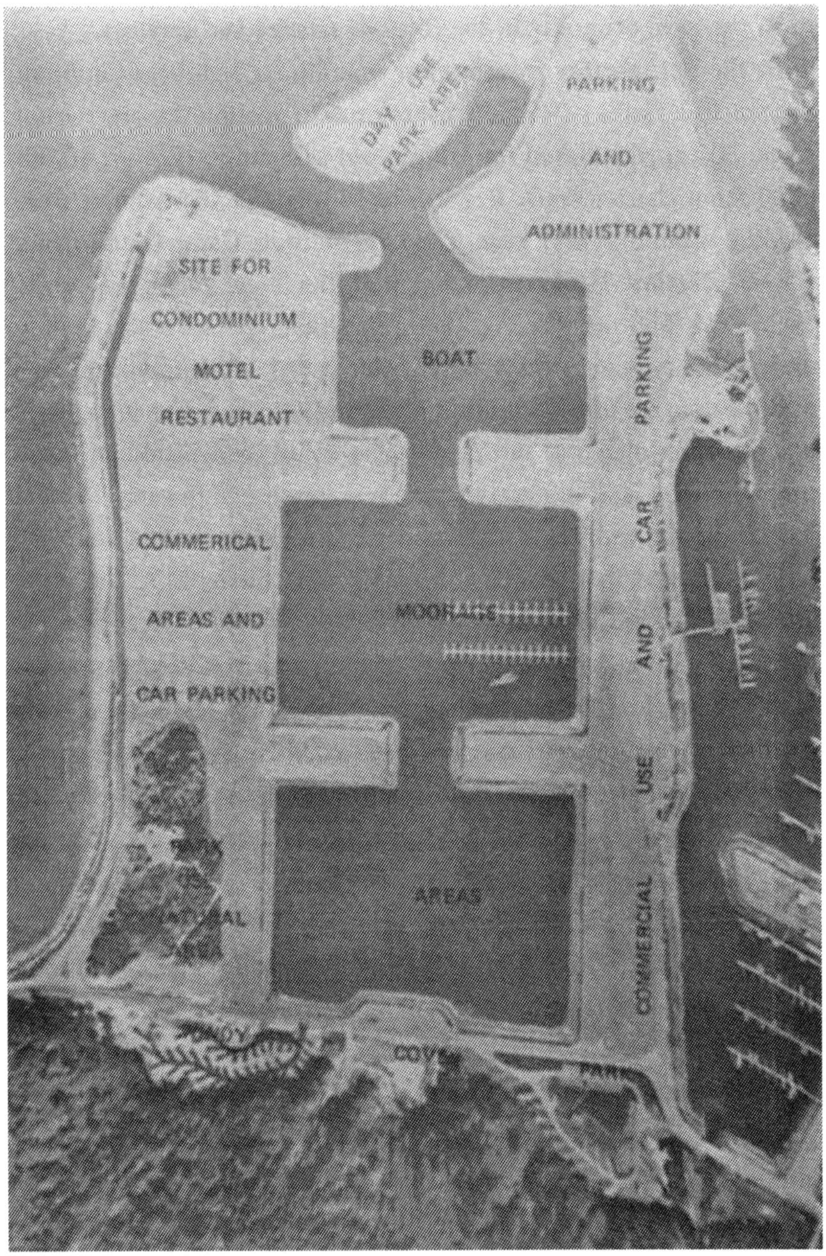

Figure 163. Land use proposal for the Salmon Harbor expansion project

period ending in 1970 totaled $439,513 and the net receipts were $177,581. The difference was used for maintenance and operating expenditures. A substantial part of this amount was reintroduced into the local economy in the form of salaries and services.

In addition to the net receipts from harbor operations, significant indirect benefits have accrued to the area as a result of salmon fishing made possible by the harbor. In 1969, for instance, about 60,500 anglers fished for salmon out of Salmon Harbor, 40,000 from private craft, and 20,500 from charter craft. Records show that 63,454 salmon were brought into the harbor by sport fishermen that year. The Oregon State Game Commission, in Technical Bulletin No. 78, September 1964, estimated that the average cost of catching salmon on sport-fishing tackle was $63 per fish. Assuming that this represents the value placed on such fishing by anglers (the money would have been spent on something else), an indirect economic impact on the area of $3.5 million for the 1969 catch was indicated.

The present expansion plan was initiated primarily in recognition of these large indirect benefits. A comprehensive land-use master plan was developed in early 1970 with the intent of integrating the existing facilities with proposed developments in a manner that would be feasible and functional, and that would optimize project benefits to the area. The resulting land-use proposal is shown in Figure 163.

Sixty-two acres of usable land have been created as a result of the expansion project. Most of this land will be devoted to recreation and ancillary facilities. The small peninsula protecting the entrance to the new basin has been set aside for a day park, providing over 3 acres of picnic area. A *nature* area and additional camping area are planned for the western arm of the project. These two areas are a natural extension of the existing park facilities at Windy Cove.

Commercial development will be confined primarily to the smaller peninsula areas that jut into and partition the three boat-berthing areas. The proposed commercial developments include: (a) A condominium-motel restaurant complex, (b) a boat and motor sales and service facility, (c) dry-storage facilities, (d) an undersea garden aquarium, (e) automatic vending facilities for around-the-clock sales of tackle and bait, (f) a floating restaurant, (g) gift shops and boutiques, (h) a bar and cocktail lounge, and (i) fish receiving and processing facilities.

The berthing capacity of the harbor will more than double, with 900 additional slips. Both private and public conveniences will be added, including expanded charter-boat service, more bait and tackle shops, a coffee shop, and marine fuel facilities.

This case study exemplifies a well planned expansion program that has considered future opportunities as well as existing facilities and operational procedures. Also, the execution is well coordinated, with the least possible disruption of ongoing operations.

b. Management.

(1) Bahia Mar. Bahia Mar is a 350-slip open-berth marina situated on the Intracoastal Waterway in Fort Lauderdale, Florida. The 40-acre complex is about evenly divided

between water and land area. It was built in 1950 under a design-and-construction contract on Federal surplus land purchased by the city of Fort Lauderdale with proceeds from bond sales. The city subsequently operated the marina, mostly at a loss, until 1962 when the entire property was leased to private enterprise. That lease is now owned by MCD Enterprises, Inc., a Maryland real estate development corporation, which has added several ancillary facilities. Bahia Mar now includes a marine fuel service depot, a large restaurant, a marine hardware store, a 115-room motel, and a shopping mall with spaces leased to boutiques and other businesses. Proceeds from these facilities supplement slip-rental income to make the total operation economically profitable. The motel operation has been so successful that the company is building a 195-room addition.

Project depth throughout the marina is 10 feet below mean low water. The berthing facilities are fixed, with concrete decks supported by concrete piles. The support piles project upward through the deck to about 2 feet above deck level, and the projections support and protect the utility risers (Fig. 164). Both 120- and 240-volt power outlets are provided at each slip without metering as part of the rental fee. A local ordinance makes it illegal to resell power.

The marina has no public address system because phone jacks are provided at each slip. The phone service is monitored through a central switchboard, and a minimum fee of 25 cents is charged for each outgoing local call, with straight rates being charged for long-distance calls. Public pay phones are also provided at the head of each pier. The marina has no sanitary pumpout facility, but one is planned in the near future. Internal trash collection is accomplished by marina personnel, and the collected trash is picked up, compacted, and hauled away by a sanitation contracting company.

Preferential parking is provided in a card-key lot for 152 cars at a rate of $50 per season. Parking spaces for the remaining slips are provided at a ratio of three spaces for every four slips on a first-come-first-served basis. Overflow from this lot is accommodated at a shopping center lot.

The pier configuration is shown in Figure 165. Breakdown of rental berths by lengths is as follows:

Slip or Dock Length (feet)	Quantity
30	210
40	28
50	20
65	7
90	15
100	5

The remaining 60+ slips are provided for commercial fishing and excursion craft.

Figure 164. Above-deck extensions of pier piles support and protect utility risers at Bahia Mar Marina, Fort Lauderdale, Florida.

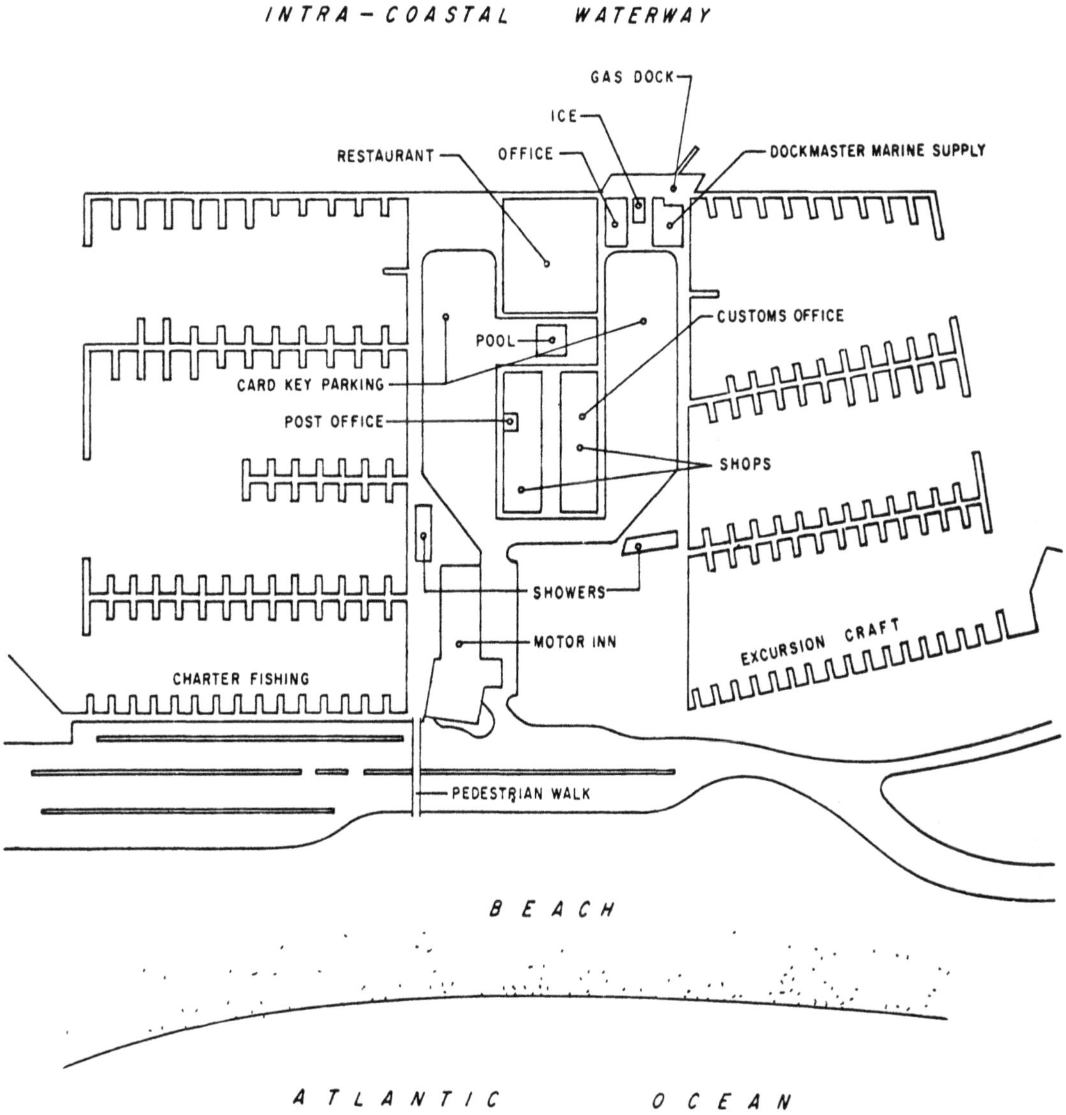

Figure 165. Bahia Mar Marina layout plan.

Located in a large boating-tourist area, the Bahia Mar facilities are designed primarily to serve transient powerboaters. The peak season is from November through April. For the 1972 winter season, the daily rates for berth rentals were 30 cents per foot, with a $9 minimum. A 4-month winter contract is offered, which reduces the rate to 25 cents per foot per day. Summer (off-season) rates are 10 cents per foot per day. Annual contracts are also available, the yearly rate ranging from $2,100 to $3,000 depending on boat length.

Service and convenience facilities include showers and dressing rooms, a freshwater swimming pool, an elevated pedestrian walkway for access to the oceanside beach, car-rental services, a U.S. Post Office, and a U.S. Customs clearing station. Because these facilities must be available to the public, no security fences are possible, but a 24-hour roving security guard is provided, and during the winter season all vehicles must pass through entrance control gates.

The only troublesome construction features are a few of the perimeter walls, which were not provided with adequate filter courses. Inadequate embedment of rebars has resulted in some spalling of the concrete. Wave and surge action has apparently pumped fill material from behind and underneath the perimeter wall into the berthing basins in some areas, leaving sinkholes behind the wall.

Bahia Mar is a good example of a small-craft harbor that has been converted from a financial burden on a community to a successful business enterprise by the addition of ancillary facilities financed by the private sector. Success is mostly attributable to an accurate assessment by the new owners of additional services desired by harbor patrons, and the investment of private capital in facilities needed to provide these services. Good management of the facilities and vigorous advertising to make the boating public aware of their availability is also recognized.

(2) Lighthouse Bay Marina. Lighthouse Bay Marina (Fig. 166) is on the shore of Pamona Lake near Quenemo, Kansas. This 200-slip marina is privately owned and operated by the Aquamarine Corporation of Quenemo. Funds for financing the project were obtained through the local bank.

The marina is managed by one of the three equal partners who own Aquamarine Corporation. The operation supports four other full-time employees year round; a mechanic, two general maintenance personnel, and a bookkeeper. During the peak summer months, April through October, the marina remains open 24 hours per day and the personnel requirements increase to 14.

The lake on which the marina is situated is regulated to a *conservation level.* In the 9 years of pool operation the lake level reached a high of 14 feet above conservation level twice and dropped once to 4.5 feet below level. The average yearly lake-level fluctuation is 5 feet. This history coupled with the average basin depth of 24 feet required a floating-dock system, anchored by cable and shore-connected, hinged struts.

Figure 166. Lighthouse Bay Marina, Quenemo, Kansas

Eighty percent of the marina patrons own conventional-hull, open cockpit powerboats and prefer covered slips. The marina obtained the slips from two different manufacturers and is satisfied with the performance of both proprietary systems. The floating docks are assembled into units referred to as *houses* by the marina personnel; each unit consists of about 12 berths (number of slips is variable) arranged abreast under a common roof. The Lighthouse Bay management decided to purchase the dock materials in bulk and do the assembling themselves so that the maintenance personnel obtain a better working knowledge of the docks.

The most recently installed house was a 20-slip unit, and the inplace cost was about $1,000 per slip. The maintenance costs on this system have been running about $3 per slip per year on 200 slips. One of the biggest maintenance problems is the need to move the shoreside anchor points frequently as the lake level varies. Each house is secured to the bank by two cables, a hinged strut at one end of the dock unit and a hinged-gangway strut at the opposite end. The time required to adjust all the 17 units is about 25 man-hours.

The only other major maintenance problem is that caused by winter ice damage. Although the lake remains frozen for about 6 weeks each winter, the resultant damage is not serious enough to warrant the difficult task of dismantling and removing the systems for the winter. Drifting ice floes or windrows have not been a problem at Lighthouse Bay; good natural protection is afforded by the cove site and by the abundance of trees along the shoreline, which break the wind.

Some of the dock flotation has been in use for 8 years and is still in serviceable condition. About 2 percent of the flotation units are replaced each year. In some cases more flotation is added to a unit instead of replacing the old sections.

The slip occupancy at Lighthouse Bay has been 100 percent for the past 3 years. The management attributes this partially to the marina's close proximity to a metropolitan area. The rental season is from 1 April to 31 October. The summer clientel are encouraged to winter-store their craft at the marina, but is not demanded in the seasonal contract. In an attempt to generate more winter activity, the marina management has completed the installation of a heated fishing dock facility.

Presently, 75 boats are stored under cover during the winter, and 20 are stored outside. The covered storage facility consists of a 90- by 200-foot building with a heated shop in one corner and a concrete floor throughout. Access to the winter storage area is controlled, and marina personnel must be in the area before an owner is allowed to inspect his craft. Practically all boats are winter-stored on trailers for the multihulled craft, which are usually placed on blocks.

The marina provides all eight of the ancillary facilities listed in the marina questionnaire (App. K) as well as tent rentals and camper hookups. The present docking arrangement provides each slip tenant a camping area immediately shoreward of his assigned boathouse. This arrangement is apparently agreeable to the patrons, as no one has suggested changing it.

Generally, this is a fine example of a well managed and efficiently operated lakeshore marina.

c. Design.

(1) Cincinnati's Ohio River Project. Through the years, the city of Cincinnati has utilized the Ohio River as a vital natural resource to enhance commercial and recreational growth. To meet present-day needs, the city began, in the late 1960's, to develop the riverfront into a modern attractive recreational area to satisfy the ever-growing demands of the general public. Within a few years, the area previously used for automobile parking, commercial boat docking, and miscellaneous storage has grown into an area that now consists of a new major sports arena, public landing area for commercial boat docking, parking facilities, river promenade, and a riverfront park facility under construction.

Land transportation is adequate, but increased boating activities have influenced officials to provide a small-craft docking facility to meet the need for water access to the area. However, due to river characteristics of frequent flooding, heavy floating debris, high river velocities, and lack of a protected basin, development of a marina in this area required some unique planning and design considerations. A stringent set of operating procedures to safeguard floating docks from damage by varying river conditions was also necessary.

This case study is based on the present (1972) status of improvement. Final design is complete and construction has started. The marina, when operational, will serve transient boaters to the use-area rather than provide permanent slip rentals by lease agreement. The marina design layout is shown in Figure 167.

The Ohio River experiences large water level fluctuations annually. The high velocity river current increases the hazard of floating debris, which requires special debris deflectors.

Because pleasure boating is not recommended on the Ohio River under high water conditions, the floating docks are designed only to operate during safe boating levels, allowing for a water fluctuation of about 13 feet above the normal pool elevation of 455 feet. To reduce marina damage, facilities must be removed from the immediate area to a safe location for storage during the winter and early spring seasons and during rises in summer river levels higher than 13 feet above normal pool.

Anchorages must withstand both lateral and longitudinal river and wind forces for unusually heavy loads due to high currents and floating debris. Lateral anchorage and restriction to upstream movements are accomplished by use of spud pipes, which stabilize the floating docks. This system permits the docks to move vertically up and down along the spud pipes maintaining uniform freeboard for boaters as water levels change. Site characteristics and river depth are not satisfactory to securely hold the bottom of the spud pipe. Therefore, a special spud-anchorage system (Fig. 168) will be constructed near the river bottom. The anchor consists of a cluster of four battered piles encased with an underwater poured-in-place concrete anchor unit. The spud pipe is seated into a pipe sleeve located in the center of each concrete unit and extends through a sleeve in the floating dock.

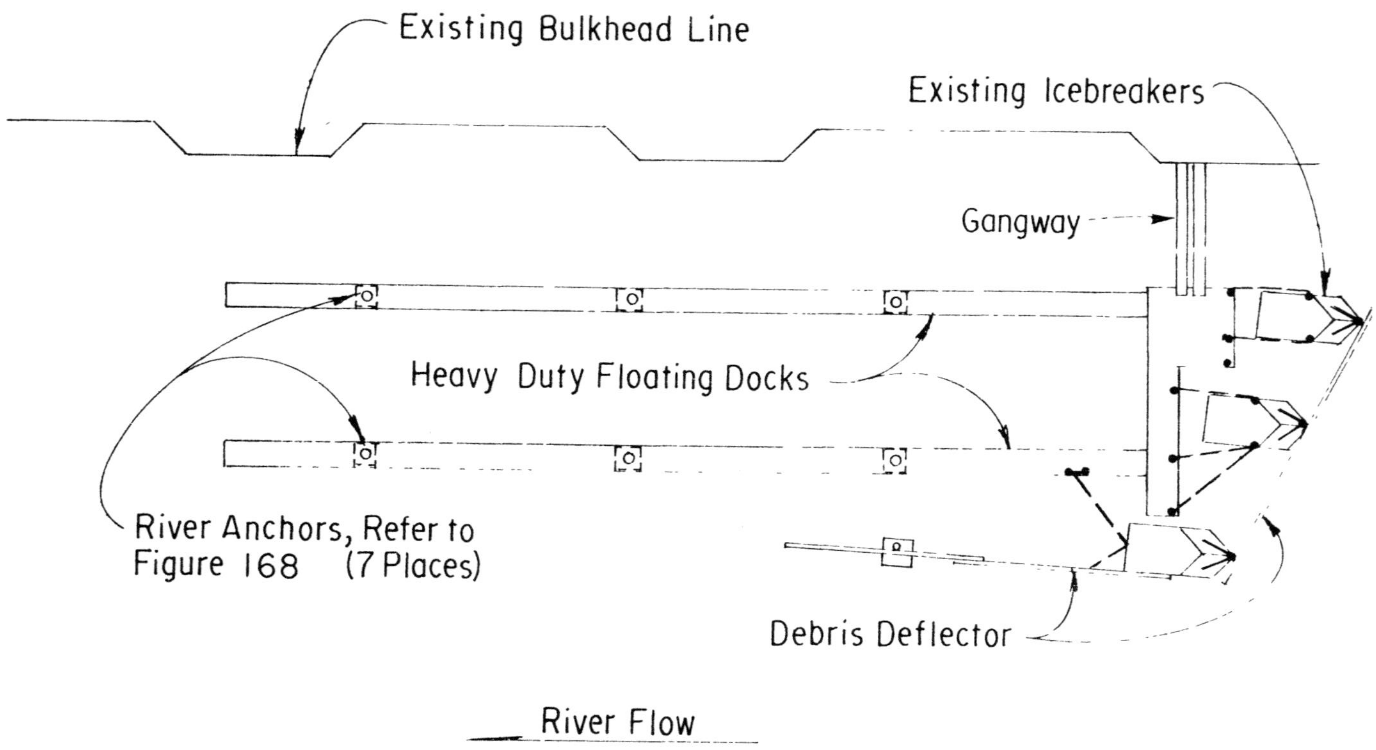

Figure 167. Layout, Cincinnati's transient marina, Ohio.

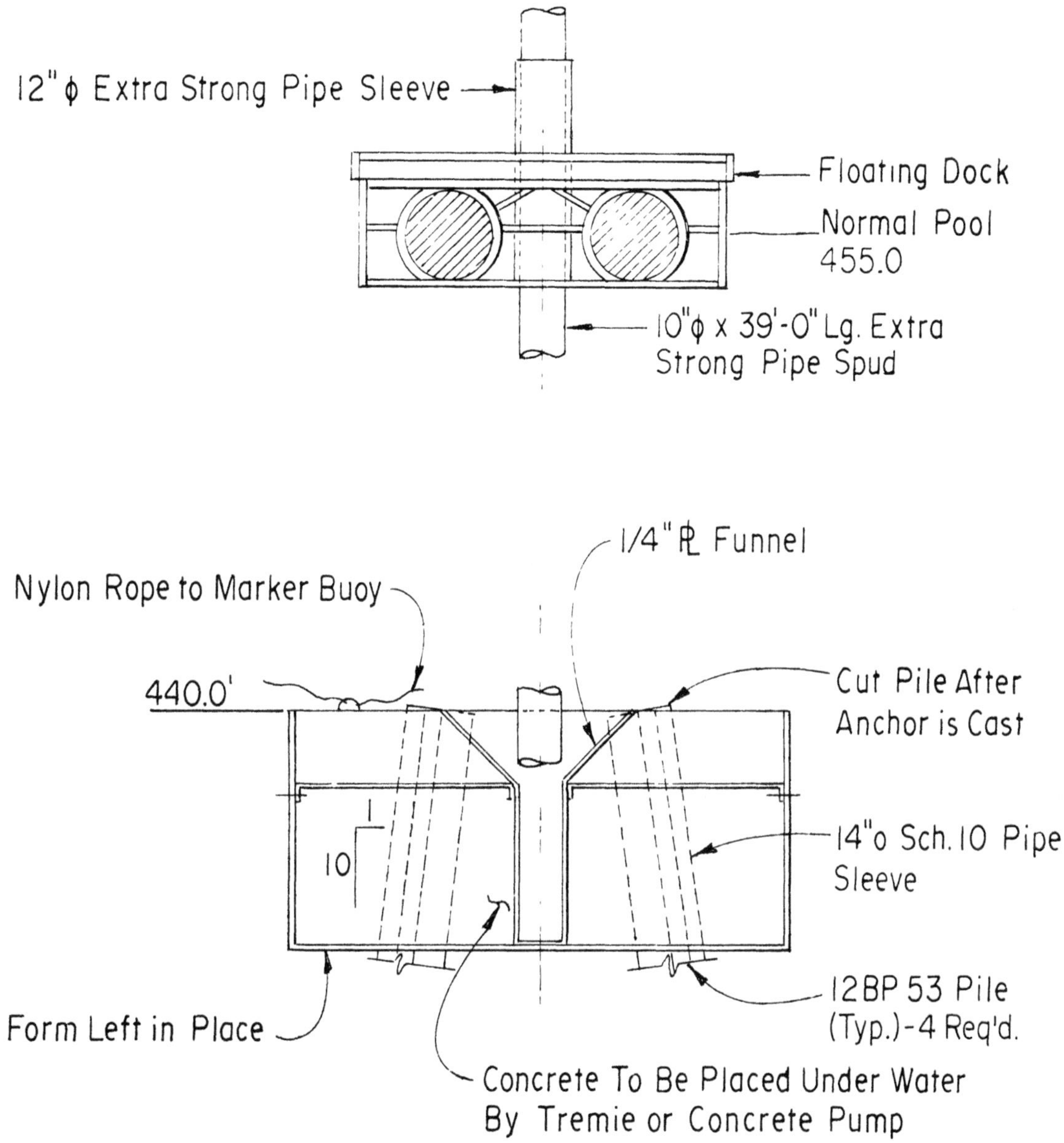

Figure 168. River anchor for floating dock on the Ohio River.

Longitudinal forces are withstood by using the existing concrete ice breakers located upriver from the docks. Connecting cables between the floating docks and the ice breakers hold the floating docks in position. Winches located on the floating docks are used to adjust the connecting cables as water fluctuations occur.

To safeguard against heavy floating debris, a *protective float* is to be located upstream from the docks so that debris will deflect around the docks to reduce maintenance and protect the docks and boats. The guard will consist of a flotation unit of three foam-filled steel pipes stacked vertically and connected to a frame braced against the upstream face of the ice breakers. Flotation units will be located as shown in Figure 167. The protective float, about 4 feet deep, will be constructed at an angle with the river flow and floating docks so that the river currents will assist in moving debris around the marina and minimize debris accumulation against the upstream face of the docks. However, operating personnel must regularly clean away debris collecting at the facility to eliminate any large buildup of debris.

The city of Cincinnati is considering operating the facility without the protective float on a trial basis.

(2) East 55th Street Marina. Cleveland's East 55th Street Marina is situated on Lake Erie. A study of the boatowner requirements for this area showed the need for a large marina with accommodations for craft up to 60 feet long. Initially, 274 berths were constructed; the facility can eventually be expanded to accommodate 426 berths. The proposed site offered both advantages and disadvantages. The nearly adequate protection afforded by existing breakwaters and sufficient depth for a small-boat harbor provided the pluses. Also, the site offered convenient highway access.

Disadvantages included weak soils in the basin bottom and rubble fill in the area selected for building and utility services. Additionally, two major combined sewers (storm and sanitary) discharged into the protected basin near its entrance. Although these factors increased development costs, they did not overshadow the harbor and location advantages.

Extreme variations in Lake Erie levels previously recorded at Cleveland were about 5 feet. Soil borings taken in the marina bottom showed loose deposits of fine materials to depths up to 70 feet below average water level. With these conditions, both convenience and economy dictated the use of floating piers. The structures provide uniform levels with respect to boat decks regardless of water elevation. They also permit economical anchorage, which does not require high-strength foundation soils.

Because of the weak soils an underwater system of mooring cables and anchors holds the floating piers in position. The galvanized coated cables are the same as those used for bridge construction. More than 12,000 feet of zinc-coated bridge rope was used on this project.

Seven cable lines, spaced to transmit about equal loads, connect all piers to the east and west anchors. Each east-west cable line is arranged in three lengths: one between the west anchor and west pier, one between the two end piers, and one between the east pier and east anchor. A wench for each cable length permits adjustment of pier position and cable tension. North-south positioning is accomplished by cables connected to anchors located beyond the north and south ends of each pier. The north cable lines also have wenches for adjustment. The seven east anchors, the nine north anchors, and five of the seven west anchors consist of concrete blocks resting on the marina bottom. Several tons of rock piled in front of the anchors provide additional resistance. Two west anchors connect to the new steel breakwater extension. Anchors located on the shoreline consist of concrete blocks weighted with earthfill. These blocks also support the bridge ramps between the shore and floating piers.

Cables are pretensioned sufficiently to hold approximate horizontal positions. The system will accommodate changes in water level up to about 5 feet without wenching adjustments. Connecting the several piers has a calming effect on the marina basin, tending to dampen waves and reduce movements during storms.

This unique anchorage system combined with the existing site-feature advantages have provided the Cleveland area with a safe full-service marina facility since 1969.

(3) Galveston Yacht Basin. Probably the most uniquely engineered small-craft installation encountered during the marina survey was the Galveston Yacht Basin (Fig. 169). This full-service marina is located only five blocks from downtown Galveston and is sited on Galveston Channel, which provides access to both the Intracoastal Waterway and the Gulf of Mexico.

This installation offers covered berths for 416 power cruisers, pigeonhole dry-storage parking for 180 boats, and open slips for 150 sailboats. The layout plan is shown in Figure 170. In the north basin, shown at the top of the figure, are four long, parallel-covered piers. The three piers labeled B, C, and D each shelter 116 craft; pier A shelters only 46 because of the land interface on its south side. These piers are protected by two moles and a breakwater.

The mole on the north side also provides protection for 50 open sailboat slips that are located in its lee. The southwest mole supports one of the two fuel stations. The south basin contains the remaining sailboat slips, a smaller covered row of 17 slips, the launching facilities, and a second fuel dock. Both basins are 12 feet deep at low water.

The most unusual engineering feature of the marina is the structural design of piers B, C and D (with pier A differing only because of land-water interface). Each main walk is of the usual reinforced concrete slab-on-stringers design, but alternate finger piers of massive concrete construction form the bases for structural steel roof-support columns about 10 feet out from each edge of the main walk. From that point outward to ends of fingers and edge of roof, both the concrete finger piers and the steel-framed roof are cantilevered. The entire concrete floor system with superimposed roofload is supported on timber piles, pressure-treated with coal-tar creosote. Intermediate finger piers between the roof-support fingers are of very narrow timber construction on creosoted timber pile supports and are intended for fendering and for mooring ties rather than for the boarding of passengers.

Figure 169. Galveston Yacht Basin, Galveston, Texas

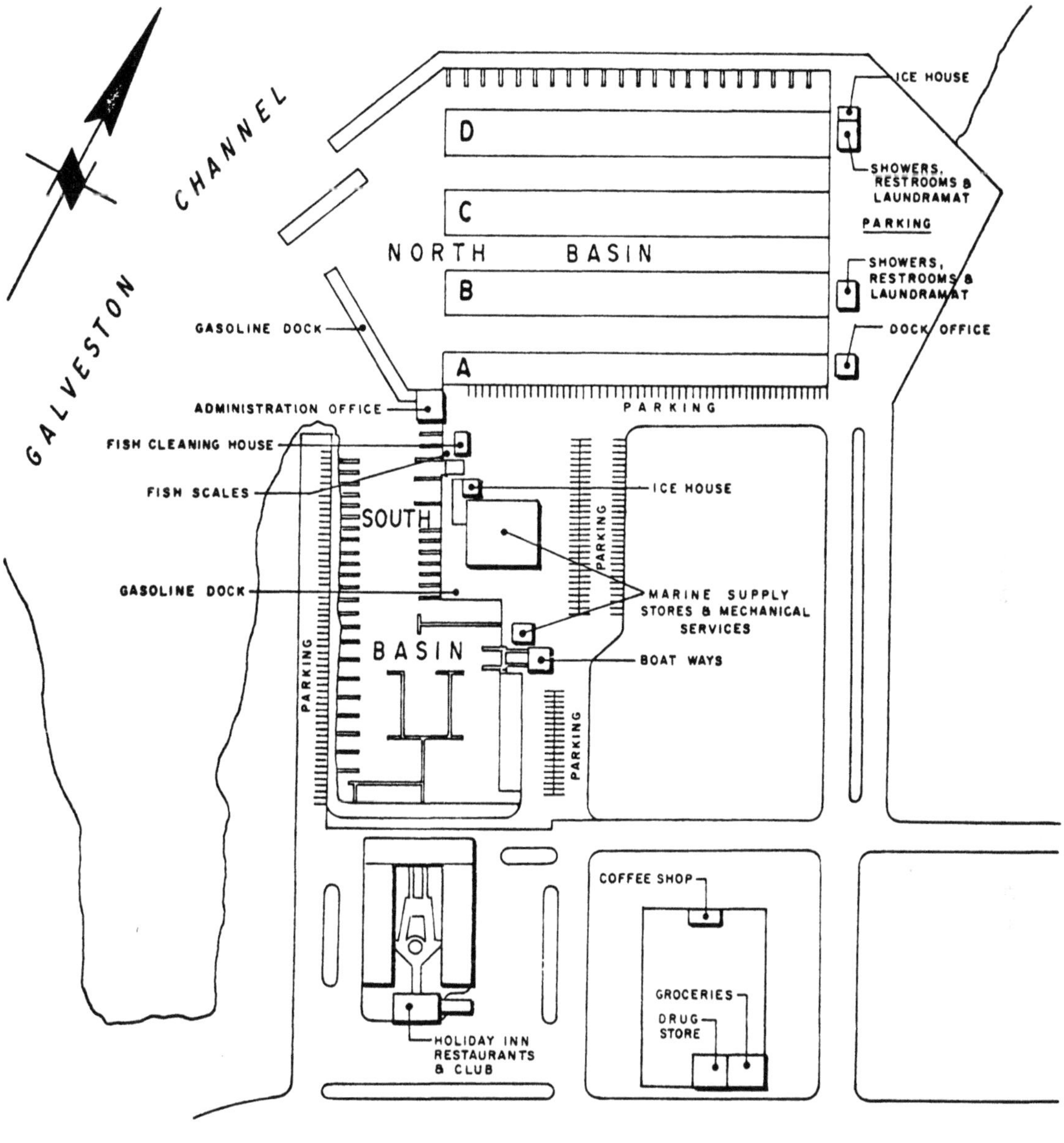

Figure 170. Galveston Yacht Basin layout plan.

Although the normal tide range at this site seldom exceeds 1 foot, the maximum hurricane-generated surge is estimated at several feet. The local building code requires houses near the coast to be built with floors at least 13 feet above mean water level, and a wind setup of 9 feet has been recorded at Galveston. For this reason the clearance between mean water level and the roof structure is 24 feet, and main roof beams are designed for the installation of a lift-out boat hoist for each slip if desired by the slip renter. Thus far, hoists have been installed in about 20 percent of the slips, mainly for use in hull maintenance and minor repairs. In the event of a hurricane warning, however, each boat under a hoist would be lifted well up toward the roof and lashed securely in place.

The slips are rented on a daily, monthly, or yearly schedule. Each slip is provided with a water connection, 120- and 240-volt electrical power, and a private full-height, walk-in locker, all included in the slip rental. A private telephone installation is also available at each slip through arrangements made directly with the phone company by the slip tenant. Trash collections are daily, and the marina is in the process (1972) of installing a sanitary pumpout facility. Paved parking spaces are available to meet the requirements of all patrons and one of the services provided by the marina management is electric cart transport from parking lot to boat berth (Figs. 169 and 170). The two service centers, located at the heads of piers B and D, offer slip tenants the use of shower baths, laundromats, air-conditioned restrooms, vending machines, and ice dispensers. A bait camp, fish-cleaning house, outboard motor sales and service, marine supply store, and yacht repair and mechanical service facilities are also available within the complex.

The launching facilities, all located in the south basin, consist of a ramp 40 feet wide for handling boats up to 40 feet long, two forklift trucks, used primarily in conjunction with the dry-storage operation, and a mobile hoist that travels on parallel piers and can launch boats up to 65 feet long.

Immediately adjacent to the marina complex are a drugstore, grocery store, coffee shop, motel with a swimming pool, restaurant, and a nightclub.

Since the weather plays a particularly important role in boating activities in this region, the marina management subscribes to the weather wire service and posts 2-hour radar reports. A 24-hour security patrol is also provided. Considering all the planning and design that have gone into this installation and the many amenities provided for tenants, the Galveston Yacht Basin represents the elite in luxury boating services.

(4) Spruce Run Reservoir. The New Jersey Green Acres Land Acquisition Act, 1961, gave the State the power to acquire, preserve, and develop parklands and recreation areas for the people. In a highly developed area like New Jersey, the competition for available lands is very keen, and land costs are high. For this reason, it is important to develop existing resources to the fullest extent possible.

The Spruce Run Reservoir located about 16 miles east of Phillipsburg, impounds the excess waters of two tributary streams and figures prominently in the present and future water supply programs of the State. To maximize benefits from this project, the State has utilized the 1961 Act to provide recreation facilities for this site consisting primarily of a launching facility for trailered small craft, an all-weather fishing pier, and a small marina installation to be leased to a concessionaire for berthing rental boats.

During periods of drought, when water demands are at peak level, the reservoir experiences an average drawdown of about 10 feet. This was a major problem in project design, but was not the only problem. Most of the reservoir site is underlain by limestone with varying thicknesses of overburden. The discovery of many sinkholes in the adjacent land area motivated a decision not to drive pilings of any kind for fear of piercing the mantle. Not only would this be unsafe for structural support, but could lead to the loss of storage water through hidden sinkholes.

The problem of variable drawdown was solved at the marina installation by articulating the landing ramps and floating dock units. Piano hinges fabricated from heavy pipe sections allowed the necessary relative movement between floating and grounded units, at the same time keeping the sections together. An elevation view showing the landing configuration is presented in Figure 171.

The floating-dock sections were fabricated from 24-inch-diameter steel pipe with seal-welded end plates. As a further precaution against vandalism, the pipes were filled with polyurethane foam. The entire floating system is fixed in place with hand-adjustable cables and anchors. At the landward end, near the paved slope areas, the guide cables were rigidly attached to the concrete pavement slab.

The launching ramp facility consists of two adjacent ramps separated by one boarding dock and flanked by two others. The ramps are of reinforced concrete slabs on compacted subgrade, with a 12 percent lakeward slope. The pontoon units that make up the 132-foot-long boarding docks are hinged together in the same fashion as the marina's floating docks, but are guided by reinforced concrete guideposts that were integrally poured with the concrete ramp (Fig. 172).

The nearby fishing pier is semicircular in plan, with two shore ends about 50 feet apart. Trestle supports of the pier are kept ice free during the winter months by a compressed-air bubbling system, which was installed so that the facility could be used year round. The submerged bubble line is made of 0.75-inch-diameter perforated polyethylene tubing that follows each side of the concrete pad. Air is supplied by a 1-inch polyethylene line that extends underground from a landside valve pit until it emerges through the bank slope about 4 feet below the high water line.

These three facilities are well designed and well integrated into a single installation that should be a high-use recreation area.

(5) Still Waters Marina. Lake Martin is a landlocked lake which, like many other lakes and watercourses in central Alabama, is well fished, cruised, and used for water skiing. Typically, the craft using these inland waters are small inboard runabouts, outboard cruisers,

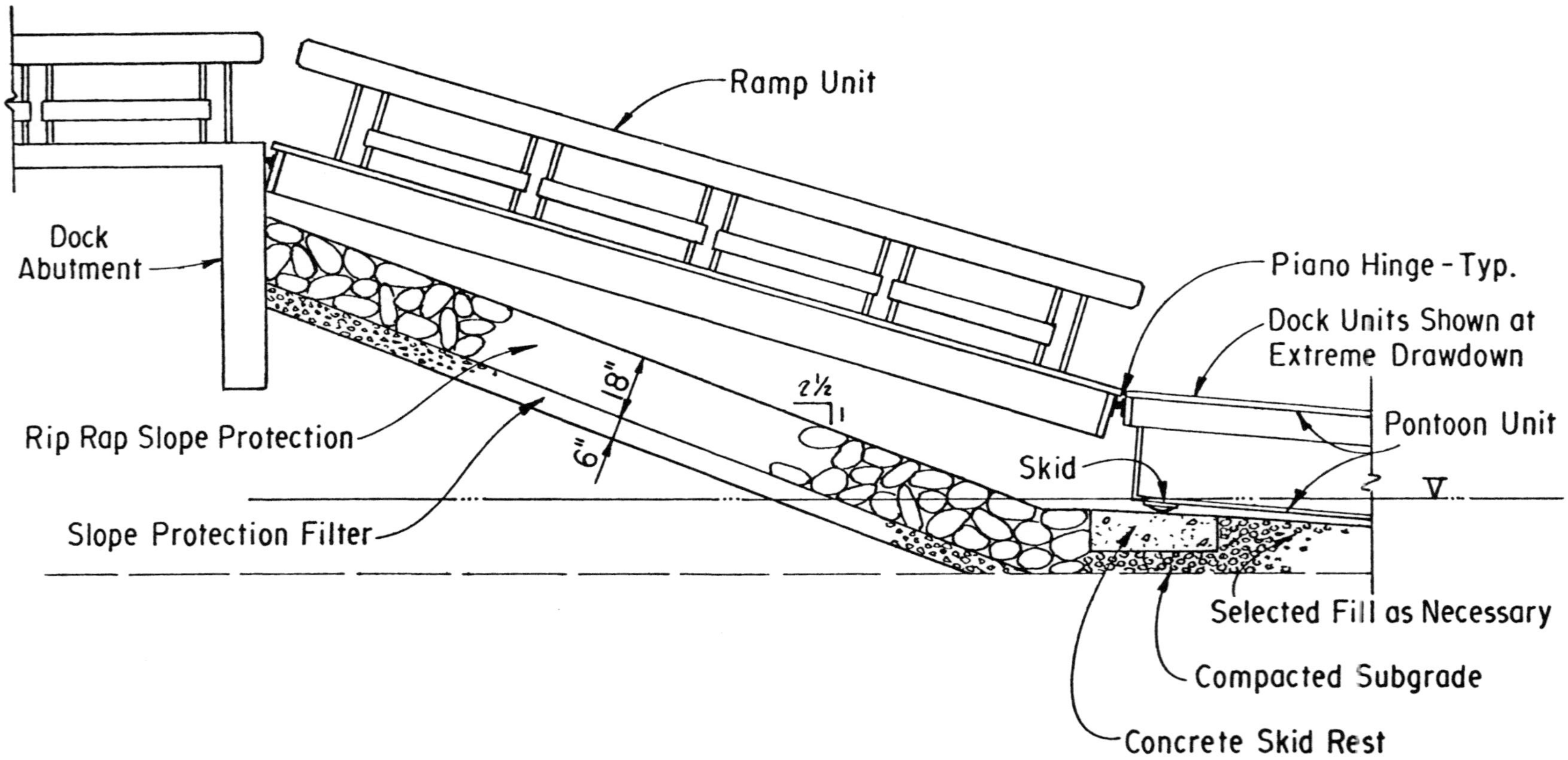

Figure 171. Landing design for floating docks, Spruce Run Reservoir, New Jersey (Courtesy of Edwards and Kelcey, Incorporated).

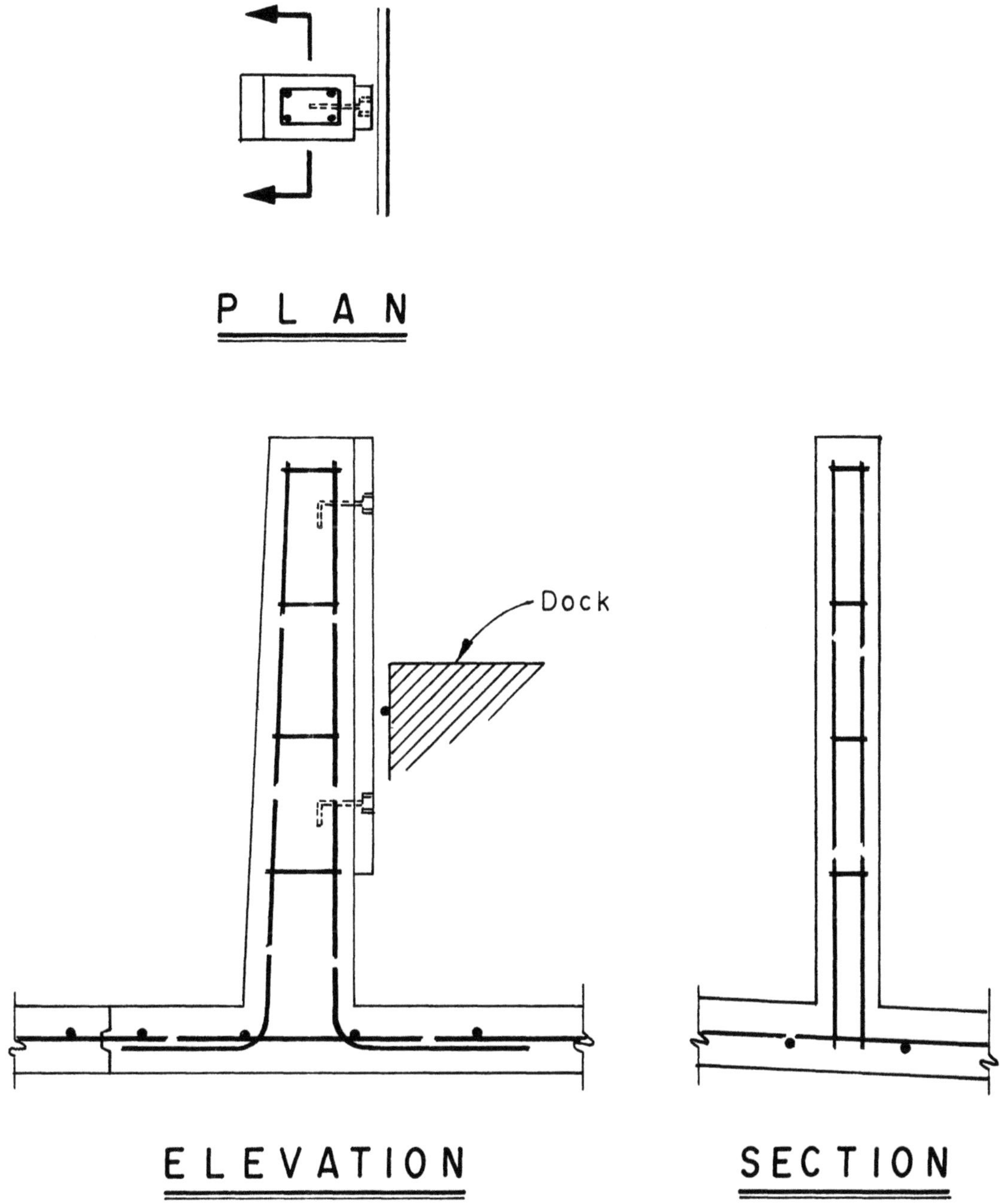

Figure 172. Boarding ramp guide posts poured integrally with concrete launching ramp Spruce Run Reservoir, New Jersey.

and inboard-outboard cruisers that are easily transported by trailer. Recognizing this local trend toward use of small craft, designers of the Still Waters Marina specifically designed the installation to accommodate them.

One of the marina's main features is a dry-storage facility. The operation is housed in a steel warehouse-type building, which is sheathed and roofed with corrugated sheet metal and has a concrete floor. One hundred boats up to 20 feet long can be dry-stored in the multilevel, well padded racks of this building (Fig. 173). A boat that is to be launched is lifted from the rack with a forklift truck, transported to the lake on the two 12-foot padded forks, and lowered into the water. The 12,000-pound-capacity lift has the capability of hyperextending its forks down over a specially designed bulkhead. Retrieval is accomplished by reversing the cycle. With this operation, the boats are kept clean and safe inside the dry-storage building until launching is requested. The dry berth rental rate is $1.10 per foot length of boat per month, and includes launching and retrieval as often as once a day if desired.

Although the dry-storage facility is limited to boats of 20 feet or less, the marina has wet berths available for craft up to 50 feet long. The seasonal water level fluctuation of Lake Martin averages 12 feet, a factor that dictated the choice of a floating rather than a fixed berthing system. A standard floating-dock system is used, but the means of anchoring the floating piers and the 26- by 26-foot floating fuel island is rather unusual. A number of vertical pipe sleeves were first bolted to the outboard edges of the dock structures and then long lengths of 4-inch-diameter pipe were dropped through the sleeves to act as guide piles. Because the marina is favorably sited, experiences little current, and is exposed to minimal wind forces and wave action, the lateral loading on the docks is small. Apparently the pipes penetrated the bottom silt to a depth that is adequate to develop the necessary passive soil resistance to prevent lateral displacement of the docks. In any event, the management reports that the system has worked well.

Covered dry-storage racks used in conjunction with an operational launching enterprise is not new or unusual, but may be a viable alternative to wet storage for smaller craft where berthing space is limited and demand is great enough. Because of the relatively high cost of the launching facilities and operating personnel required, this system can only be successful at installations serving more than 100 boats. An efficient operational system that does not keep a client waiting long for his boat to be launched is also required. Unsatisfactory experiences with such operations have had a deterrent effect on the dry-storage market in many areas in the past, but with improved equipment, new handling systems, and a dwindling inventory of good undeveloped wet berthing space, dry storage should expand significantly in the future. Still Waters Marina is a good example of a successful combination of both types of small-craft berthings.

(6) Wahweap and Rainbow Bridge Marinas. Lake Powell, on the Arizona-Utah boundary, was formed on 21 January 1963, when the control gates of the newly completed

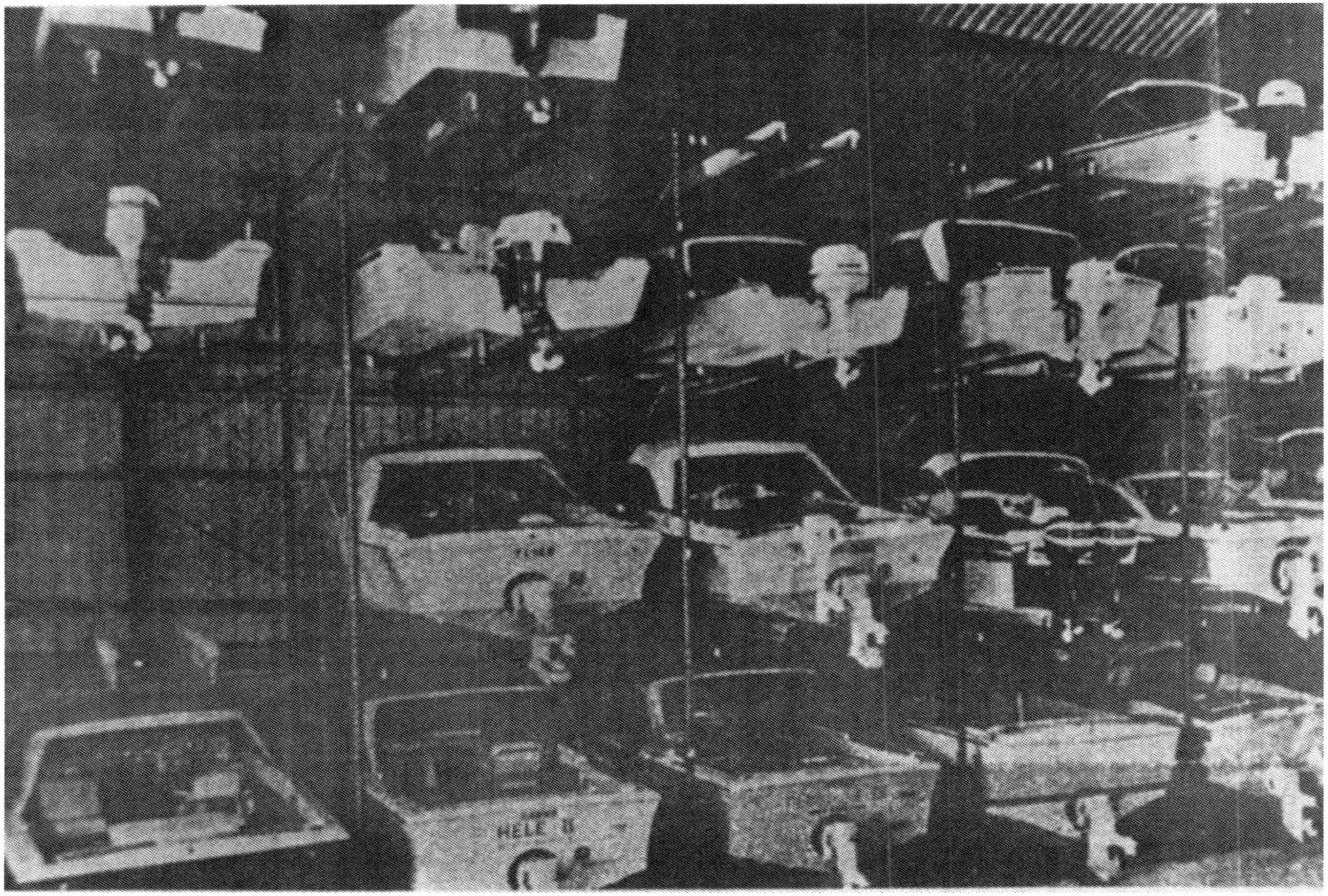

Figure 173. Small-craft dry storage facility, Still Waters Marina, Dadeville, Alabama

Glen Canyon Dam were closed. By the summer of 1972 the lake, fed by the Colorado River, was nearly 400 feet deep at the dam and extended 289 miles upstream with over 1,800 miles of serpentine finger shoreline. During 1972 over 770,000 tourists visited the Lake Powell recreation area. Many of these visitors took advantage of the marina facilities at Wahweap.

Wahweap Lodge and Marina north of Page, Arizona, is a total recreation complex operated by Canyon Tours, Inc., under exclusive contract to the U.S. National Park Service. The marina has an ultimate capacity of 300 slips and is located near the lodge parking lot (Fig. 174). Several unique design features are incorporated into this marina. The dock is a floating system anchored by cable, dictated by the great depth of water and the annual water level fluctuation of about 18 feet. The original installation consisted of 32 slips of wood and polystyrene construction, but the newer docks are of steel construction. The flotation units consist of steel plates rolled into 42-inch-diameter by 8-foot-long tubes with sealed ends. The tubes are bolted together with steel straps and angle iron braces, which act as supports for the metal framework in which the 2- by 4-foot steel deck panels rest. The main walks are constructed in 400-foot lengths interconnected with hinge-and-swivel joints. The finger piers are also connected to the main walks with hinge-and-swivel joints to accommodate wave action without overstressing the system.

The floating slip is moored in place by a web network under the docks to which are attached 1.125-inch-diameter wire rope mooring lines leading to 5,000-pound bottom anchors. The system is adjusted laterally as the water surface fluctuates by hand winches; land access is achieved by adding or subtracting floating walkway sections.

The berths are protected by a unique floating breakwater designed by the U.S. National Park Service. The breakwater is constructed of cellular concrete containing polystyrene flotation elements. Manufactured sections are about 8 feet wide and 10 feet high. These sections float with a freeboard of about 3 feet and a *keel* depth of 7 feet. The sections are interconnected with heavy chain, and the entire breakwater is moored in position with 1-inch-diameter wire rope lines attached to 20 (U.S. Navy-type) 1,500-pound anchors. A system of hand winches adjusts to the varying lake levels. The breakwater is at the lower left in Figure 175.

Marina occupancy is averaging 70 percent during the winter months and exceeds capacity during the summer season. During the 1972 summer season, 210 private power craft, 8 private sailboats, 12 commercial tour boats, and 105 rental boats were berthed at the marina. Slip-rental rates were $1.75 per foot per month, or $2.50 to $3.50 per day when available. The marina manager estimates the major activities of the patrons to be cruising (40 percent), fishing (35 percent), and skiing (25 percent).

Electrical power (120 volts) for lighting is supplied to the docks and the charge is included in the slip-rental fee. The dock lighting is a high-level system. Water is supplied free, there is a public address system to the docks, and four phones are strategically located along the piers. The marina has a sanitary holding tank, pumpout facilities, and daily trash collection.

Figure 174. Wahweap Lodge and Marina complex, Lake Powell, Arizona (Courtesy of Canyon Tours, Incorporated)

Figure 175. Rainbow Bridge Marina, Lake Powell, Arizona. This "way-station" marina is dwarfed by the sheer cliffs of the canyon in which it is anchored.

A ramp, 1,800 feet long and 200 feet wide, extends at a 9 percent slope into the lake and functions both as a launching ramp and a parking area. The ramp can handle craft up to 65 feet long. Launchings are free, and there are no hoists at the marina.

The ancillary facilities are extensive. Three two-story buildings now comprise the lodge; construction on a fourth is scheduled for 1973. When completed, a total of 320 rooms will be available. The lodge complex also includes dining rooms, a cocktail lounge, and a convention center. Cabins, a trailer park, and campgrounds are also available.

Rentals available include boats for water skiing, fishing, and houseboats. In addition, guided tours by boat and vehicle are offered.

One of the most unusual of all the ancillary facilities encountered during the marina survey was Wahweap's sister-installation, the Rainbow Bridge Marina, a floating marina with no land interface. Modules were constructed at Wahweap and floated in sections for 50 miles to the present anchorage (Fig. 175). Apartments for operating personnel are an integral part of this small isolated marina, which functions primarily as a *halfway house*, providing fuel, food, and general supplies for boaters touring the lake. The fueling facilities consist of 18 pumps, which pump up to 10,000 gallons per day. The station is supplied by a small tanker.

LITERATURE CITED

AMERICAN ASSOCIATION OF STATE HIGHWAY OFFICIALS, "Standard Specifications for Highway Bridges," 10*th* ed., Washington, D.C., 1970.

AMERICAN CONCRETE INSTITUTE, "Recommended Practice for Selecting Proportions for Normal and Heavyweight Concrete," ACI Standard 211.1-74, Detroit, Mich., Jan. 1974.

AMERICAN INSTITUTE OF TIMBER CONSTRUCTION, *Timber Construction Manual,* 2*nd* ed., Wiley, New York, 1966.

AMERICAN SOCIETY OF CIVIL ENGINEERS, "Wind Forces on Structures," *Transactions of the American Society of Civil Engineers,* Vol. 125, Part II, Paper No. 3269, 1961, pp. 1124–1197.

AMERICAN SOCIETY OF CIVIL ENGINEERS, "Small-Craft Harbors," Manuals and Reports on Engineering Practice–No. 50, ASCE, New York, N.Y., 1969.

AMERICAN SOCIETY FOR TESTING AND MATERIALS, "Applicable ASTM Standards," Philadelphia, Pa., 1971 (periodically updated).

AMERICAN WOOD-PRESERVER'S ASSOCIATION, "Standards of the AWPA," Standards C-1 and C-18, Washington, D.C., 1971 (periodically updated).

AMERICAN WOOD-PRESERVER'S INSTITUTE, "Bulkheads: Design and Construction," AWPI Technical Guidelines for Pressure Treated Wood, Parts I, II, and III, Washington, D.C., 1970.

BARRETT, R. J., "Use of Plastic Filters in Coastal Structures," *Proceedings of the 10th Conference on Coastal Engineering,* ASCE, Vol. 2, 1966, pp. 1048–1067.

BOWLES, J. E., *Foundation Analysis and Design,* McGraw-Hill, New York, 1968.

CALHOUN, C. C., Jr., "Development of Design Criteria and Acceptance Specifications for Plastic Filter Cloths," Technical Report S-72-7, U.S. Army Engineer Waterways Experiment Station, Vicksburg, Miss., 1972.

CHANEY, C. A., "Marinas–Recommendations for Design, Construction and Maintenance," National Association of Engine and Boat Manufacturers, Inc., Greenwich, Conn., 1961.

CHELLIS, R. D., *Pile Foundations,* McGraw-Hill, New York, 1961.

ELY, A. L., "Congestion in Small-Craft Harbor Entrance Channels," *National Transportation Engineering Meeting,* ASCE, Preprint, 1972.

JAMES, W., "Spectral Response of Harbor Resonator Configurations," *Abstracts of the 12th Coastal Engineering Conference,* ASCE, No. 51, 1970, pp. 151–154.

KAMEL, A. M., and DAVIDSON, D. D. "Hydraulic Characteristics of Mobile Breakwaters Composed of Tires or Spheres," Technical Report H-68-2, U.S. Army Engineer Waterways Experiment Station, Vicksburg, Miss., June 1968.

LE MEHAUTE, B., and HWANG, LI-SAN, "Harbor Design: Scale Model or Computer?," Tetra Tech, Inc., Pasadena, Calif., 1970.

MORROW, C. T., "Diffraction of Ocean Waves About a Breakwater," *Journal of the Waterways and Harbors Division,* ASCE, Vol. 29, No. WW3, Aug. 1966, pp. 25–44.

NATIONAL ASSOCIATION OF ENGINE AND BOAT MANUFACTURERS, "Marina Operations and Service," Greenwich, Conn., 1967.

NATIONAL ASSOCIATION OF ENGINE AND BOAT MANUFACTURERS, "Marina Architects, Consultants, Designers and Engineers," Greenwich, Conn., 1967 (periodically updated).

NATIONAL ELECTRICAL CODE, National Fire Protection Association, Boston, Mass., 1971.

NATIONAL OCEANIC AND ATMOSPHERIC ADMINISTRATION, *Tide Tables,* Rockville, Md., 1974 (annually updated).

NAVY BUREAU OF YARDS AND DOCKS, "Soil Mechanics, Foundations, and Earth Structures," *Design Manual,* DM-7, NAVDOCKS (now Naval Facilities Engineering Command), Naval Supply Depot, Philadelphia, Pa., 1962 (periodically updated).

OUTBOARD BOATING CLUB OF AMERICA, *Directory of Architects and Engineers,* Chicago, Ill. (periodically updated).

PORTLAND CEMENT ASSOCIATION, "Small Boat Launching Ramps," *Concrete Information,* Water Resources Bureau, Chicago, Ill., 1965.

QUINN, A. D., *Design and Construction of Ports and Marine Structures,* 2nd ed., McGraw-Hill, New York, 1972.

SMYTHE, A., "Countering Erosion," *New Zealand Harbour & Shipping Export Review,* June 1961.

SORENSEN, R. M., "Water Waves Produced by Ships," *Journal of the Waterways, Harbors, and Coastal Engineering Division,* ASCE, Vol. 99, No. WW2, Paper No. 9754, May 1973, pp. 245–256.

STATE OF CALIFORNIA, "Corrosion Protection for Concrete," *Bridge Planning and Design Manual,* Vol. 1, "Design Specifications," Item 6-14, Department of Public Works, Division of Highways, Bridge Department, Sacramento, Calif., 1969.

STATE OF CALIFORNIA, "California Boating Law," Department of Navigation and Ocean Development, Resources Agency, Sacramento, Calif., 1971.

TERZAGHI, K., "Anchored Bulkheads," *Transactions of the American Society of Civil Engineers,* ASCE, Vol. 119, 1954, pp. 1243–1324.

UNIFORM BUILDING CODE, "Uniform Building Code Standards," *International Conference of Building Officials,* Vol. 1, 1970.

U.S. ARMY, CORPS OF ENGINEERS, "Digest of Water Resources–Policies and Activities," EP 1165-1-1, Washington, D.C., Dec. 1972.

U.S. ARMY, CORPS OF ENGINEERS, COASTAL ENGINEERING RESEARCH CENTER, *Shore Protection Manual,* Vols. I, II, and III, Stock No. 0822-00077, U.S. Government Printing Office, Washington, D.C., 1973, 1,160 pp.

UHLIG, H. H., *The Corrosion Handbook,* Wiley, New York, 1953.

VAN DE KREEKE, J., "Damage Function of Rubble-Mound Breakwaters," *Journal of the Waterways and Harbors Division,* ASCE, Vol. 95, No. WW3, Aug. 1969, pp. 345–354.

WATKINS, L. L. "Corrosion Test of Steel Piling in Sea Water," *Journal of the Waterways, Harbors, and Coastal Engineering Division,* ASCE, Vol. 97, No. WW3, Aug. 1971, pp. 549–558.

APPENDIX A

GLOSSARY OF TERMS

accretion—May be either **natural** or **artificial.** Natural accretion is the buildup of land solely by the action of the forces of nature, on a **beach** by deposition of waterborne or airborne material. Artificial accretion is a similar buildup of land by reason of an act of man, such as the accretion formed by a groin, breakwater, or beach fill deposited by mechanical means.

amplitude—The magnitude of the displacement of a wave from a mean value.

anchor pile—A pile or column that is an integral part of a structure, and whose main function is to keep the structure firmly in place, having been driven into the earth for this purpose.

ancillary facilities—Installations or services provided at a harbor that complement the harbor's operations, but are not essential to harbor functioning, per se (e.g., snackbars, ice-vending machines, and sail repairs).

armor—An outer layer of large stone or concrete armor units whose function is to ensure the integrity of an embankment, jetty, or breakwater for protection against wave action or currents.

aseptic—Free from pathogenic or infecting micro-organisms.

attenuate—To lessen the amplitude of waves or surge.

auxiliary power—Any means of propelling a craft that is not the primarily designed means (e.g., an engine-driven propeller on a sailboat).

B-stone—Quarrystone sized for use in the supporting layer below the armor in a breakwater, jetty, or revetment.

bar—A submerged or emerged embankment of sand, gravel, or other unconsolidated material built in shallow water by waves and currents.

battens—Thin wood strips, often those affixed vertically around large pilings for protection.

batter—The slope or cant off the vertical of a pile or the face of a wall.

bearing boards—The wood members that transmit the deckloads to the floats in some floating-dock systems.

bedding layer—The first and lowermost layer of gravel or stone that acts as a bearing layer for larger stones or armor units placed upon it. It also functions as a filter layer for the material beneath the structure.

benefits—The dollar value placed on the overall contributions of a project such as a marina to the economy, including the indirect as well as the direct contributions (as distinguished from actual revenues).

benefit-cost ratio—The ratio of the estimated average annual benefits from a project to its estimated average annual cost, including debt-servicing, periodic replacement of major components, operating expense, and maintenance.

berm—A nearly horizontal part of the beach or backshore formed by the deposit of material by wave action. Some beaches have no berms, others have one or several.

berth—A place where a boat may be secured to a fixed or floating structure and left unattended.

berthing area—The water area in which craft are berthed.

bifurcation—A division of the main stream or channel into two branches, often caused by an island or exposed bar in such a manner that the two branches remerge into a single channel downstream.

bight—A bend in a coastline forming an open bay. A bay formed by such a bend.

boat—Any type of surface craft that may be berthed in a small-craft harbor.

boater—A person who uses any type of small craft.

breaker depth—The stillwater depth at the point where the wave breaks.

breakwater—A structure protecting a shore area, harbor, anchorage, or basin from waves.

bulkhead—A structure or partition to retain or prevent sliding of the land. A secondary purpose is to protect the upland against damage from wave action.

caisson—A hollow structure filled with selected material that is used as a breakwater or jetty section.

cantilever—A projecting beam, pile, or structural member that resists (or is capable of resisting) a transverse load applied to the part projecting beyond its last point of support.

cap—The finished structural member topping off a wall or bulkhead, providing strength, protection, and continuity to the wall or bulkhead.

channel meandering—The tendency of a sedimentary channel to change course mainly due to the changing pattern of erosion and deposition of bottom sediments.

cleat—A wooden or metal fitting usually with two projecting horns around which a rope may be tied.

core—The inner makeup or part of a rubble-mound jetty or breakwater that is beneath the armor layer or lowest armor-supporting layer.

contiguous land—Adjacent or adjoining lands.

Continental Shelf—The zone bordering a continent and extending from the low water line to the depth (usually about 100 fathoms) where there is a marked or rather steep descent toward a greater depth.

counterfort—A triangular bracing wall between a bulkhead wall and its footing.

course—In stonework, a layer one stone thick.

crowning—Warping or raising the center part of a flat surface so that it will shed water.

cruising range—Total distance over which a powered craft can travel using one store of self-contained fuel, excluding its emergency safety reserve.

cutterhead dredge—A dredge that uses a rotating, toothed head mounted on the end of a translating boom for loosening or cutting away the bottom material that is to be removed or dredged.

damping—Reducing in amplitude or intensity; attenuating.

deadman—The buried object to which bulkhead or seawall tiebacks are attached. An anchor pile may be used in lieu of a deadman.

deep water—Water so deep that surface waves are little affected by the ocean bottom. Generally, water deeper than one-half the surface wavelength is considered deep water.

delta—An alluvial deposit, usually triangular or digitate in shape, formed at a river mouth.

diaphragm breakwater—A comparatively thin, impervious wall, membrane, or structure designed to resist wave penetration, usually a secondary defense structure.

diffraction diagram—A diagram showing lines of equal wave height (and sometimes successive crest positions) after a wave of given characteristics has undergone diffraction about the end of a breakwater or through a gap.

dock—A fixed or floating decked structure against which a boat may be berthed either temporarily or indefinitely.

dolphins—A cluster of battered pilings joined at the top.

downdrift—The direction of predominant movement of littoral materials.

ebbtide—The period of tide between high water and the succeeding low water; a falling tide.

ecological preservation—Maintaining the natural balance or pattern in the relationship between marine or near-water organisms and their environment.

eddy current—A circular movement of water formed on the side of a main current. Eddies may be created at points where the main stream passes projecting obstructions or where two adjacent currents flow counter to each other.

estuary—(1) The part of a river that is affected by tides. (2) The region near a river mouth in which the freshwater of the river mixes with the saltwater of the sea.

exchange of water—The flushing or replacement of water in a basin area, usually that required to combat pollution or to ensure good water quality.

fairway—The parts of a waterway that are open and unobstructed for navigation.

fendering system—A protective bumper system designed to prevent damage to either boats or docks whenever they come in contact.

fetch—The area in which **seas** are generated by a wind having a rather constant direction and speed. Sometimes used synonymously with **fetch length.**

fetch length—The horizontal distance (in the direction of the wind) over which a wind generates **seas** or creates a **wind setup.**

filter—The underlayer of small rock or gravel that permits proper seepage and dissipation or distribution of water beneath or behind a structure wall or riprapped slope without allowing the earth or other retained material to escape. Plastic cloth may also be used as a filter.

fines—Grains of silt and clay magnitude that are carried away in suspension by waves and currents. In stonework, any material smaller than the smallest grading classification or sieve size in a given specification.

finger pier—A comparatively smaller pier structure attached (usually perpendicular) to the headwalk of a multislip pier; usually provided to facilitate access to the berthed craft.

foot-candle—A uniform measure of light intensity standardized by the Illumination Engineering Society.

formwork—The structure used to contain and shape poured concrete.

freeboard—The additional height of a structure above design high water level to prevent overflow. Also, at a given time, the vertical distance between the water level and the top of the structure. On a ship, the distance from the water line to main deck or gunwale.

french drain—A continuous run of buried gravel through which water can flow for drainage purposes.

gabion—A connected system of wicker or metal cages filled with brush or rock and used for slope protection or stabilization.

gangway—A pedestrian or handcart bridge affording access from shore or a shore-connected fixed pier to a floating structure (sometimes called brow).

grid (gridiron)—A woven or welded metal mat or framework used for ramps, slope stabilization, or erosion control.

guide piles—The piles in a floating dock system that resist the horizontal displacement of the system but allow and guide its vertical movement with changes in the level of the water surface.

gunwale—The upper edge of the side of a ship or a boat.

harbormaster—The officer who executes the regulations respecting the use of a harbor.

harbor service—The action taken and facilities provided by the harbor administration for the benefit and protection of the harbor patrons.

head—A toilet in a boat or a ship.

head walk—The main walkway on a multislip pier. Usually more substantial structurally than finger piers; supports the utility lines and may support the lighting system, firefighting equipment, and locker boxes.

heave—The rhythmic vertical displacements of an entire craft due to sea, swell, or boat waves.

impact statement—A report on the expected effects or influence any development will have on its environment.

interstices—The internal spaces or voids in a rubble-mound structure.

jetty—(1) (U.S. usage) On open seacoasts, a structure extending into a body of water, and designed to prevent shoaling of a channel by littoral materials, and to direct and confine the stream or tidal flow. Jetties are built at the mouth of a river or tidal inlet to help maintain and stabilize a channel. (2) (British usage) Jetty is synonymous with wharf or pier.

king piles—The primary structural pile members that carry or support the panels, sheating, or sheet piles of a diaphragm breakwater or bulkhead.

land access—Means provided for reaching any facility in the water from shore.

marina—A small-craft harbor complex that includes most or all of the support and ancillary facilities needed or desired by boatmen, such as launching equipment, repair facility, a fueling station, restrooms, marine hardware supply, and restaurants. This term is used to describe harbors that are intended primarily for recreational craft.

operational launching—Boat launching on a routine in-and-out basis at the boater's option, as distinguished from initial or seasonal launching.

orthogonal—On a wave-refraction diagram, a line drawn perpendicularly to the wave crests.

oscillation—A periodic motion backward and forward. To vibrate or vary above and below a mean value.

pierhead line—The limiting line to which any pier or dock structure can extend into the main channel area.

pitch—The rotational oscillations of a craft about its transverse axis under wave excitation.

prevailing winds—The characteristic and most commonly occurring winds, both in magnitude and direction, observed at a given location.

profile height—The quotient obtained by dividing the cross-sectional area of a craft's end (or side) profile above the water surface by the overall width (or length) of that craft.

project bottom depth—The design depth of a basin or channel, usually measured below a given water surface elevation such as mean low water. Authorized depth to be maintained.

protected water—Sheltered water areas, protected either by natural or manmade wave barriers, which are adequate for the safe mooring of watercraft.

pumpout station—A facility for removal of sanitary wastes from a boat's holding tank or head.

reach—A length, distance, or leg of a channel or other watercourse.

rebar—Reinforcing bar or rod imbedded in a concrete structure.

refraction diagram—A drawing showing successive positions of a wave crest and/or orthogonals in a given area for a specific deepwater wave period and direction.

residual wave action—Wave action that remains after an incident wave has been partially attenuated, reflected, absorbed, or otherwise modified by any natural or manmade obstacle.

resonance—The phenomenon of amplification of a free wave or oscillation of a system by a forced wave or oscillation of exactly equal period. The forced wave may arise from an impressed force upon the system or from boundary conditions.

revetment—A facing of stone, concrete, or other material, built to protect a scarp, embankment, or shore structure against erosion by wave action or currents.

riprap—A layer, facing, or protective mound of stones randomly placed to prevent erosion, scour, or sloughing of a structure or embankment. Also the stone so used.

riser—A vertical pipe. post, or support extending above deck level to support utility outlets and other facilities.

roadstead—(Nautical) A sheltered area of water near shore where vessels may anchor in relative safety. Also road.

rock—Stone in the mass, as in a quarry or rock stratum.

roll—The rotational oscillations of a craft about its longitudinal axis under wave excitation.

rubble mound—A mound of random-shaped and random-placed stones protected with a cover layer of selected stones or specially shaped concrete armor units. (Armor units in primary cover layer may be placed in orderly manner or dumped at random.)

rub strake—A longitudinal rib or protective strip running along the hull of a craft to function as a bumper.

scend—The sinkage of a craft in the trough of a wave.

scour—Removal of underwater material by waves and currents, especially at the base or toe of a shore structure.

screw—A propeller on any type of craft.

sea—(1) An ocean, or alternatively a large body of (usually) saltwater smaller than an ocean. (2) Waves caused by wind at the place and time of observation. (3) State of the ocean or lake surface in regard to waves.

section modulus—A relative measure of the ability of a cross-sectional configuration to resist bending forces. Technically, the maximum allowable bending movement of a homogenous beam divided by its maximum allowable fiber stress.

seiche—A standing wave oscillation of an enclosed water body that continues, pendulum fashion, after the cessation of the originating force.

sheet pile—A pile with a generally flat cross section to be driven into the ground or seabed and meshed or interlocked with like members to form a diaphragm, wall, or bulkhead.

shingle—(1) Loosely and commonly, any beach material coarser than ordinary gravel, especially any having flat or flattish pebbles. (2) Strictly and accurately, beach material of smooth, well-rounded pebbles that are roughly the same size. The spaces between pebbles are not filled with finer materials. Shingle often gives out a musical note when stepped on.

shoal—(verb) (1) To *become* shallow gradually. (2) To *cause* to become shallow. (3) To *proceed* from a greater to a lesser depth of water.

shoal—(noun) A detached elevation of the sea bottom comprised of any material except rock or coral, and which may endanger surface navigation.

significant wave height—The average height of the one-third highest waves of a given wave group. Note that the composition of the highest waves depends upon the extent to which the lower waves are considered. In wave record analysis, the average height of the highest one-third of a selected number of waves, this number being determined by dividing the time of record by the significant period.

slip—A berthing space between two finger piers.

sonic probe—Investigation or exploration of depths by use of radio or sound pulse echos.

spalls—Small fragments or chips of stone.

squat—The vertical downward displacement of a craft under power with respect to its position in the water when not underway.

stone—Quarried rock. Also, any individual piece of rock broken away from its original mass.

stringers—The relatively long, main, horizontal beams that support the deck of a fixed pier or dock between bearing points. In a floating structure, the continuous beams (usually along the sides) that join a series of floating modules.

submarine canyon—A cut or gorge in the Continental Shelf.

substrata—The layers of material beneath the surface soil.

support facilities—Installations or services needed to support the functions, such as utility services, fueling stations, repair and launching facilities, the harbor headquarters, and restrooms.

surcharge loading—Any load applied to the surface area immediately behind a retaining wall, bulkhead, or embankment.

surge—The name applied to wave motion with a period intermediate between that of the ordinary wind wave and that of the tide, say from 30 seconds to 60 minutes. It is of low height; usually less than 0.3 foot.

swell—Wind-generated waves that have traveled out of their generating area.

temperature rebars—Reinforcing bars provided primarily to resist temperature and shrinkage stresses in concrete.

tidal prism—The total amount of water that flows into a harbor or estuary and out again with movement of the tide, excluding any freshwater flow.

trailing floating slips—Floating (usually multiboat) slips that aline themselves with the prevailing river current. Although they must be secured against downstream displacement, they do not require guide piles.

training wall—A wall or jetty designed to direct or stabilize current flow.

transient boater—A boater within the confines of a harbor that is not his home base facility.

tremie-poured concrete—Concrete poured under water through a tube (tremie) and cured below the water surface. The placement process systematically displaces a given water volume with concrete, with very little dilution of the fresh concrete by water penetration.

tsunami—A long-period wave caused by an underwater disturbance such as volcanic eruption or earthquake. Commonly miscalled tidal wave.

unwater—To pump the water out of an enclosed basin to permit construction in the dry of any facilities to be built within the enclosure.

updrift—The direction opposite that of the predominant movement of littoral materials.

wale—A strake, rib, ridge, or raised reinforcing strip along any structure. A horizontal stringer along a sheet-pile bulkhead to which tie rods are attached or against which king piles bear.

water exchange—Cyclic replacement from outside sources of water in a basin. This may occur as a result of tidal action, river currents, forced circulation, or other means.

wave attenuation—Reduction of wave height or amplitude for any reason as the wave is propagated from one area to another.

wave breaker—A device that absorbs a large amount of wave energy with respect to the amount it reflects or transmits.

wave convergence—(1) In refraction phenomena, the decreasing of the distance between orthogonals. Denotes an area of increasing wave height and energy concentration. (2) In wind setup phenomena, the increase in setup observed over that which would occur in an equivalent rectangular basin of uniform depth, caused by changes in planform or depth; also the decrease in basin width or depth causing such increase.

wave energy—The theoretical capacity of a wave to do work.

wave hindcasts—The calculation from historic synoptic wind charts of the characteristics of waves that probably occurred at some past time.

wave period—The time for a wave crest to traverse a distance equal to one wavelength. The time for two successive wave crests to pass a fixed point.

wave propagation—Transmission of waves through water.

wave refraction—(1) The process by which the direction of a wave moving in shallow water at an angle to the contours is changed. The part of the wave advancing in shallower water moves more slowly than that part still advancing in deeper water, causing the wave crest to bend toward alinement with the underwater contours. (2) The bending of wave crests by currents.

wave steepness—The ratio of the wave height to the wavelength.

weep hole—A drainage or pressure relief opening in an otherwise watertight structure.

wind setup–(1) The vertical rise in the stillwater level on the leeward side of a body of water caused by wind stresses on the surface of the water. (2) The difference in stillwater levels on the windward and the leeward sides of a body of water caused by wind stresses on the surface of the water. (3) Synonymous with **wind tide** and **storm surge.** Storm surge is usually reserved for use on the ocean and large bodies of water. Wind setup is usually reserved for use on reservoirs and smaller bodies of water.

APPENDIX B

AWPI GUIDELINES FOR PRESSURE TREATED WOOD

APPENDIX B

AWPI Technical Guidelines

For Pressure Treated Wood

S2
1970

Bulkheads: Design and Construction—Part I

SUMMARY

This bulletin describes several variations of bulkhead construction, including two principal types of anchorage systems The reader is given summaries of both the proper construction sequence and the design procedures The step-by-step descriptions will be discussed in Parts II and III and will assist the designer in completing the design and specification of treated timber bulkheads

INTRODUCTION

There are many approaches to bulkhead design and construction Three typical systems are shown in Figs 1, 2 and 3 These systems illustrate the differences that may be expected under varying site conditions, particularly with respect to variations in water levels An engineer or architect who is not experienced in the design of bulkheads may want to rely upon consultants for the design of critical bulkheads and seawalls, the latter being a bulkhead exposed to more severe forces

The principal difference between a bulkhead and a retaining or crib wall is that a bulkhead is built near water, adding an additional design factor In designing bulkheads, the architect and engineer can utilize their experiences with design of similar earth-retaining structures The main question that may arise is how to cope with the presence of water Therefore, the purpose of this bulletin is to show where information about the effects of water can be found and to express, in concise terms, the procedures that bulkhead designers have consistently used

The designs in this bulletin are relatively conservative Where existing soil conditions are superior to those assumed in these design procedures, the thicknesses of sheet pilings and the sizes and spacings of tie rods and anchor systems can be reduced Yet, precisely because these procedures are somewhat conservative, they may be used with confidence

TYPICAL TIMBER BULKHEADS

Figure 1 shows a conservatively-designed, low bulkhead for installation where the existing grade along the sheet piles is somewhat higher than the low-water level Figure 2 shows an intermediate bulkhead suitable for retaining fill at the site of a marina or for providing a finished waterfront in a housing development If these bulkheads are located further in shore, or if the outside water level variations are less than shown in the figures, the heights of bulkheads and lengths of sheet piles may be reduced

The anchorage systems shown in these two figures depend upon the passive resistance of a mound of earth immediately around the anchor post and wales The theoretical mound required to develop this passive resistance is shown by a dotted line If backfill were placed directly against the sheet piles before this mound of earth is compacted around the anchor system, the resulting forces would displace the bulkhead or even push it over, resulting in a costly and disastrous failure After earth is compacted over the anchorage system, backfill can be deposited against the sheet piles by a drag line, truck dumping, hydraulic pipe line, or other suitable method

Figure 3 shows a bulkhead suitable for the deep-water portions of marinas, or for locations where the existing water depths are 6 to 8 feet and extensive land fills are desirable The anchorage system is a self-supporting A-frame that does not depend on passive earth resistance The anchor system is particularly adaptable to filling by the hydraulic method because the backfill can be raised behind the sheet piles without regard for the placement of backfill at the anchorage location

To install the A-frame anchorage, a pile-driving rig is required to drive the piles to specified bearing capacity At locations with less water-level variation than the 4 feet shown, the height of finished grade may be lowered proportionally For an increase in water-level variation, a similar increase in height of finished grade can be made with a corresponding reduction in water depth The 5-foot vertical distance from finished grade to tie rod level should be maintained

CONSTRUCTION PROCEDURES

The proper sequence for bulkhead construction is

1 Drive all round timber piles, both vertical and battered, set or drive all posts
2 Using bolts, attach the horizontal wales for the sheet piles and the anchorage system
3 Drive sheet piling
4 Complete all bolted connections and install tie rods

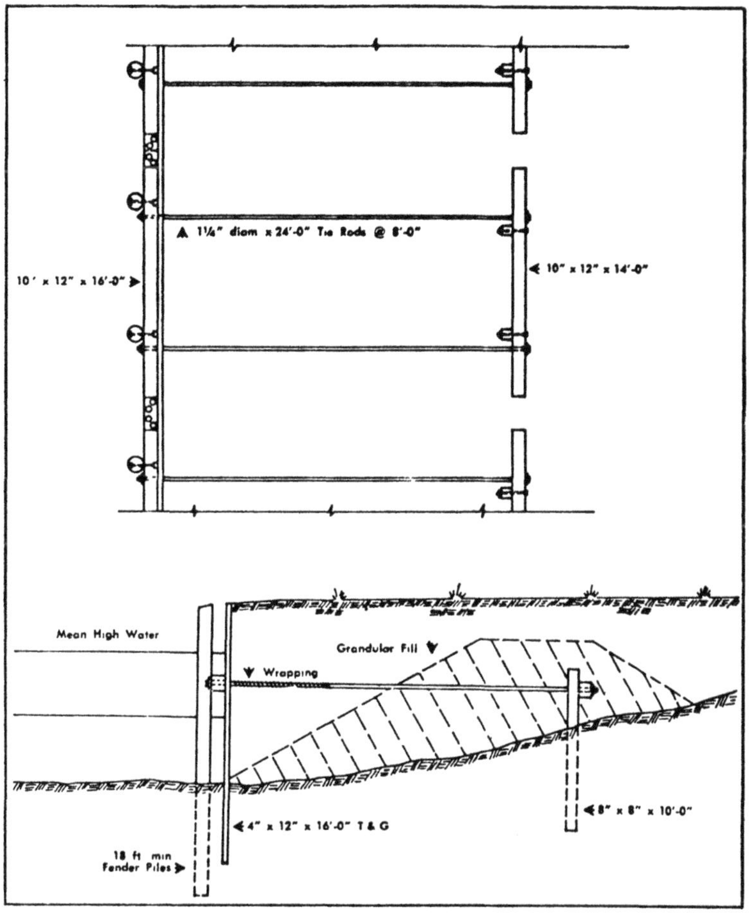

Fig. 2: Bulkhead design, four-foot water depth.

from the standpoint of foundation support for shore structures

If the hydraulic method is used for backfilling, provide sufficient drainage to permit rapid escape of water at the ends of the construction area, both to prevent formation of pools and to maintain as low a free-water level in the backfill as possible Although the bulkhead is designed to hold earth, it may not be designed to resist water pressures that can be generated during hydraulic filling

One method of facilitating drainage is to provide openings through the sheet piling above the level of the outside wale Space these openings at intervals of about 60 feet to supplement the escape of drainage water The final three feet of backfill adjacent to the sheet piles should be put into place by earth moving equipment to avoid hydraulic pressures at the upper portions of the bulkhead The hydraulic discharge line should be parallel to the bulkhead alignment, not directed at it, and should be located at least 100 feet behind the bulkhead sheet piling

DESIGN STEPS

At a construction site, the natural conditions that exert the most influence on the design of any waterfront structure are water level variation, wave action, and type of soil Ice conditions are a special consideration for locations subject to the effects of solid ice sheets, floating ice fields or large ice packs

Removal Of Poor Quality Soils

Often the natural soil is capable of providing the necessary resistance for the lower ends of the sheet piling in seawalls, bulkheads, and groins However, if the bottom soil is soft silt, mud, or soft clay, it should be removed and replaced with granular materials

In most cases, earth fill is required above the existing ground line for some distance shoreward from the face of the sheet piling The filling material for a sufficient width to encompass the anchor system should be predominantly granular in nature, even though it may be necessary to transport it from a considerable distance **(continued, part II)**

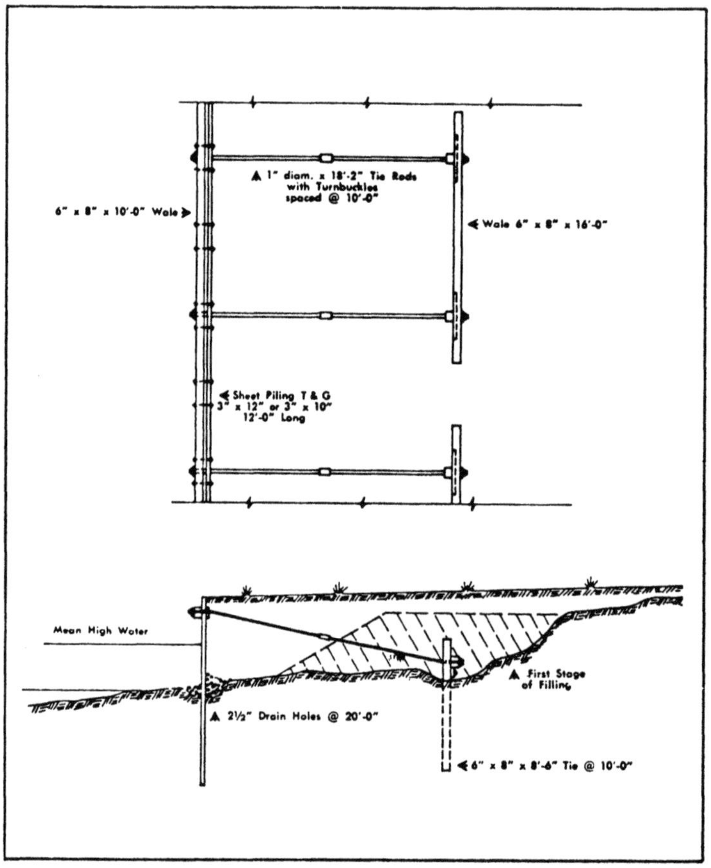

Fig. 1: Bulkhead design, zero-foot water depth.

5. Place the backfill. Where passive resistance anchorage systems are used, be certain to place fill over the anchors before backfilling behind the sheet-pile wall.

Establish accurate survey lines as a control for the construction of water front and shore-protection structures and anchorage systems. Exercise care in locating positions of round piles and posts that support horizontal timbers that will be used as driving guides for sheet piles. In all cases, use a driving guide for sheet piles, preferably the permanent horizontal wale that is attached to the vertical round piling or posts. (See Figs 1, 2, and 3.)

It is particularly important to drive the first few sheet piles accurately vertical in all directions. The wall must be plumb and the sheet piles must not be inclined within the plane of the wall. One of the common problems facing piling contractors is "creep," the tendency for successive sheet piles to lean more and more in the direction of construction of the wall. The wall can be perfectly plumb, yet piles can lean; this error in alignment tends to accumulate and, if left uncorrected, can create considerable difficulties in driving successive piles. A discussion of this situation and how to correct it is given in Ref. 1, pp. 332 and 333.

As sheet-pile driving proceeds, place the tongue of each new sheet in the forward position and the groove in tight contact with the tongue of the sheet previously driven. Keep the joints between piles as tight as practicable. Remember that the maximum allowable opening at joints is ½ inch for splined and Wakefield piling and ¼ inch for tongue-and-groove piling. If wider joints appear after the sheet piles have been spiked to the outer wale, cover them with treated timber lath to prevent the backfill material from gradually filtering through the cracks and being lost.

Wherever passive-resistance anchorage systems are used, the anchorage must be well covered with a compacted mound of earth before backfill material is deposited to any appreciable depth against the piling. Otherwise, the pressures generated by the backfill may disrupt the sheet piling and the anchorage system.

Use a predominately granular material for backfill adjacent to the sheet piling and over the anchorage system. Shoreward of the anchorage, a poorer quality filling material may be used unless it is objectionable

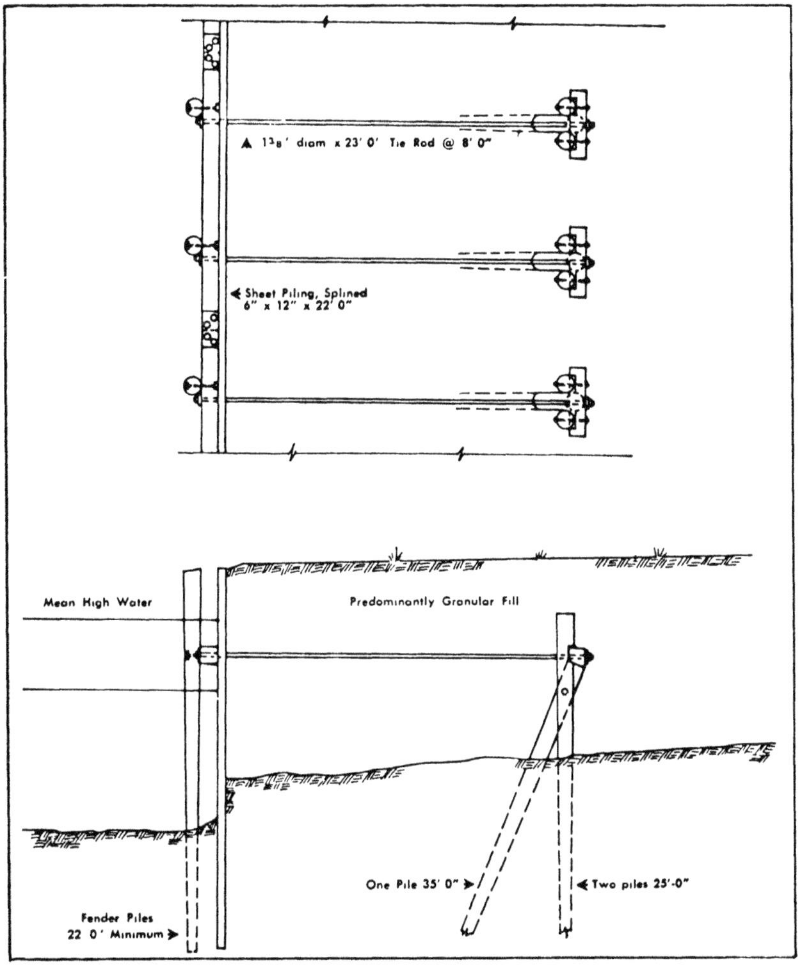

Fig. 3: Bulkhead design, eight-foot water depth.

AWPI Technical Guidelines

For Pressure Treated Wood

S3
1970

Bulkheads: Design and Construction—Part II

SUMMARY

This bulletin discusses the nine (9) procedures which are required in the design for anchored bulkhead construction. The step-by-step descriptions which include some basic information are presented in detail from steps Ia through Ig. Descriptions from step 2 through step 9 will be discussed in later bulletins.

Design Steps

Several steps are required in the design of an anchored bulkhead as follows:

1. Determine the following basic information: (See Fig. 4.)
   a. Water depth required (by owner).
   b. Water level variation in front of sheet piling. (From Tide Tables.)
   c. Effects of scour (elevation of stable mudline).
   d. Ground water level behind bulkhead at time of low-water level in front.
   e. Level of finished grade behind bulkhead.
   f. Types of soil available for backfill and for resisting movement of lower ends of sheet piling; unit weights of moist and submerged soils. (See Table 2.)
   g. Amount of vertical surcharge loads (if any) anticipated on ground behind bulkhead (determined from proposed use of site).
2. Prepare earth-pressure diagrams for inner and outer faces of sheet piling to obtain resultant pressure diagram. (See Figs. 6 and 7, Part III.)
3. Determine depth of sheet pile penetration.
4. Determine pull in tie rods.
5. Compute bending moment in sheet piling.
6. Determine required thickness of sheet piling.
7. Determine size and spacing of tie rods.
8. Determine bending moment and required size of wales.
9. Design anchorage to resist for tie-rod pull.

Each of these steps is described in detail and in sequence in the following paragraphs.

WATER DEPTH REQUIRED (Step Ia)

The depth of water against the face of a timber bulkhead must be determined before the bulkhead can be designed. If the bulkhead is simply a "retaining wall" for shore protection of a building site, the depth of water may simply be whatever naturally results when the bulkhead is constructed along the desired shoreline.

Elsewhere, in a marina for example, the depth of water may depend upon the types of boats that may be brought adjacent to the bulkhead. The property owner, or the developer of the marina, may establish minimum requirements for water depth. Some dredging may be required to achieve these depths, although, generally, the least costly solution is to place the bulkhead farther from shore in deeper water.

WATER LEVEL VARIATION (Step 1b)

At coastal locations, the water level variation on the exposed face of the sheet piling may be obtained from tide tables published by the U. S. Coast and Geodetic Survey and by the Navy Oceanographic Office. At inland locations adjacent to lakes and rivers, seasonal records of high- and low-water levels may be used. These latter data are available from offices of the Corps of Engineers, state or local conservation agencies, city or county engineers, and building officials.

The "design value" used for the "high-water level" requires some judgment. This value need not be the highest water level ever recorded in the vicinity, but should be a reasonably high value that, statistically, is expected to occur within the design life of the structure.

DEPTH OF SCOUR (Step 1c)

The construction of a bulkhead may deflect wave energy in such a way that it erodes the bottom material adjacent to the sheet piling. This is most likely to occur in shallow depths, especially if mean low water is less than two feet. It is advisable to assume that the energy deflection, occurring at the time of intermediate and higher tide stages, will erode the bottom material to a depth of 2 or 3 feet below mean low water.

Accordingly, when lacking more definitive data, assume a minimum low-water depth of 2 to 3 feet. This will result in a requirement for slightly longer sheet piling, but will prevent the failure that can occur if the bottom material is eroded to the extent that the sheet piling may be shoved out of place.

WATER LEVEL IN THE BACKFILL (Step 1d)

The ground-water level in the backfill will rise and fall as the tide rises and falls. This is caused mostly by natural percolation of water through the soil, behind the lower portions of the bulkhead wall, rather than

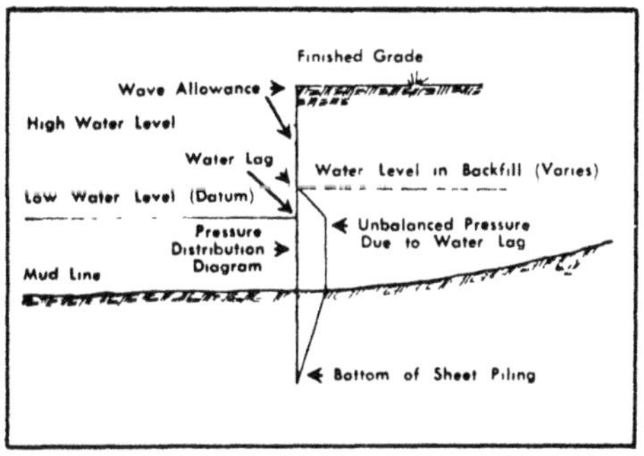

Fig. 4: Bulkhead or seawall, basic vertical dimensions.

by passage of water through cracks in the bulkhead Because the soil slows down the flow of water somewhat, the ground-water level behind the bulkhead is seldom the same as the free-water level on the seaward side As the tide rises, the ground water level rises, but at a slower rate Similarly, as the tide recedes, the ground-water level falls also at a slower rate The *distance* between these two water levels is called the *water lag* (Fig 4)

When the free water is higer than the ground water there is no particular design problem The water pressure is resisted by the soil backfill But when the ground-water level is higher than the free water, there is an outward pressure that is resisted only by the wall and its anchorage system The net effect of the ground-water pressure is similar to that of the backfill against the bulkhead The outward pressure of both water and soil must be considered in the design of a bulkhead wall

To determine the ground-water level, dig a hole behind the sheet piles of any nearby bulkhead Water will rise and fall to measurable levels If no bulkheads exist in the vicinity use a minimum value of one foot, or a maximum of one-half the tidal variation, for the water lag

Table 1 Estimated Wave/Bulkhead Heights

Wave Height Crest to Trough	Minimum Height of Bulkhead above high water level, as allowance for wave action
1 to 2 ft	3 ft
2 to 3 ft	4 ft

As the water level recedes during ebb tide, the water in the ground behind the sheeting will escape through any openings left in the wall, carrying away the finer particles of the soil backfill During numerous tidal interchanges, large quantities of backfill can escape in this manner, unless tight joints are provided between sheet piles Figure 5 shows joints used to minimize losses

FINISHED ELEVATIONS (Step 1e)

Figure 4 indicates the criteria used to determine the basic vertical dimensions of a bulkhead The finished elevation of the bulkhead and backfill were established by allowing for wave action above the high water level

Wave height is a function of wind velocity and duration, of expanse of water (fetch) over which the wind blows, and of water depth Reference 2 describes methods for estimating wave height and period under various site conditions

In protected areas, where treated timber bulkheads have their best application, the length of fetch and water depth are limited Consequently in most cases the expected wave heights are in the range of 1 to 2 feet

Use Table 1 to estimate wave action, where a more precise calculation is not required By observing the heights of existing structures in a particular locality, experience can often be used as a guide in arriving at the elevation of the finished grade

AVAILABLE SOIL MATERIALS (Step 1f)

Granular materials, such as sand and gravel are preferred for backfill behind bulkheads and for base

Table 2 Unit Weights of Soils and Coefficients of Earth Pressure

Type of Soil	Unit Weight of Soil Pounds per cubic foot				Active Earth Pressure			Passive Earth Pressure	
	Moist		Submerged		Coefficient K_a		Angles of Friction, Degrees	Coefficient, K_p for Soils in Place	Angles of Friction, Degrees
	Min	Max	Min	Max	Backfill	Soils in Place			
Clean Sand									
Dense	110	140	65	78		0 20	38 20	9 0	38 25
Medium	110	130	60	68		0 25	34 17	7 0	34 23
Loose	90	125	56	63	0 35	0 30	30 15	5 0	30 20
Silty Sand									
Dense	110	150	70	88		0 25		7 0	
Medium	95	130	60	68		0 30		5 0	
Loose	80	125	50	63	0 50	0 35		3 0	

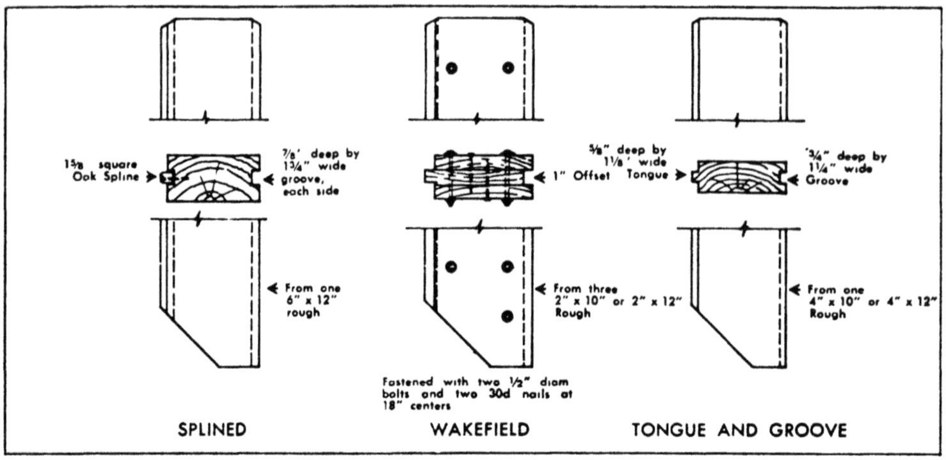

Fig. 5: Types of sheet piling.

materials into which sheet piles are driven Granular materials have several characteristics that make them desirable The active pressures they exert on bulkheads are considerably less than those exerted by silts and clays or even by sands that are "contaminated" with silt or clay Granular materials can develop a higher passive resistance, permitting the use of shorter sheet piles

It is often preferable to bring granular materials to the construction site by truck or other conveyance rather than to backfill with silt or clay In the final analysis, a judgment is required A designer must determine whether it is more economical or otherwise more desirable to increase the thicknesses of materials in the bulkhead rather than to transport granular materials from far distances Fortunately, granular materials are quite often readily available near bodies of water

Granular material also promotes drainage behind the bulkhead and is less likely to be eroded away through minor fissures in the bulkhead as the water level rises and falls The drainage is advantageous wherever the backfill is to be used for roadways, sidewalks, parks, yards, and so on Grass and plants will grow better in a soil that is moderately well drained, the granular material will not shrink and swell with alternate drying and wetting to the same degree that clay will The backfilled area is not likely to resemble a marshy bog where drainage is provided through granular materials Where lawns and other plantings are required, a relatively thin layer of top soil can be placed over the granular material

Where the material at the shoreline is silt or clay, the passive resistance will be very low This will require lengthening the sheet piling to obtain sufficient passive resistance to avert a failure The greater length of sheet piling induces a higher degree of bending in the piling, requiring a thicker piling than would otherwise be needed

Under these circumstances, the designer should consider removing the poor soil and replacing it with a granular material, such as clean sand The cost is often far less than that of providing thicker materials and more elaborate anchorage systems for the bulkhead

SURCHARGE (Step 1g)

Where the backfill behind a bulkhead will be used for supporting loads above and beyond that of the earth itself, for example a road or street subject to automobile traffic, the effect of these additional forces must be considered in the design of the bulkhead To simplify the design, lighter loadings are translated into surcharges for convenience in design Surcharges are uniform loadings applied to the surface of the finished elevation Lateral earth pressures resulting from surcharges are calculated in the same manner as the lateral earth pressure of the soil behind the bulkhead itself

The magnitude of the surcharge depends upon the use to which the land adjacent to the bulkhead will be put Only a very small surcharge would be required for a footpath, heavier surcharges for light vehicles Where automobile and truck loadings will occur, refer to authoritative texts, such as the *AASHO Bridge Specifications*, for load data **(continued, part III)**

REFERENCES

1 Chellis, Robert D , *Pile Foundations*, McGraw-Hill, 1961

2 U S Army—Coastal Engineering Research Center, *Shore Protection, Planning and Design, Technical Report No 4*, Third Edition, 1966

3 Terzaghi, Karl, "Anchored Bulkheads," *Transactions, ASCE* 1954 (Vol 119) pp 1243 to 1324

AWPI Technical Guidelines

For Pressure Treated Wood

S4
1970

Bulkheads: Design and Construction—Part III

SUMMARY

This bulletin discusses the remaining eight (8) procedures which are required in the design for anchored bulkhead construction. Part II of the series on Bulkheads Design and Construction, presented nine (9) design steps and discussed Steps 1a through 1g. The step-by-step descriptions* which include some basic information are presented in detail from steps two (2) through nine (9)

EARTH PRESSURE DIAGRAMS (Design Step 2)

The sheet piles are driven into the ground to hold earth on one side of the wall at a higher level than on the other. The pressure exerted against the sheet piles by the retained earth is called *active earth pressure*. The pressure exerted by the earth on the low side in resistance to lateral movement of the sheet piling is known as the *passive earth pressure*. Because cohesive soils can resist higher forces than they can create, the allowable passive pressure is virtually always higher than the active pressure occasionally nearly ten times as much.

The magnitude and distribution of active pressure against a sheet pile wall depends on a number of factors. These factors include the physical properties of the retained earth, the friction between the earth and the sheet pile wall, the amount of deflection of the wall, and the flexibility of the sheet piles. Similar factors influence the magnitude and distribution of the passive earth pressure. For a discussion of the behavior of sheetpile walls and the influence of various factors on the active and passive earth pressures, see Ref. 1.

Customarily the lateral earth pressures are proportional to the vertical pressure at any given level. If the symbol P_v designates the vertical pressure at any given level (weight of overlying soil) the lateral pressures are expressed as K_AP_V for active pressure and as K_PP_V for passive. The symbols K_A and K_P, are known as coefficients of earth pressure. In determining the vertical pressure P_V at any level in the soil for bulkhead design, the unit weight is that of moist earth above and that of submerged earth below, the free water surface.

Sands are classified as dense, medium or loose, to approximately describe the density, and as clean or silty to indicate the absence or presence of fine materials. These physical properties influence both the unit weights of the material and the earth pressure coefficients. Table 1 showing some physical properties for clean and silty sand, is from Ref. 1, p. 1268.

For designing timber sheet pile walls backfilled with predominately grandular materials and driven into natural undisturbed deposits, the following average values may be used for the unit weights of sand and for the earth pressure coefficients.

Table 1 Unit Weights of Soils and Coefficients of Earth Pressure

Type of Soil	Unit Weight of Soil Pounds per cubic foot				Active Earth Pressure			Passive Earth Pressure	
	Moist		Submerged		Coefficient, K_A		Angles of Friction Degrees	Coefficient, K_A for Soils in Place	Angles of Friction Degrees
	Min	Max	Min	Max	Backfill	Soils in Place			
Clean Sand									
Dense	110	140	65	78		0 20	38 20	9 0	38 25
Medium	110	130	60	68		0 25	34 17	7 0	34 23
Loose	90	125	56	63	0 35	0 30	30 15	5 0	30 20
Silty Sand									
Dense	110	150	70	88		0 25		7 0	
Medium	95	130	60	68		0 30		5 0	
Loose	80	125	50	63	0 50	0 35		3 0	

Unit weight of moist sand 100 lb per cu ft
Unit weight of submerged sand 60 lb per cu ft
$K_P = 5.0$
$K_A = 3.0$
For sand fills $K_P = 3.0$

Preparation of Earth Pressure Diagrams

Figure 7 shows a typical combined lateral pressure diagram for the bulkhead illustrated in Fig 2, part 1 (S2) Granular materials are assumed throughout, with the material below the outside bottom assumed as undisturbed soil The combined lateral pressure diagram is obtained from the diagrams on Fig 6 showing the separate effects of active earth pressure, water lag, and passive earth pressure The active earth pressure increases at a rate of K_AW_e, where K_A is taken as 0 3 and W_e is effective unit weight of earth W_e is taken as 100 pounds per cubic foot (pcf) above the free water surface within the backfill and as 60 pcf below the free water surface Hence the active earth pressure increases at a rate of 30 pcf above the free water surface and at 18 pcf for the submerged earth

The available passive earth pressure increases at a rate of K_PW_e but, to allow for a factor of safety (Fs) against the outward movement of the lower ends of the sheet piles, the rate of increase in the passive pressure is taken as K_PW_e/F_s The factor of safety against the outward movement of the sheet piles should be in the range of 1 5 to 2 0 Hence, if Fs is taken as 1 67 and K_P is 5 0, the rate of increase for the passive pressure, including allowance for the factor of safety is

(5 0/1 67) X 60 pcf (submerged earth) = 180 pcf

The free water surface behind the sheet piling for the diagrams in Fig 6 is taken as 1 0 foot above the outside water level, which for the purpose of analysis is taken as MLW because the maximum outward forces occur at low water level The unbalanced water pressure due to this one-foot water lag augments the outward pressures of the earth

Assuming that sea water weighs 64 pcf, the water lag pressure has a constant value of 64 pounds per square foot (psf) from MLW level to the level of the outside bottom at elevation minus four feet (-4 0) Although the water lag pressure probably decreases from its value of 64 psf at the level of the outside bottom (-4 0) to a zero value at the bottom of the sheet piles, the assumption is often made that the water-lag pressure continues at a constant value to the bottom of the sheet piles as shown by the solid line in Fig 6b and 6d This assumption simplifies the computation process in arriving at the combined lateral pressure diagram, because the required depth of penetration for the sheet piling remains to be determined at the time that the construction of the pressure diagrams is in progress

Thus the slope, or rate of increase of the combined pressure diagram below elevation 4 0 in Fig 7 is simply the difference between the value of 180 pcf for the passive pressure and 18 pcf for the active pressure, giving 162 pcf if the water-lag pressure remains at its constant value of 64 psf to the bottom of the sheet piles

The diagram for active earth pressure plus water lag shown in Fig 6d is obtained by adding the pressure diagrams shown in figs 6a and 6b The combined pressure diagram shown in Fig 7 results from adding the pressure diagrams shown on the two sides of the sheet piles in Fig 6c and 6d

DEPTH OF PENETRATION (Design Step 3)

The sheet piling must extend below the outside bottom to a depth such that the total resultant force developed by the passive earth pressure will be of sufficient magnitude and so located that, together with the tie rod reaction, it will equalize the effects of the summation of outward loads produced by the combined active earth pressure and water lag pressure This requirement is known as the equilibrium of moments condition about the tie rod reaction

To determine the required depth of penetration, the forces for the several pressure areas behind the bulkhead, and the positions of their centers of gravity with respect to the tie rod level are determined as shown in Fig 7 The moments of these forces about the tie

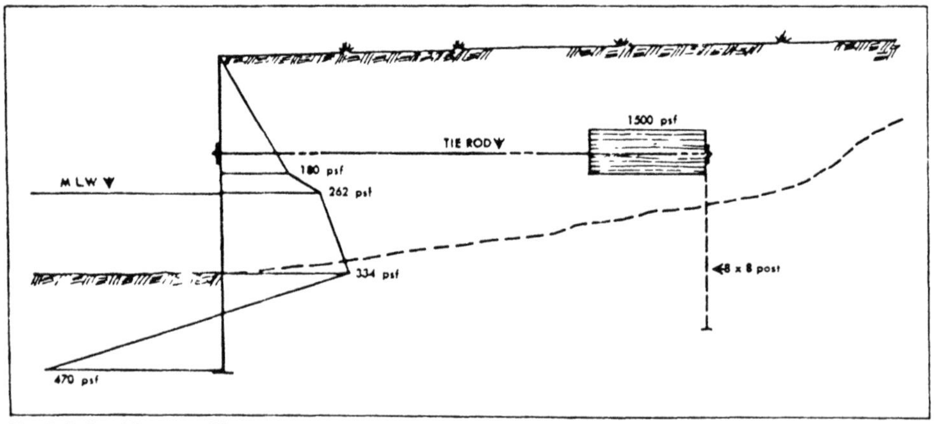

Fig. 6: Lateral Pressure Diagrams

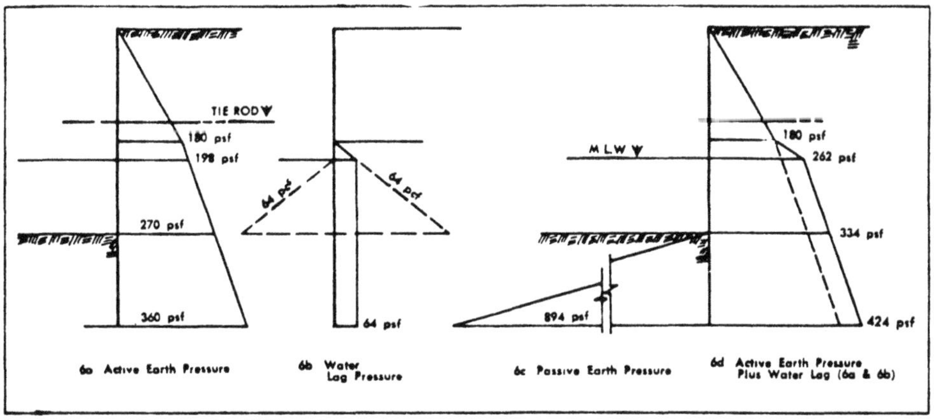

Fig. 7: Pressure Diagram For Bulkhead.

rod level are computed as shown in the following tabulation

Force	Arm	Moment
540 pounds	-1.0 Ft	-540 Pound Ft
220	1 52	340
1,190	4 08	4,850
340	6 69	2,280
2,290		6,930

The location of the zero value (point "0" in Fig 7) for the combined pressure diagram is found to be at a distance below the outside bottom of

334/162 = 2 06 ft

The required penetration below point "0" is designated by the letter "d" Then the equilibrium of moments condition is expressed by the equation

$(162d^2/2)(8\ 06 + 2d/3) = 6{,}930$

This equation is best solved by trial, whence d = 2 92 ft The total depth of penetration below outside bottom is then 2 06 + 2 92 = 4 98 ft The residual passive pressure intensity at the bottom of the sheet piles is computed as

162 X 2 92 = 470 psf

The area of the residual passive pressure triangle is

470 X 2 92 X 1/2 = 690 pounds per lineal foot of wall

TIE ROD REACTION (Design Step 4)

The tie reaction then becomes

2,290 − 690 = 1,600 pounds per ft of wall

BENDING MOMENT IN SHEET PILING (Design Step 5)

Bending moment in sheet piling Maximum moment occurs at elevation of zero shear

Let "Z" = distance below MLW to plane of zero shear

Then in the case of Fig 3 (Part 1, S2) Z is found as follows

Shear at MLW = 1,600 −540 −220 = 840 lbs

$262Z + 18Z^2/2 = 840$ lbs

Z = 2 92 ft

Moments of forces above plane of zero shear taken with respect to plane of zero shear are shown in the following tabulation

Force	Arm	Moment Clockwise	Moment Counter-clockwise
1,600	4.92	7,870	
540	5 92		3,200
220	3 39		750
763	1 46		1,110
77	0 97		75
1,600			5,135

Bending moment in sheet piling per foot of wall length = 7870 − 5135 = 2735 ft-lbs This is conventionally called the "free earth support" moment

The "free earth support" moment may be reduced by a factor of 30% to allow for beneficial effects related to flexibility of sheet piling (see Ref 1, p 1261) Hence, Design Bending Moment = 0 7 X 2,735 = 1,910 lbs per foot

SHEET PILING THICKNESS (Design Step 6)

Required thickness of sheet piling

Try 4 X 12 sheeting, actual thickness = 3⅝ inch

Section modulus = $12 \times 3\ 625^2/6 = 26\ 3$ inch^3

Bending stress = 1,910 X 12/26 3 = 870 psi

Try 3 X 12 sheeting, actual thickness = 2⅝ inch

Section modulus = $12 \times 2\ 625^2/6 = 13\ 8$ inch^3

Bending stress = (1910 X 12)/13 8 = 1,660 psi

For sheet piling of common commercial grade, the 4-inch thickness should be used

TIE ROD SIZE AND SPACING (Design Step 7)

The size and spacing of tie rods must be compatible with the design of the wales and anchorage A spacing of eight feet will be selected

Tie Rod Pull = 8 X 1,600 = 12,800 lbs

For 1¼-inch diameter tie rod, area at root of thread = 0 89 square inches

Tensile stress = 12,900/0 89 = 14,400 psi

This conservative unit stress is desirable to allow for effects of corrosion, surcharges, unequal yield of anchorages and other variables

WALE SIZE (Design Step 8)

Bending moment in wales:

a. Front Wale

Max moment over support

$Wl^2/10 = 1{,}600 \times 8^2/10 = 10\,200$ lbs per foot

Deduct 2 inch width for tie rod hole

For 12 X 10 wale, width = 11.5 − 2.0 = 9.5 depth = 9.5

Section modulus = $9.5 \times 9.5^2/6 = 143$ inches^3

Bending stress = (10,200 X 12)/143 = 850 psi

Horizontal shear = $\frac{3}{2} \times \frac{1{,}600 \times 4.0}{9.5 \times 9.5} = 107$ psi

b. Anchor Wale

Max moment over support =

$Wa^2/3 = 1{,}600 \times 3^2/2 = 7{,}200$ lbs. per foot

For 12 X 10 wale with 2 inch width squared deducted for tie rod hole, bending stress = (7200 X 12)/143 = 600 psi

Max moment at L span = $Wl^2/8 - Wa^2/2 = 1{,}600 \times 8^2/8 - 7{,}200 = 5{,}600$

Section modulus (no hole) = 11.5 X 9.3/6 = 173 inch^3

Bending stress = 5 600 X 12/173 = 390 psi

PASSIVE RESISTANCE ANCHORAGE (Design Step 9)

The anchorage for the bulkhead shown on Fig. 3 (Part 1) relies on passive resistance. To locate this anchorage at a safe distance behind the bulkhead, a slope line is drawn from the bottom of the sheet piles to the finished grade at an angle Ø to the horizontal, where Ø is the angle of internal friction of the material. For the sand backfill assumed, Ø is taken as 30 degrees. The upper edge of the anchorage should be located beneath this slope line.

For a continuous anchorage, like that for the bulkhead shown on Fig. 2 (Part 1) in which the height of bearing contact is only one foot, with center of bearing contact located 5.5 feet below the ground surface, the ultimate resistance is approximately equal to the bearing capacity of a continuous strip of footing of one foot width located at a depth of 5.5 feet. See Ref. 2, p. 231. The overburden pressure is 5.5 feet X 100 pounds per cubic foot which is 550 pounds per square foot.

Then by Ref. 2, p. 125:

Ultimate Resistance = 550 N_q, where N_q is a bearing capacity factor to be taken from curve. N_q is read as 7.0.

Ultimate Resistance = 550 X 7.0 = 3 850 psf

For 7 foot length of anchor wale to each tie rod, the ultimate resistance of the anchor wale is 7 x 1.0 x 3,850 = 26,950 pounds per tie rod. With a tie rod reaction of 1,600 pounds per foot, x 8 feet = 12,800 pounds, the factor of safety is 25,950/12,800 = 2.1. The resistance of the anchor post would increase this safety factor somewhat. A value of 2.0 is considered adequate. If an arbitrary limit of 3,000 pounds per square foot (1.5 tons) is assumed for the ultimate resistance as shown in Fig. 7, the total resistance including the anchor post will still provide a safety factor of 2.0.

REFERENCES

1. Chellis, Robert D., *Pile Foundations*, McGraw-Hill, 1961

2. U.S. Army—Coastal Engineering Research Center, *Shore Protection, Planning and Design, Technical Report No. 4*, Third Edition, 1966.

3. Terzaghi, Karl, "Anchored Bulkheads," *Transactions, ASCE* 1954 (Vol. 119) pp. 1243 to 1324

APPENDIX C

ACI STANDARD
RECOMMENDED PRACTICE FOR SELECTING PROPORTIONS FOR NORMAL AND HEAVYWEIGHT CONCRETE

(ACI 211.1–74)

APPENDIX C

ACI Standard

Recommended Practice for Selecting Proportions for Normal and Heavyweight Concrete (ACI 211.1-74)*

Reported by ACI Committee 211

JOHN R. WILSON
Chairman, Committee 211

EDWARD A. ABDUN-NUR, ROBERT A. BURMEISTER, WILLIAM A. CORDON, CLAYTON L. DAVIS, EDWIN A. DECKER, DONALD E. DIXON, FRANK G. ERSKINE, H. P. FAUERBY, RICHARD J. FRAZIER, A. T. HERSEY, WILLIAM W. HOTALING, JR., EDWARD J. HYLAND, HECTOR I. KING, PAUL KLIEGER, FRANK J. LAHM, RICHARD C. MEININGER, JOHN T. MOLNAR, AUSTIN H. MORGAN, JR., J. NEIL MUSTARD, FRANK P. NICHOLS, JR., JOHN E. PALO, SANDOR POPOVICS, JOHN M. SCANLON, JR., GEORGE W. WASHA, CECIL H. WILLETTS, CEDRIC WILLSON, JOHN C. WYCOFF

Describes, with examples, two methods for selecting and adjusting proportions for normal weight concrete. One method is based on an estimated weight of the concrete per unit volume; the other is based on calculations of the absolute volume occupied by the concrete ingredients. The procedures take into consideration the requirements for placeability, consistency, strength, and durability. Example calculations are shown for both methods, including adjustments based on the characteristics of the first trial batch.

The proportioning of heavyweight concrete for such purposes as radiation shielding and bridge counterweight structures is described in an appendix. This appendix uses the absolute volume method which is generally accepted and is more convenient for heavyweight concrete.

Keywords: absorption; aggregates; air entrainment; cement content; coarse aggregates; concrete durability; **concretes;** consistency; durability; fine aggregates; heavyweight aggregates; **heavyweight concretes; mix proportioning;** quality control; radiation shielding; slump tests; volume; water-cement ratio.

1. SCOPE

1.1—This recommended practice describes methods for selecting proportions for concrete made with aggregates of normal and high density (as distinguished from lightweight and special high density aggregates) and of workability suitable for usual cast-in-place construction (as distinguished from special mixtures for concrete products manufacture).

1.2—The methods provide a first approximation of proportions intended to be checked by trial batches in the laboratory or field and adjusted, as necessary, to produce the desired characteristics of the concrete.

1.3—U. S. customary units are used in the main body of the text. Adaptation for the metric system is provided in Appendix 1, and demonstrated in an example problem in Appendix 2.

1.4—Test methods mentioned in the text are listed in Appendix 3.

2. INTRODUCTION

2.1—Concrete is composed principally of cement, aggregates, and water. It will contain some amount of entrapped air and may also contain purposely entrained air obtained by use of an admixture or air-entraining cement. Admixtures are also frequently used for other purposes such as to accelerate, retard, improve workability, reduce mixing water requirement, increase strength, or alter other properties of the concrete.

2.2—The selection of concrete proportions involves a balance between reasonable economy and requirements for placeability, strength, durability, density, and appearance. The required characteristics are governed by the use to which the concrete will be put and by conditions ex-

*Adopted as a standard of the American Concrete Institute May 1974, to supersede ACI 211.1-70, in accordance with the Institute standardization procedure.

pected to be encountered at the time of placement These are often, but not always, reflected in specifications for the job.

2.3—The ability to tailor concrete properties to job needs reflects technological developments which have taken place, for the most part, since the early 1900s. The use of the water-cement ratio as a tool for estimating strength was recognized about 1918. The remarkable improvement in durability resulting from the entrainment of air was recognized in the early 1940s. These two significant developments in concrete technology have been augmented by extensive research and development in many related areas, including the use of admixtures to counteract possible deficiencies, develop special properties, or achieve economy.* It is beyond the scope of this discussion to review the theories of concrete proportioning which have provided the background and sound technical basis for the relatively simple methods of this recommended practice. More detailed information can be obtained from the list of references.

2.4—Proportions calculated by any method must always be considered subject to revision on the basis of experience with trial batches. Depending on circumstances, the trial mixes may be prepared in a laboratory or, perhaps preferably, as full-size field batches. The latter procedure, when feasible, avoids possible pitfalls of assuming that data from small batches mixed in a laboratory environment will predict performance under field conditions. Trial batch procedures and background testing are described in Appendix 3.

3. BASIC RELATIONSHIP

3.1—Concrete proportions must be selected to provide necessary placeability, strength, durability, and density for the particular application. Well established relationships governing these properties are discussed briefly below.

3.2—*Placeability* (including satisfactory finishing properties) encompasses traits loosely accumulated in the terms "workability" and "consistency." For the purpose of this discussion, *workability* is considered to be that property of concrete which determines its capacity to be placed and consolidated properly and to be finished without harmful segregation. It embodies such concepts as moldability, cohesiveness, and compactability. It is affected by the grading, particle shape and proportions of aggregate, the amount of cement, the presence of entrained air, admixtures, and the consistency of the mixture. Procedures in this recommended practice permit these factors to be taken into account to achieve satisfactory placeability economically.

3.3—*Consistency,* loosely defined, is the wetness of the concrete mixture. It is measured in terms of slump—the higher the slump the wetter the mixture—and it affects the ease with which the concrete will flow during placement. It is related to but not synonymous with workability. In properly proportioned concrete, the unit water content required to produce a given slump will depend on several factors. Water requirement increases as aggregates become more angular and rough textured (but this disadvantage may be offset by improvements in other characteristics such as bond to cement paste). Required mxing water decreases as the maximum size of well graded aggregate is increased. It also decreases with the entrainment of air. Mixing water requirement may often be significantly reduced by certain admixtures.

3.4—*Strength.* Strength is an important characteristic of concrete, but other characteristics such as durability, permeability, and wear resistance are often equally or more important. These may be related to strength in a general way but are also affected by factors not significantly associated with strength. For a given set of materials and conditions, concrete strength is determined by the net quantity of water used per unit quantity of cement. The net water content excludes water absorbed by the aggregates. Differences in strength for a given water-cement ratio may result from changes in: maximum size of aggregate; grading, surface texture, shape, strength, and stiffness of aggregate particles; differences in cement types and sources; air content; and the use of admixtures which affect the cement hydration process or develop cementitious properties themselves. To the extent that these effects are predictable in the general sense, they are taken into account in this recommended practice. However, in view of their number and complexity, it should be obvious that accurate predictions of strength must be based on trial batches or experience with the materials to be used.

3.5—*Durability.* Concrete must be able to endure those exposures which may deprive it of its serviceability—freezing and thawing, wetting and drying, heating and cooling, chemicals, deicing agents, and the like. Resistance to some of these may be enhanced by use of special ingredients: low-alkali cement, pozzolans, or selected aggregate to prevent harmful expansion due to the alkali-aggregate reaction which occurs in some areas when concrete is exposed in a moist environment: sulfate resisting cement or pozzolans for concrete exposed to seawater or sulfate-

*See ACI Committee 212, "Admixtures for Concrete," ACI JOURNAL, *Proceedings* V. 60, No. 11, Nov. 1963, pp. 1535-1534.

bearing soils, or aggregate free of excessive soft particles where resistance to surface abrasion is required Use of a low water cement ratio will prolong the life of concrete by reducing the penetration of aggressive liquids Resistance to severe weathering, particularly freezing and thawing, and to salts used for ice removal is greatly improved by incorporation of a proper distribution of entrained air Entrained air should be used in all exposed concrete in climates where freezing occurs.*

3.6—*Density* For certain applications concrete may be used primarily for its weight characteristic Examples of applications are counterweights on lift bridges, weights for sinking oil pipelines under water, shielding from radiation, and for insulation from sound By using special aggregates, placeable concrete of densities as high as 350 lb per cu ft can be obtained—see Appendix 4

4. BACKGROUND DATA

4.1—To the extent possible selection of concrete proportions should be based on test data or experience with the materials actually to be used Where such background is limited or not available, estimates given in this recommended practice may be employed.

4 2—The following information for available materials will be useful:

4 2.1 Sieve analyses of fine and coarse aggregates

4.2.2 Unit weight of coarse aggregate

4.2.3 Bulk specific gravities and absorptions of aggregates

4.2.4 Mixing water requirements of concrete developed from experience with available aggregates

4.2.5 Relationships between strength and water-cement ratio for available combinations of cement and aggregate

4.3—Estimates from Tables 5 3 3 and 5 3 4, respectively, may be used when the last two items of information are not available As will be shown, proportions can be estimated without the knowledge of aggregate specific gravity and absorption, Item 4 2 3.

5. PROCEDURE

5 1—The procedure for selection of mix proportions given in this section is applicable to normal weight concrete Although the same basic data and procedures can be used in proportioning heavyweight concrete, additional information as well as sample computations for this type of concrete are given in Appendix 4

5.2—Estimating the required batch weights for the concrete involves a sequence of logical, straightforward steps which, in effect, fit the characteristics of the available materials into a mixture suitable for the work The question of suitability is frequently not left to the individual selecting the proportions The job specifications may dictate some or all of the following:

5.2.1 Maximum water-cement ratio

5.2 2 Minimum cement content

5.2.3 Air content

5.2.4 Slump

5.2.5 Maximum size of aggregate

5.2.6 Strength

5 2.7 Other requirements relating to such things as strength overdesign, admixtures, and special types of cement or aggregate.

5.3—Regardless of whether the concrete characteristics are prescribed by the specifications or are left to the individual selecting the proportions, establishment of batch weights per cubic yard of concrete can best be accomplished in the following sequence:

5.3.1 *Step 1 Choice of slump* If slump is not specified, a value appropriate for the work can be selected from Table 5 3 1. The slump ranges shown apply when vibration is used to consolidate the concrete Mixes of the stiffest consistency that can be placed efficiently should be used.

TABLE 5.3 1—RECOMMENDED SLUMPS FOR VARIOUS TYPES OF CONSTRUCTION

Types of construction	Slump, in	
	Maximum*	Minimum
Reinforced foundation walls and footings	3	1
Plain footings, caissons, and substructure walls	3	1
Beams and reinforced walls	4	1
Building columns	4	1
Pavements and slabs	3	1
Mass concrete	2	1

*May be increased 1 in for methods of consolidation other than vibration

5.3.2 *Step 2 Choice of maximum size of aggregate.* Large maximum sizes of well graded aggregates have less voids than smaller sizes. Hence, concretes with the larger-sized aggregates require less mortar per unit volume of concrete Generally, the maximum size of aggregate should be the largest that is economically available and consistent with dimensions of the structure In no event should the maximum size exceed one-fifth of the narrowest dimension between sides of forms, one-third the depth of slabs, nor three-fourths of the minimum clear spacing between individual reinforcing bars, bundles of bars, or pretensioning strands These limitations are sometimes waived if workability and methods of consolidation are such that the concrete can be

*For further details, see ACI Committee 201, "Durability of Concrete in Service" ACI JOURNAL, *Proceedings* V 59, No 12, Dec 1962, pp 1771-1820

placed without honeycomb or void. When high strength concrete is desired, best results may be obtained with reduced maximum sizes of aggregate since these produce higher strengths at a given water-cement ratio.

5.3.3 *Step 3. Estimation of mixing water and air content.* The quantity of water per unit volume of concrete required to produce a given slump is dependent on the maximum size, particle shape and grading of the aggregates, and on the amount of entrained air. It is not greatly affected by the quantity of cement. Table 5 3 3 provides estimates of required mixing water for concretes made with various maximum sizes of aggregate, with and without air entrainment. Depending on aggregate texture and shape, mixing water requirements may be somewhat above or below the tabulated values, but they are sufficiently accurate for the first estimate. Such differences in water demand are not necessarily reflected in strength since other compensating factors may be involved. For example, a rounded and an angular coarse aggregate, both well and similarly graded and of good quality, can be expected to produce concrete of about the same compressive strength for the same cement factor in spite of differences in water-cement ratio resulting from the different mixing water requirements. Particle shape per se is not an indicator that an aggregate will be either above or below average in its strength-producing capacity

TABLE 5.3.3—APPROXIMATE MIXING WATER AND AIR CONTENT REQUIREMENTS FOR DIFFERENT SLUMPS AND MAXIMUM SIZES OF AGGREGATES*

Slump, in	Water, lb per cu yd of concrete for indicated maximum sizes of aggregate							
	⅜ in	½ in	¾ in	1 in	1½ in	2 in †	3 in †	6 in †
	Non-air-entrained concrete							
1 to 2	350	335	315	300	275	260	240	210
3 to 4	385	365	340	325	300	285	265	230
6 to 7	410	385	360	340	315	300	285	—
Approximate amount of entrapped air in non-air-entrained concrete, percent	3	2 5	2	1.5	1	0.5	0.3	0.2
	Air-entrained concrete							
1 to 2	305	295	280	270	250	240	225	200
3 to 4	340	325	305	295	275	265	250	220
6 to 7	365	345	325	310	290	280	270	—
Recommended average total air content, percent	8	7	6	5	4.5	4	3.5	3

*These quantities of mixing water are for use in computing cement factors for trial batches They are maxima for reasonably well-shaped angular coarse aggregates graded within limits of accepted specifications

†The slump values for concrete containing aggregate larger than 1½ in are based on slump tests made after removal of particles larger than 1½ in by wet-screening

Table 5.3.3 indicates the approximate amount of entrapped air to be expected in non-air-entrained concrete, and shows the recommended levels of average air content for concrete in which air is to be purposely entrained for durability. Air-entrained concrete should always be used for structures which will be exposed to freezing and thawing, and generally for structures exposed to sea water or sulfates. When severe exposure is not anticipated, beneficial effects of air entrainment on concrete workability and cohesiveness can be achieved at air content levels approximately half those shown for air-entrained concrete.

When trial batches are used to establish strength relationships or verify strength-producing capability of a mixture, the least favorable combination of mixing water and air content should be used. This is, the air content should be the maximum permitted or likely to occur, and the concrete should be gaged to the highest permissable slump. This will avoid developing an over-optimistic estimate of strength on the assumption that average rather than extreme conditions will prevail in the field. For information on air content recommendations, see ACI 201, 301, and 302.

5.3.4 *Step 4. Selection of water-cement ratio.* The required water-cement ratio is determined not only by strength requirements but also by factors such as durability and finishing proper ties Since different aggregates and cements generally produce different strengths at the same water-cement ratio, it is highly desirable to have or develop the relationship between strength and water-cement ratio for the materials actually to be used In the absence of such data, approximate and relatively conservative values for concrete containing Type I portland cement can be taken from Table 5.3.4(a). With typical materials, the tabulated water-cement ratios should produce the strengths shown, based on 28-day tests of specimens cured under standard laboratory conditions The average strength selected must, of course, exceed the specified strength by a sufficient margin to keep the number of low tests within specified limits.*

For severe conditions of exposure, the water-cement ratio should be kept low even though strength requirements may be met with a higher value Table 5 3 4(b) gives limiting values.

5.3.5 *Step 5 Calculation of cement content.* The amount of cement per unit volume of concrete is fixed by the determinations made in Steps 3 and 4 above The required cement is equal to the estimated mixing water content (Step 3) divided by the water-cement ratio (Step 4) If, however, the specification includes a separate minimum limit on cement in addition to

*See "Recommended Practice for Evaluation of Compression Test Results of Field Concrete (ACI 214-65)."

TABLE 5.3.4(a)—RELATIONSHIPS BETWEEN WATER-CEMENT RATIO AND COMPRESSIVE STRENGTH OF CONCRETE

Compressive strength at 28 days, psi*	Water-cement ratio, by weight	
	Non-air-entrained concrete	Air-entrained concrete
6000	0.41	—
5000	0.48	0.40
4000	0.57	0.48
3000	0.68	0.59
2000	0.82	0.74

*Values are estimated average strengths for concrete containing not more than the percentage of air shown in Table 5.3.3 For a constant water-cement ratio, the strength of concrete is reduced as the air content is increased
Strength is based on 6 x 12 in cylinders moist-cured 28 days at 73.4 ± 3 F (23 ± 1.7 C) in accordance with Section 9(b) of ASTM C 31 for Making and Curing Concrete Compression and Flexure Test Specimens in the Field
Relationship assumes maximum size of aggregate about ¾ to 1 in, for a given source, strength produced for a given water-cement ratio will increase as maximum size of aggregate decreases, see Sections 3.4 and 5.3.2

TABLE 5.3.4(b)—MAXIMUM PERMISSIBLE WATER-CEMENT RATIOS FOR CONCRETE IN SEVERE EXPOSURES*

Type of structure	Structure wet continuously or frequently and exposed to freezing and thawing†	Structure exposed to sea water or sulfates
Thin sections (railings, curbs, sills, ledges, ornamental work) and sections with less than 1 in cover over steel	0.45	0.40‡
All other structures	0.50	0.45‡

*Based on report of ACI Committee 201, "Durability of Concrete in Service," previously cited
†Concrete should also be air-entrained
‡If sulfate resisting cement (Type II or Type V of ASTM C 150) is used, permissible water-cement ratio may be increased by 0.05

requirements for strength and durability, the mixture must be based on whichever criterion leads to the larger amount of cement.

The use of pozzolanic or chemical admixtures will affect properties of both the fresh and hardened concrete.*

5.3.6 *Step 6 Estimation of coarse aggregate content.* Aggregates of essentially the same maximum size and grading will produce concrete of satisfactory workability when a given volume of coarse aggregate, on a dry-rodded basis, is used per unit volume of concrete Appropriate values for this aggregate volume are given in Table 5.3.6 It can be seen that, for equal workability, the volume of coarse aggregate in a unit volume of concrete is dependent only on its maximum size and the fineness modulus of the fine aggregate. Differences in the amount of mortar required for workability with different aggregates, due to differences in particle shape and grading, are compensated for automatically by differences in dry-rodded void content.

The volume of aggregate, in cubic feet, on a dry-rodded basis, for a cubic yard of concrete is equal to the value from Table 5.3.6 multiplied by 27 This volume is converted to dry weight of coarse aggregate required in a cubic yard of concrete by multiplying it by the dry-rodded weight per cubic foot of the coarse aggregate.

TABLE 5.3.6—VOLUME OF COARSE AGGREGATE PER UNIT OF VOLUME OF CONCRETE

Maximum size of aggregate, in	Volume of dry-rodded coarse aggregate* per unit volume of concrete for different fineness moduli of sand			
	2.40	2.60	2.80	3.00
⅜	0.50	0.48	0.46	0.44
½	0.59	0.57	0.55	0.53
¾	0.66	0.64	0.62	0.60
1	0.71	0.69	0.67	0.65
1½	0.75	0.73	0.71	0.69
2	0.78	0.76	0.74	0.72
3	0.82	0.80	0.78	0.76
6	0.87	0.85	0.83	0.81

*Volumes are based on aggregates in dry-rodded condition as described in ASTM C 29 for Unit Weight of Aggregate
These volumes are selected from empirical relationships to produce concrete with a degree of workability suitable for usual reinforced construction For less workable concrete such as required for concrete pavement construction they may be increased about 10 percent For more workable concrete, such as may sometimes be required when placement is to be by pumping, they may be reduced up to 10 percent

5.3.7 *Step 7. Estimation of fine aggregate content.* At completion of Step 6, all ingredients of the concrete have been estimated except the fine aggregate. Its quantity is determined by difference. Either of two procedures may be employed: the "weight" method (Section 5.3.7.1) or the "absolute volume" method (Section 5.3.7.2).

TABLE 5.3.7.1—FIRST ESTIMATE OF WEIGHT OF FRESH CONCRETE

Maximum size of aggregate, in	First estimate of concrete weight, lb per cu yd*	
	Non-air-entrained concrete	Air-entrained concrete
⅜	3840	3690
½	3890	3760
¾	3960	3840
1	4010	3900
1½	4070	3960
2	4120	4000
3	4160	4040
6	4230	4120

*Values calculated by Eq (5-1) for concrete of medium richness (550 lb of cement per cu yd) and medium slump with aggregate specific gravity of 2.7 Water requirements based on values for 3 to 4 in slump in Tables 5.3.3 If desired, the estimated weight may be refined as follows if necessary information is available for each 10 lb difference in mixing water from the Table 5.3.3 values for 3 to 4 in slump, correct the weight per cu yd 15 lb in the opposite direction, for each 100 lb difference in cement content from 550 lb correct the weight per cu yd 15 lb in the same direction, for each 0.1 by which aggregate specific gravity deviates from 2.7, correct the concrete weight 100 lb in the same direction

5.3.7.1 If the weight of the concrete per unit volume is assumed or can be estimated from experience, the required weight of fine aggregate is simply the difference between the weight of fresh concrete and the total weight of the other ingredients. Often the unit weight of concrete is known with reasonable accuracy from previous experience with the materials. In the absence of such information, Table 5.3.7.1 can be used to make a first estimate Even if the estimate of concrete weight per cubic yard is rough, mixture proportions will be sufficiently accurate to per-

*See report of ACI Committee 212 "Admixtures for Concrete," ACI Journal, *Proceedings* V 60, No 11, Nov 1963, pp. 1481-1524

mit easy adjustment on the basis of trial batches as will be shown in the examples.

If a theoretically exact calculation of fresh concrete weight per cubic yard is desired, the following formula can be used:

$$U = 16.85\, G_a (100 - A) + C(1 - G_a/G_c) - W(G_a - 1) \qquad (5\text{-}1)$$

where

U = weight of fresh concrete per cubic yard, lb
G_a = weighted average specific gravity of combined fine and coarse aggregate, bulk SSD*
G_c = specific gravity of cement (generally 3.15)
A = air content, percent
W = mixing water requirement, lb per cu yd
C = cement requirement, lb per cu yd

5.3.7.2 A more exact procedure for calculating the required amount of fine aggregate involves the use of volumes displaced by the ingredients. In this case, the total volume displaced by the known ingredients—water, air, cement, and coarse aggregate—is subtracted from the unit volume of concrete to obtain the required volume of fine aggregate. The volume occupied in concrete by any ingredient is equal to its weight divided by the density of that material (the latter being the product of the unit weight of water and the specific gravity of the material).

5.3.8 *Step 8 Adjustments for aggregate moisture.* The aggregate quantities actually to be weighed out for the concrete must allow for moisture in the aggregates. Generally, the aggregates will be moist and their dry weights should be increased by the percentage of water they contain, both absorbed and surface. The mixing water added to the batch must be reduced by an amount equal to the free moisture contributed by the aggregate—i e, total moisture minus absorption.

5.3.9 *Step 9. Trial batch adjustments.* The calculated mixture proportions should be checked by means of trial batches prepared and tested in accordance with ASTM C 192, "Making and Curing Concrete Compression and Flexure Test Specimens in the Laboratory," or full-sized field batches. Only sufficient water should be used to produce the required slump regardless of the amount assumed in selecting the trial proportions. The concrete should be checked for unit weight and yield (ASTM C 138) and for air content (ASTM C 138, C 173, or C 231). It should also be carefully observed for proper workability, freedom from segregation, and finishing properties. Appropriate adjustments should be made in the proportions for subsequent batches in accordance with the following procedure.

5.3.9.1 Re-estimate the required mixing water per cubic yard of concrete by multiplying the net mixing water content of the trial batch by 27 and dividing the product by the yield of the trial batch in cubic feet. If the slump of the trial batch was not correct, increase or decrease the re-estimated amount of water by 10 lb for each required increase or decrease of 1 in. in slump.

5.3.9.2 If the desired air content (for air-entrained concrete) was not achieved, re-estimate the admixture content required for proper air content and reduce or increase the mixing water content of paragraph 5.3.9.1 by 5 lb for each 1 percent by which the air content is to be increased or decreased from that of the previous trial batch.

5.3.9.3 If estimated weight per cubic yard of fresh concrete is the basis for proportioning, re-estimate that weight by multiplying the unit weight in pounds per cubic foot of the trial batch by 27 and reducing or increasing the result by the anticipated percentage increase or decrease in air content of the adjusted batch from the first trial batch.

5.3.9.4 Calculate new batch weights starting with Step 4 (Paragraph 5.3.4), modifying the volume of coarse aggregate from Table 5.3.6 if necessary to provide proper workability.

6. SAMPLE COMPUTATIONS

6.1—Two example problems will be used to illustrate application of the proportioning procedures The following conditions are assumed:

6.1.1 Type I non-air-entraining cement will be used and its specific gravity is assumed to be 3 1 5.†

6.1.2 Coarse and fine aggregates in each case are of satisfactory quality and are graded within limits of generally accepted specifications.‡

6.1.3 The coarse aggregate has a bulk specific gravity of 2 68† and an absorption of 0 5 percent.

6.1.4 The fine aggregate has a bulk specific gravity of 2 64,† an absorption of 0 7 percent, and fineness modulus of 2.8.

6.2—*Example 1* Concrete is required for a portion of a structure which will be below ground level in a location where it will not be exposed to severe weathering or sulfate attack. Structural considerations require it to have an average 28-day compressive strength of 3500 psi.§ On the basis of information in Table 5.3.1, as well as

---

*SSD indicates saturated-surface-dry basis used in considering aggregate displacement The aggregate specific gravity used in calculations must be consistent with the moisture condition assumed in the basic aggregate batch weights—i e bulk dry if aggregate weights are stated on a dry basis, and bulk SSD if weights are stated on a saturated-surface-dry basis

†The specific gravity values are not used if proportions are selected to provide a weight of concrete assumed to occupy 1 cu yd

‡Such as the "Specifications for Concrete Aggregates," (ASTM C 33)

§This is not the specified strength used for structural design but a higher figure expected to be produced on the average For the method of determining the amount by which average strength should exceed design strength see "Recommended Practice for Evaluation of Compression Test Results of Field Concrete (ACI 214-65)."

previous experience, it is determined that under the conditions of placement to be employed, a slump of 3 to 4 in. should be used and that the available No 4 to 1½-in. coarse aggregate will be suitable. The dry-rodded weight of coarse aggregate is found to be 100 lb per cu ft Employing the sequence outlined in Section 5, the quantities of ingredients per cubic yard of concrete are calculated as follows:

6.2.1 *Step 1.* As indicated above, the desired slump is 3 to 4 in.

6.2.2 *Step 2.* The locally available aggregate, graded from No. 4 to 1½ in., has been indicated as suitable.

6.2.3 *Step 3.* Since the structure will not be exposed to severe weathering, non-air-entrained concrete will be used. The approximate amount of mixing water to produce a 3- to 4-in. slump in non-air-entrained concrete with 1½-in. aggregate is found from Table 5 3 3 to be 300 lb per cu yd. Estimated entrapped air is shown as 1 percent.

6.2.4 *Step 4* From Table 5.3 4(a), the water-cement ratio needed to produce a strength of 3500 psi in non-air-entrained concrete is found to be about 0.62.

6.2.5 *Step 5.* From the information derived in Steps 3 and 4, the required cement content is found to be 300/0 62 = 484 lb per cu yd.

6.2.6 *Step 6* The quantity of coarse aggregate is estimated from Table 5 3 6. For a fine aggregate having a fineness modulus of 2 8 and a 1½ in maximum size of coarse aggregate, the table indicates that 0.71 cu ft of coarse aggregate, on a dry-rodded basis, may be used in each cubic foot of concrete. For a cubic yard, therefore, the coarse aggregate will be 27 × 0.71 = 19 17 cu ft Since it weighs 100 lb per cu ft, the dry weight of coarse aggregate is 1917 lb.

6.2.7 *Step 7.* With the quantities of water, cement, and coarse aggregate established, the remaining material comprising the cubic yard of concrete must consist of sand and whatever air will be entrapped. The required sand may be determined on the basis of either weight or absolute volume as shown below:

6.2.7.1 *Weight basis* From Table 5 3.7.1, the weight of a cubic yard of non-air-entrained concrete made with aggregate having a maximum size of 1½ in. is estimated to be 4070 lb. (For a first trial batch, exact adjustments of this value for usual differences in slump, cement factor, and aggregate specific gravity are not critical.) Weights already known are:

Water (net mixing)	300 lb
Cement	484 lb
Coarse aggregate	1917 lb (dry)*
Total	2701 lb

The weight of sand, therefore, is estimated to be

4070 − 2701 = 1369 lb (dry)*

6 2.7 2 *Absolute volume basis.* With the quantities of cement, water, and coarse aggregate established, and the approximate entrapped air content (as opposed to purposely entrained air) taken from Table 5.3 3, the sand content can be calculated as follows:

Volume of water	$= \frac{300}{62.4}$	= 4 81 cu ft
Solid volume of cement	$= \frac{484}{3.15 \times 62.4}$	= 2 46 cu ft
Solid volume of coarse aggregate	$= \frac{1917}{2.68 \times 62.4}$	= 11.46 cu ft
Volume of entrapped air	= 0 01 × 27	= 0 27 cu ft
Total solid volume of ingredients except sand		= 19.00 cu ft
Solid volume of sand required	= 27 − 19 00	= 8.00 cu ft
Required weight of dry sand	= 8.00 × 2 64 × 62 4	= 1318 lb

6.2.7.3 Batch weights per cubic yard of concrete calculated on the two bases are compared below:

	Based on estimated concrete weight, lb	Based on absolute volume of ingredients, lb
Water (net mixing)	300	300
Cement	484	484
Coarse aggregate (dry)	1917	1917
Sand (dry)	1369	1318

6.2.8 *Step 8.* Tests indicate total moisture of 2 percent in the coarse aggregate and 6 percent in the fine aggregate. If the trial batch proportions based on assumed concrete weight are used, the adjusted aggregate weights become

Coarse aggregate (wet) = 1917 (1.02) = 1955 lb
Fine aggregate (wet) = 1369 (1.06) = 1451 lb

Absorbed water does not become part of the mixing water and must be excluded from the adjustment in added water. Thus, surface water contributed by the coarse aggregate amounts to 2 − 0.5 = 1.5 percent; by the fine aggregate 6 − 0.7 = 5 3 percent. The estimated requirement for added water, therefore, becomes

300 − 1917(0.015) − 1369 (0.053) = 199 lb

The estimated batch weights for a cubic yard of concrete are:

*Aggregate absorption is disregarded since its magnitude is inconsequential in relation to other approximations.

Water (to be added)	199 lb
Cement	484 lb
Coarse aggregate (wet)	1955 lb
Fine aggregate (wet)	1451 lb

6.2.9 *Step 9.* For the laboratory trial batch, it is found convenient to scale the weights down to produce 0 03 cu yd or 0 81 cu ft of concrete. Although the calculated quantity of water to be added was 5 97 lb, the amount actually used in an effort to obtain the desired 3 to 4 in slump is 7 00 lb The batch as mixed therefore, consists of

Water (added)	7 00 lb
Cement	14 52 lb
Coarse aggregate (wet)	58 65 lb
Fine aggregate (wet)	43 53 lb
Total	123 70 lb

The concrete has a measured slump of 2 in and unit weight of 149 0 lb per cu ft It is judged to be satisfactory from the standpoint of workability and finishing properties To provide proper yield and other characteristics for future batches, the following adjustments are made:

6.2.9.1 Since the yield of the trial batch was

$$123\ 70/149\ 0 = 0\ 830 \text{ cu ft}$$

and the mixing water content was 7 00 (added) + 0 86 on coarse aggregate + 2 18 on fine aggregate = 10 04 lb, the mixing water required for a cubic yard of concrete with the same slump as the trial batch should be

$$\frac{10\ 04 \times 27}{0\ 830} = 327 \text{ lb}$$

As indicated in Paragraph 5 3 9.1, this amount must be increased another 15 lb to raise the slump from the measured 2 in to the desired 3 to 4 in range, bringing the net mixing water to 342 lb.

6.2.9.2 With the increased mixing water, additional cement will be required to provide the desired water-cement ratio of 0 62 The new cement content becomes

$$342/0\ 62 = 552 \text{ lb}$$

6.2.9.3 Since workability was found to be satisfactory, the quantity of coarse aggregate per unit volume of concrete will be maintained the same as in the trial batch The amount of coarse aggregate per cubic yard becomes

$$\frac{58.65}{0\ 83} \times 27 = 1908 \text{ lb wet}$$

which is

$$\frac{1908}{1.02} = 1871 \text{ lb dry}$$

and

$$1871\ (1.005) = 1880 \text{ SSD*}$$

6.2.9.4 The new estimate for the weight of a cubic yard of concrete is $149\ 0 \times 27 = 4023$ lb. The amount of sand required is, therefore,

$$4023 - (342 + 552 + 1880) = 1249 \text{ lb SSD}$$

or

$$1249/1.007 = 1240 \text{ lb dry}$$

The adjusted basic batch weights per cubic yard of concrete are

Water (net mixing)	342 lb
Cement	552 lb
Coarse aggregate (dry)	1871 lb
Fine aggregate (dry)	1240 lb

6.2.10 Adjustments of proportions determined on an absolute volume basis follow a procedure similar to that just outlined. The steps will be given without detailed explanation:

6.2.10.1 Quantities used in nominal 0 81 cu ft batch are

Water (added)	7 00 lb
Cement	14 52 lb
Coarse aggregate (wet)	58 65 lb
Fine aggregate (wet)	41 91 lb
Total	122 08 lb

Measured slump 2 in ; unit weight 149 0 lb per cu ft, yield 122 08/149 0 = 0 819 cu ft; workability o k

6.2.10.2 Re-estimated water for same slump as trial batch:

$$\frac{27\,(7.00 + 0.86 + 2.09)}{0.819} = 328 \text{ lb}$$

Mixing water required for slump of 3 to 4 in :

$$328 + 15 = 343 \text{ lb}$$

6.2.10.3 Adjusted cement content for increased water:

$$343/0\ 62 = 553 \text{ lb}$$

6.2.10.4 Adjusted coarse aggregate requirement:

$$\frac{58\ 65}{0\ 819} \times 27 = 1934 \text{ lb wet}$$

or

$$1934/1\ 02 = 1896 \text{ lb dry}$$

6.2.10.5 The volume of ingredients other than air in the original trial batch was

Water	$\frac{9.95}{62.4}$	= 0 159 cu ft
Cement	$\frac{14\ 52}{3.15 \times 62.4}$	= 0 074 cu ft
Coarse aggregate	$\frac{57.50}{2\ 68 \times 62.4}$	= 0.344 cu ft
Fine aggregate	$\frac{39.54}{2\ 64 \times 62.4}$	= 0 240 cu ft
Total		= 0.817 cu ft

*Saturated-surface-dry

Since the yield was 0.819 cu ft, the air content was

$$\frac{0.819 - 0.817}{0.819} = 0.2 \text{ percent}$$

With the proportions of all components except fine aggregate established, the determination of adjusted cubic yard batch quantities can be completed as follows:

Volume of water	=	$\frac{343}{62.4}$	= 5.50 cu ft
Volume of cement	=	$\frac{553}{3.15 \times 62.4}$	= 2.81 cu ft
Volume of air	=	0.002×27	= 0.05 cu ft
Volume of coarse aggregate	=	$\frac{1896}{2.68 \times 62.4}$	= 11.34 cu ft
Total volume exclusive of fine aggregate			= 19.70 cu ft
Volume of fine aggregate required	=	$27 - 19.70$	= 7.30 cu ft
Weight of fine aggregate (dry basis)	=	$7.30 \times 2.64 \times 62.4$	= 1203 lb

The adjusted basic batch weights per cubic yard of concrete, then, are:

Water (net mixing)	343 lb
Cement	553 lb
Coarse aggregate (dry)	1896 lb
Fine aggregate (dry)	1203 lb

These differ only slightly from those given in Paragraph 6.2.9.4 for the method of assumed concrete weight. Further trials or experience might indicate small additional adjustments for either method.

6.3—*Example 2.* Concrete is required for a heavy bridge pier which will be exposed to fresh water in a severe climate. An average 28-day compressive strength of 3000 psi will be required. Placement conditions permit a slump of 1 to 2 in. and the use of large aggregate, but the only economically available coarse aggregate of satisfactory quality is graded from No. 4 to 1 in. and this will be used. Its dry-rodded weight is found to be 95 lb per cu ft. Other characteristics are as indicated in Section 6.1.

The calculations will be shown in skeleton form only. Note that confusion is avoided if all steps of Section 5 are followed even when they appear repetitive of specified requirements.

6.3.1 *Step 1.* The desired slump is 1 to 2 in.

6.3.2 *Step 2.* The locally available aggregate, graded from No. 4 to 1 in., will be used.

6.3.3 *Step 3.* Since the structure will be exposed to severe weathering, air-entrained concrete will be used. The approximate amount of mixing water to produce a 1 to 2-in slump in air-entrained concrete with 1-in. aggregate is found from Table 5.3.3 to be 270 lb per cu yd. The recommended air content is 5 percent.

6.3.4 *Step 4* From Table 5.3.4(a), the water-cement ratio needed to produce a strength of 3000 psi in air-entrained concrete is estimated to be about 0.59. However, reference to Table 5.3.4(b) reveals that, for the severe weathering exposure anticipated, the water-cement ratio should not exceed 0.50 This lower figure must govern and will be used in the calculations.

6.3.5 *Step 5.* From the information derived in Steps 3 and 4, the required cement content is found to be 270/0.50 = 540 lb per cu yd.

6.3.6. *Step 6.* The quantity of coarse aggregate is estimated from Table 5.3.6. With a fine aggregate having a fineness modulus of 2.8 and a 1 in. maximum size of coarse aggregate, the table indicates that 0.67 cu ft of coarse aggregate, on a dry-rodded basis, may be used in each cubic foot of concrete. For a cubic yard, therefore, the coarse aggregate will be $27 \times 0.67 = 18.09$ cu ft. Since it weighs 95 lb per cu ft, the dry weight of coarse aggregate is $18.09 \times 95 = 1719$ lb.

6.3.7 *Step 7.* With the quantities of water, cement and coarse aggregate established, the remaining material comprising the cubic yard of concrete must consist of sand and air. The required sand may be determined on the basis of either weight or absolute volume as shown below.

6.3.7.1 *Weight basis.* From Table 5.3.7.1, the weight of a cubic yard of air-entrained concrete made with aggregate of 1 in. maximum size is estimated to be 3900 lb. (For a first trial batch, exact adjustments of this value for differences in slump, cement factor, and aggregate specific gravity are not critical.) Weights already known are:

Water (net mixing)	270 lb
Cement	540 lb
Coarse aggregate (dry)	1719 lb
Total	2529 lb

The weight of sand, therefore, is estimated to be

$$3900 - 2529 = 1371 \text{ lb (dry)}$$

6.3.7.2 ***Absolute volume basis.*** With the quantities of cement, water, air, and coarse aggregate established, the sand content can be calculated as follows:

Volume of water	$= \frac{270}{62.4}$	= 4.33 cu ft
Solid volume of cement	$= \frac{540}{3.15 \times 62.4}$	= 2.75 cu ft

Solid volume of coarse aggregate	$= \frac{1719}{2.68 \times 62.4}$	= 10.28 cu ft
Volume of air	= 0.05 × 27	= 1.35 cu ft
Total volume of ingredients except sand		= 18.71 cu ft
Solid volume of sand required	= 27 − 18.71	= 8.29 cu ft
Required weight of dry sand	= 8.29 × 2.64 × 62.4 = 1366 lb	

6.3.7.3 Batch weights per cubic yard of concrete calculated on the two bases are compared below:

	Based on estimated concrete weight, lb	Based on absolute volume of ingredients, lb
Water (net mixing)	270	270
Cement	540	540
Coarse aggregate (dry)	1719	1719
Sand (dry)	1371	1366

6.3.8 *Step 8.* Tests indicate total moisture of 3 percent in the coarse aggregate and 5 percent in the fine aggregate. If the trial batch proportions based on assumed concrete weight are used, the adjusted aggregate weights become

Coarse aggregate (wet) = 1719 (1.03) = 1771 lb
Fine aggregate (wet) = 1371 (1.05) = 1440 lb

Absorbed water does not become part of the mixing water and must be excluded from the adjustment in added water. Thus, surface water contributed by the coarse aggregate amounts to 3 − 0.5 = 2.5 percent, by the fine aggregate 5 − 0.7 = 4.3 percent. The estimated requirement for added water, therefore, becomes

$$270 - 1719\,(0.025) - 1371\,(0.043) = 168 \text{ lb}$$

The estimated batch weights for a cubic yard of concrete are:

Water (to be added)	168 lb
Cement	540 lb
Coarse aggregate (wet)	1771 lb
Fine aggregate (wet)	1440 lb
Total	3919 lb

6.3.9 *Step 9.* For the laboratory trial batch, the weights are scaled down to produce 0.03 cu yd or 0.81 cu ft of concrete. Although the calculated quantity of water to be added was 5.04 lb the amount actually used is in an effort to obtain the desired 1 to 2-in slump is 4.50 lb. The batch as mixed, therefore, consists of

Water (added)	4.50 lb
Cement	16.20 lb
Coarse aggregate (wet)	53.13 lb
Fine aggregate (wet)	43.20 lb
Total	117.03 lb

The concrete has a measured slump of 2 in., unit weight of 141.8 lb per cu ft, and air content of 6.5 percent. It is judged to be slightly oversanded for the easy placement condition involved. To provide proper yield and other characteristics for future batches, the following adjustments are made:

6.3.9.1 Since the yield of the trial batch was

$$117.03/141.8 = 0.825 \text{ cu ft}$$

and the mixing water content was 4.50 (added) + 1.29 on coarse aggregate + 1.77 on fine aggregate = 7.56 lb the mixing water required for a cubic yard of concrete with the same slump as the trial batch should be

$$\frac{7.56 \times 27}{0.825} = 247 \text{ lb}$$

The slump was satisfactory but, since the air content was too high by 1.5 percent, more water will be needed for proper slump when the air content is corrected. As indicated in Paragraph 5.3.9.2, the mixing water should be increased roughly 5 × 1.5 or about 8 lb, bringing the new estimate to 255 lb per cu yd.

6.3.9.2 With the decreased mixing water, less cement will be required to provide the desired water-cement ratio of 0.5. The new cement content becomes

$$255/0.5 = 510 \text{ lb}$$

6.3.9.3 Since the concrete was found to be oversanded, the quantity of coarse aggregate per unit volume will be increased 10 percent, to 0.74, in an effort to correct the condition. The amount of coarse aggregate per cubic yard becomes

$$0.74 \times 27 \times 95 = 1898 \text{ lb dry}$$

or

$$1898 \times 1.03 = 1955 \text{ lb wet}$$

and

$$1898 \times 1.005 = 1907 \text{ lb SSD*}$$

6.3.9.4 The new estimate for the weight of the concrete with 1.5 percent less air is 141.8/0.985 = 144.0 lb per cu ft or 144.0 × 27 = 3888 lb per cu yd. The weight of sand, therefore, is

$$3888 - (255 + 510 + 1907) = 1216 \text{ lb SSD*}$$

or

$$1216/1.007 = 1208 \text{ lb dry}$$

*Saturated-surface-dry

The adjusted basic batch weights per cubic yard of concrete are

Water (net mixing)	255 lb
Cement	510 lb
Coarse aggregate (dry)	1898 lb
Fine aggregate (dry)	1208 lb

Admixture dosage must be reduced to provide the desired air content.

6.3.10 Adjustments of proportions determined on an absolute volume basis would follow the procedure outlined in Paragraph 6 2 10 which will not be repeated for this example.

REFERENCES

1. Fuller, William B , and Thompson, Sanford E , "The Laws of Proportioning Concrete," *Transactions*, ASCE, V 59, Dec 1907, pp. 67-143.

2. Abrams, Duff A., "Design of Concrete Mixtures," *Bulletin* No. 1, Structural Materials Research Laboratory, Lewis Institute, Chicago, 1918

3. Edwards, L N , "Proportioning the Materials of Mortars and Concretes by Surface Areas of Aggregates," *Proceedings*, ASTM, V 18, Part 2, 1918, p. 235.

4. Young, R B , "Some Theoretical Studies on Proportioning Concrete by the Method of Surface Area of Aggregate," *Proceedings*, ASTM, V 19, Part 2, 1919, p. 444.

5. Talbot, A N , "A Proposed Method of Estimating the Density and Strength of Concrete and of Proportioning the Materials by Experimental and Analytical Consideration of the Voids in Mortar and Concrete," *Proceedings*, ASTM, V 21, 1921, p 940.

6 Weymouth, C A G , "A Study of Fine Aggregate in Freshly Mixed Mortars and Concretes," *Proceedings*, ASTM, V 38, Part 2, 1938, pp 354-372

7 Dunagan, W M , "The Application of Some of the Newer Concepts to the Design of Concrete Mixes," ACI JOURNAL, *Proceedings* V 36, No 6, June 1940, pp 649-684.

8 Goldbeck, A T, and Gray, J E, "A Method of Proportioning Concrete for Strength, Workability, and Durability," *Bulletin* No 11, National Crushed Stone Association, Dec 1942, 30 pp (Revised 1953 and 1956)

9 Swayze, M A , and Gruenwald, E , "Concrete Mix Design—Modification of Fineness Modulus Method," ACI JOURNAL, *Proceedings* V 43, No 7, Mar 1947, pp. 829-844.

10 Discussion of "Concrete Mix Design—A Modification of the Fineness Modulus Method" by Stanton Walker and Fred F Bartel, ACI JOURNAL, *Proceedings* V 43, Part 2, Dec 1947, pp 844-1—844-17.

11 Henrie, James O., "Properties of Nuclear Shielding Concrete," ACI JOURNAL, *Proceedings* V. 56, No. 1, July 1959, pp 37-46

12 Mather, Katharine, "High Strength, High Density Concrete," ACI JOURNAL, *Proceedings* V 62, No 8, Aug 1965, pp 951-960

13 Clendenning, T G , Kellam, B , and MacInnis, C , "Hydrogen Evolution from Ferrophosphorous Aggregate in Portland Cement Concrete," ACI JOURNAL, *Proceedings* V 65, No 12, Dec 1968, pp 1021-1028

14 Popovics, Sandor, "Estimating Proportions for Structural Concrete Mixtures," ACI JOURNAL, *Proceedings* V 65, No 2, Feb 1968, pp 143-150.

15 "Tentative Specification for Aggregates for Radiation-Shielding Concrete," (ASTM C 637), American Society for Testing and Materials, Philadelphia

16 Davis, H S , "Aggregates for Radiation Shielding Concrete," *Materials Research and Standards*, V. 7, No 11, Nov 1967, pp 494-501

17 *Concrete for Nuclear Reactors*, SP-34, American Concrete Institute, Detroit, 1972, 1736 pp

*Saturated-surface-dry

APPENDICES

APPENDIX 1—METRIC SYSTEM ADAPTATION

A1.1—Procedures outlined in this recommended practice have been presented using British (United States customary) units of measurement. The principles are equally applicable in the metric system with proper adaptation of units This Appendix provides all of the information necessary to apply the proportioning procedure using International SI (metric) measurements Table A1.1 gives relevant conversion factors A numerical example is presented in Appendix 2.

A1.2—For convenience of reference, numbering of subsequent paragraphs in this Appendix corresponds to the body of the report except that the designation "A1" is prefixed. All tables have been converted and reproduced. Descriptive portions are included only where use of the metric system requires a change in a procedure or formula. To the extent practicable, conversions to metric units have been made in such a way that values are realistic in terms of usual practice and significance of numbers For example, aggregate and sieve sizes in the metric tables are ones commonly used in Europe. Thus, there is not always a precise mathematical correspondence between British and metric values in corresponding tables.

TABLE A1.1—CONVERSION FACTORS, BRITISH TO METRIC UNITS*

Quantity	British (U S customary) unit	SI† (Metric) unit	Conversion factor (Ratio British/SI)
Length	inch (in)	centimeter (cm)	2 540
	inch (in)	millimeter (mm)	25 40
Volume	cubic foot (ft^3)	cubic meter (m^3)	0 02832
	cubic yard (yd^3)	cubic meter (m^3)	0.7646
Mass	pound (lb)	kilogram (kg)	0 4536
Stress	pounds per square inch (psi)	kilograms force per square centimeter (kgf/cm^2)	0 0703
Density	pounds per cubic foot (lb/ft^3)	kilograms per cubic meter (kg/m^3)	16 02
	pounds per cubic yard (lb/yd^3)	kilograms per cubic meter (kg/m^3)	0.5933
Temperature	degrees Fahrenheit (F)	degrees Centigrade (C)	‡

*Gives names (and abbreviations) of measurement units in the British (U.S. customary) system as used in the body of this report and in the S.I. (metric) system, along with multipliers for converting the former to the latter. From "ASTM Metric Practice Guide" (2nd Edition, 1966).

†Système International d'Unités

‡C = (F − 32)/1.8.

A1.5.2 *Steps in calculating proportions.* Except as discussed below, the methods for arriving at quantities of ingredients for a unit volume of concrete are essentially the same when metric units are employed as when British units are employed. The main difference is that the unit volume of concrete becomes the cubic meter and numerical values must be taken from the proper "A1" table instead of the one referred to in the text.

A1.5.2.1 *Step 1. Choice of slump* See Table A1.5 2.1.

A1.5.2.2 *Step 2. Choice of maximum size of aggregate*

A1.5.2.3 *Step 3. Estimation of mixing water and air content.* See Table A1.5.2.3.

A1.5.2.4 *Step 4. Selection of water-cement ratio.* See Table A1.5.2.4.

A1.5.2.5 *Step 5. Calculation of cement content.*

A1.5.2.6 *Step 6 Estimation of coarse aggregate content.* The dry weight of coarse aggregate required for a cubic meter of concrete is equal to the value from Table A1.5 2.6 multiplied by the dry-rodded unit weight of the aggregate in kilograms per cubic meter.

TABLE A1.5.2.1—RECOMMENDED SLUMPS FOR VARIOUS TYPES OF CONSTRUCTION (METRIC)

Types of construction	Slump, cm	
	Maximum*	Minimum
Reinforced foundation walls and footings	8	2
Plain footings, caissons, and substructure walls	8	2
Beams and reinforced walls	10	2
Building columns	10	2
Pavements and slabs	8	2
Heavy mass concrete	8	2

*May be increased 2 cm for methods of consolidation other than vibration.

TABLE A1.5.2.3—APPROXIMATE MIXING WATER REQUIREMENTS FOR DIFFERENT SLUMPS AND MAXIMUM SIZES OF AGGREGATES (METRIC)*

Slump, cm	Water, kg/m³ of concrete for indicated maximum sizes of aggregate in mm							
	10	12 5	20	25	40	50†	70†	150†
Non-air-entrained concrete								
3 to 5	205	200	185	180	160	155	145	125
8 to 10	225	215	200	195	175	170	160	140
15 to 18	240	230	210	205	185	180	170	—
Approximate amount of entrapped air in non-air-entrained concrete, percent	3	2 5	2	1 5	1	0 5	0.3	0.2
Air-entrained concrete								
3 to 5	180	175	165	160	145	140	135	120
8 to 10	200	190	180	175	160	155	150	135
15 to 18	215	205	190	185	170	165	160	—
Recommended average total air content, percent	8	7	6	5	4 5	4	3 5	3

*These quantities of mixing water are for use in computing cement factors for trial batches They are maxima for reasonably well-shaped angular coarse aggregates graded within limits of accepted specifications

†The slump values for concrete containing aggregate larger than 40 mm are based on slump tests after removal of particles larger than 40 mm by wet-screening

TABLE A1.5.2.4(a)—RELATIONSHIPS BETWEEN WATER-CEMENT RATIO AND COMPRESSIVE STRENGTH OF CONCRETE (METRIC)

Compressive strength at 28 days, kgf/cm²*	Water-cement ratio, by weight	
	Non-air-entrained concrete	Air-entrained concrete
450	0 38	—
400	0 43	—
350	0 48	0 40
300	0 55	0 46
250	0 62	0 53
200	0 70	0 61
150	0 80	0 71

*Values are estimated average strengths for concrete containing not more than the percentage of air shown in Table A1 5 2 3 For a constant water-cement ratio, the strength of concrete is reduced as the air content is increased

Strength is based on 15 x 30 cm cylinders moist-cured 28 days at 23 ± 1 7 C in accordance with Section 9(b) of ASTM C 31 for Making and Curing Concrete Compression and Flexure Test Specimens in the Field Cube strengths will be higher by approximately 20 percent

Relationship assumes maximum size of aggregate about 20 to 30 mm, for a given source, strength produced by a given water-cement ratio will increase as maximum size decreases see Sections 3 4 and 5 3.2

TABLE A1.5.2.4(b)—MAXIMUM PERMISSIBLE WATER-CEMENT RATIOS FOR CONCRETE IN SEVERE EXPOSURES (METRIC)*

Type of Structure	Structure wet continuously or frequently and exposed to freezing and thawing†	Structure exposed to sea water or sulfates
Thin sections (railings, curbs, sills, ledges, ornamental work) and sections with less than 3 cm cover over steel	0 45	0 40‡
All other structures	0 50	0 45‡

*Based on the report of ACI Committee 201, "Durability of Concrete in Service," previously cited

†Concrete should also be air-entrained

‡If sulfate resisting cement (Type II or Type V of ASTM C 150) is used, permissible water-cement ratio may be increased by 0.05.

A1.5.2 7 *Step 7. Estimation of fine aggregate content* In the metric system, the formula for calculation of fresh concrete weight per cubic meter is:

$$U_M = 10G_a(100 - A) + C_M(1 - G_a/G_c) - W_M(G_a - 1)$$

where

U_M = weight of fresh concrete, kg/m³
G_a = weighted average specific gravity of combined fine and coarse aggregate, bulk, SSD
G_c = specific gravity of cement (generally 3.15)
A = air content, percent
W_M = mixing water requirement, kg/m³
C_M = cement requirement, kg/m³

A1.5.2.9 *Trial batch adjustments.* The following "rules of thumb" may be used to arrive at closer approximations of unit batch quantities based on results for a trial batch:

A1 5.2.9.1 The estimated mixing water to produce the same slump as the trial batch will be equal to the net amount of mixing water used divided by the yield of the trial batch in m³. If slump of the trial batch was not correct, increase or decrease the re-estimated water content by 2 kg/m³ of concrete for each increase or decrease of 1 cm in slump desired.

A1.5 2.9.2 To adjust for the effect of incorrect air content in a trial batch of air-entrained concrete on slump, reduce or increase the mixing water

TABLE A1.5.2.6—VOLUME OF COARSE AGGREGATE PER UNIT OF VOLUME OF CONCRETE (METRIC)

Maximum size of aggregate, mm	Volume of dry-rodded coarse aggregate* per unit volume of concrete for different fineness moduli† of sand			
	2 40	2 60	2 80	3 00
10	0.50	0 48	0 46	0 44
12 5	0 59	0 57	0 55	0 53
20	0 66	0 64	0 62	0 60
25	0 71	0 69	0 67	0 65
40	0 76	0 74	0 72	0 70
50	0 78	0 76	0 74	0 72
70	0 81	0 79	0 77	0 75
150	0 87	0 85	0 83	0 81

*Volumes are based on aggregates in dry-rodded condition as described in ASTM C 29 for Unit Weight of Aggregate

These volumes are selected from empirical relationships to produce concrete with a degree of workability suitable for usual reinforced construction For less workable concrete such as required for concrete pavement construction they may be increased about 10 percent For more workable concrete, such as may sometimes be required when placement is to be by pumping, they may be reduced up to 10 percent

†Fineness modulus of sand = sum of ratios (cumulative) retained on sieves with square openings of 0 149, 0.297, 0.595, 1 19, 2 38, and 4 76 mm

TABLE A1.5.2.7.1—FIRST ESTIMATE OF WEIGHT OF FRESH CONCRETE (METRIC)

Maximum size of aggregate, mm	First estimate of concrete weight, kg/m³*	
	Non-air-entrained concrete	Air-entrained concrete
10	2285	2190
12 5	2315	2235
20	2355	2280
25	2375	2315
40	2420	2355
50	2445	2375
70	2465	2400
150	2505	2435

*Values calculated by Eq (A1 5 2 7) for concrete of medium richness (330 kg of cement per m³) and medium slump with aggregate specific gravity of 2 7 Water requirements based on values for 8 to 10 cm slump in Table A1 5 2 3 If desired, the estimate of weight may be refined as follows if necessary information is available for each 5 kg difference in mixing water from the Table A1 5 2.3 values for 8 to 10 cm slump, correct the weight per m³ 8 kg in the opposite direction; for each 20 kg difference in cement content from 330 kg, correct the weight per m³ 3 kg in the same direction, for each 0 1 by which aggregate specific gravity deviates from 2 7, correct the concrete weight 70 kg in the same direction.

content of A1 5 2 9 1 by 3 kg/m³ of concrete for each 1 percent by which the air content is to be increased or decreased from that of the trial batch.

A1.5.2.9.3 The re-estimated unit weight of the fresh concrete for adjustment of trial batch proportions is equal to the unit weight in kg/m³ measured on the trial batch, reduced or increased by the percentage increase or decrease in air content of the adjusted batch from the first trial batch.

APPENDIX 2—

EXAMPLE PROBLEM IN METRIC SYSTEM

A2.1—Example 1. Example 1 presented in Section 6 2 will be solved here using metric units of measure. Required average strength will be 250 kgf/cm² with slump of 8 to 10 cm. The coarse aggregate has a maximum size of 40 mm and dry-rodded weight of 1600 kg/m³ As stated in Section 6 1, other properties of the ingredients are cement—Type I with specific gravity of 3 15, coarse aggregate—bulk specific gravity 2.68 and absorption 0 5 percent, fine aggregate—bulk specific gravity 2 64, absorption 0 7 percent, and fineness modulus 2 8.

A2.2— All steps of Section 5 3 should be followed in sequence to avoid confusion, even though they sometimes merely restate information already given.

A2.2.1 *Step 1.* The slump is required to be 8 to 10 cm.

A2.2.2 *Step 2* The aggregate to be used has a maximum size of 40 mm.

A2.2.3 *Step 3* The concrete will be non-air-entrained since the structure is not to be exposed to severe weathering From Table A1 5 2 3, the estimated mixing water for a slump of 8 to 10 cm in non-air-entrained concrete made with 40-mm aggregate is found to be 175 kg/m³.

A2.2.4 *Step 4* The water-cement ratio for non-air-entrained concrete with a strength of 250 kgf/cm² is found from Table A1 5 2 4(a) to be 0.62.

A2.2.5 *Step 5* From the information developed in Steps 3 and 4, the required cement content is found to be 175/0.62 = 282 kg/m³.

A2.2.6 *Step 6* The quantity of coarse aggregate is estimated from Table A1 5 2 6 For a fine aggregate having a fineness modulus of 2 8 and a 40 mm maximum size of coarse aggregate, the table indicates that 0 72 m³ of coarse aggregate, on a dry-rodded basis, may be used in each cubic meter of concrete The required dry weight is, therefore, 0 72 × 1600 = 1152 kg

A2.2.7 *Step 7* With the quantities of water, cement and coarse aggregate established, the remaining material comprising the cubic meter of concrete must consist of sand and whatever air will be entrapped The required sand may be determined on the basis of either weight or absolute volume as shown below·

A2.2.7.1 *Weight Basis* From Table A1 5 2 7 1, the weight of a cubic meter of non-air-entrained concrete made with aggregate having a maximum size of 40 mm is estimated to be 2420 kg (For a first trial batch, exact adjustments of this value for usual differences in slump, cement factor, and aggregate specific gravity are not critical) Weights already known are·

Water (net mixing)	175 kg
Cement	282 kg
Coarse aggregate	1152 kg
Total	1609 kg

The weight of sand, therefore, is estimated to be

$$2420 - 1609 = 811 \text{ kg}$$

A2.2.7 2 *Absolute volume basis* With the quantities of cement, water, and coarse aggregate established, and the approximate entrapped air content (as opposed to purposely entrained air) of 1 percent determined from Table A1 5 2 3, the sand content can be calculated as follows:

Volume of water	$= \frac{175}{1000}$	0.175 m³
Solid volume of cement	$= \frac{282}{3\ 15 \times 1000}$	0 090 m³
Solid volume of coarse aggregate	$= \frac{1152}{2\ 68 \times 1000}$	0 430 m³
Volume of entrapped air	= 0 01 × 1 000	0 010 m³
Total solid volume of ingredients except sand		0 705 m³
Solid volume of sand required	= 1 000 − 0 705	0.295 m³
Required weight of dry sand	= 0 295 × 2 64 × 1000	779 kg

A2.2.7 3 Batch weights per cubic meter of concrete calculated on the two bases are compared below:

	Based on estimated concrete weight, kg	Based on absolute volume of ingredients, kg
Water (net mixing)	175	175
Cement	282	282
Coarse aggregate (dry)	1152	1152
Sand (dry)	811	779

A2.2.8 *Step 8.* Tests indicate total moisture of 2 percent in the coarse aggregate and 6 percent in the fine aggregate If the trial batch proportions based on assumed concrete weight are used, the adjusted aggregate weights become

Coarse aggregate (wet) = 1152(1 02) = 1175 kg
Fine aggregate (wet) = 811(1 06) = 860 kg

Absorbed water does not become part of the mixing water and must be excluded from the adjustment in added water Thus, surface water contributed by the coarse aggregate amounts to 2 − 0.5 = 1.5 percent; by the fine aggregate 6 − 0 7 = 5 3 percent. The estimated requirement for added water, therefore, becomes

$$175 - 1152(0\ 015) - 811(0\ 053) = 115 \text{ kg}$$

The estimated batch weights for a cubic meter of concrete are:

Water (to be added)	115 kg
Cement	282 kg
Coarse aggregate (wet)	1175 kg
Fine aggregate (wet)	860 kg
Total	2432 kg

A2.2 9 *Step 9* For the laboratory trial batch, it is found convenient to scale the weights down to produce 0 02 m³ of concrete Although the calculated quantity of water to be added was 2 30 kg, the amount actually used in an effort to obtain the desired 8 to 10 cm slump is 2 70 kg The batch as mixed, therefore, consists of

Water (added)	2 70 kg
Cement	5 64 kg
Coarse aggregate (wet)	23.50 kg
Fine aggregate (wet)	17.20 kg
Total	49 04 kg

ACI STANDARD

The concrete has a measured slump of 5 cm and unit weight of 2390 kg/m^3 It is judged to be satisfactory from the standpoint of workability and finishing properties To provide proper yield and other characteristics for future batches, the following adjustments are made·

A2.2.9.1 Since the yield of the trial batch was

$$49\ 04/2390 = 0\ 0205\ m^3$$

and the mixing water content was 2 70 (added) + 0 34 (on coarse aggregate) + 0 86 (on fine aggregate) = 3 90 kg, the mixing water required for a cubic meter of concrete with the same slump as the trial batch should be

$$\frac{3\ 90}{0\ 0205} = 190\ kg$$

As indicated in A1 5 2 9 1, this amount must be increased another 8 kg to raise the slump from the measured 5 cm to the desired 8 to 10 cm range, bringing the total mixing water to 198 kg

A2.2.9.2 With the increased mixing water, additional cement will be required to provide the desired water-cement ratio of 0 62 The new cement content becomes

$$198/0\ 62 = 319\ kg$$

A2.2.9.3 Since workability was found to be satisfactory, the quantity of coarse aggregate per unit volume of concrete will be maintained the same as in the trial batch The amount of coarse aggregate per cubic meter becomes

$$\frac{23\ 50}{0\ 0205} = 1146\ kg\ wet$$

which is

$$\frac{1146}{1\ 02} = 1124\ kg\ dry$$

and

$$1124 \times 1\ 005 = 1130\ kg\ SSD^*$$

A2.2.9.4 The new estimate for the weight of a cubic meter of concrete is the measured unit weight of 2390 kg/m^3 The amount of sand required is, therefore

$$2390 - (198 + 319 + 1130) = 743\ kg\ SSD$$

or

$$743/1\ 007 = 738\ kg\ dry$$

The adjusted basic batch weights per cubic meter of concrete are

Water (net mixing)	198 kg
Cement	319 kg
Coarse aggregate (dry)	1124 kg
Fine aggregate (dry)	738 kg

A2.2.10 Adjustments of proportions determined on an absolute volume basis follow a procedure similar to that just outlined The steps will be given without detailed explanation

A2.2.10.1 Quantities used in the nominal 0 02 m^3 batch are

Water (added)	2 70 kg
Cement	5 64 kg
Coarse aggregate (wet)	23 50 kg
Fine aggregate (wet)	16 51 kg
Total	48 35 kg

Measured slump 5 cm, unit weight 2390 kg/m^3, yield 48 35/2390 = 0 0202 m^3, workability o.k.

A2 2.10 2 Re-estimated water for same slump as trial batch:

$$\frac{2\ 70 + 0\ 34 + 0\ 83}{0\ 0202} = 192\ kg$$

Mixing water required for slump of 8 to 10 cm

$$192 + 8 = 200\ kg$$

A2 2.10.3 Adjusted cement content for increased water:

$$200/0\ 62 = 323\ kg$$

A2.2.10.4 Adjusted coarse aggregate requirement:

$$\frac{23\ 50}{0\ 0202} = 1163\ kg\ wet$$

or

$$1163/1\ 02 = 1140\ kg\ dry$$

A2 2.10 5 The volume of ingredients other than air in the original trial batch was

Water	$\frac{3\ 87}{1000}$	= 0 0039 m^3
Cement	$\frac{5\ 64}{3\ 15 \times 1000}$	= 0 0018 m^3
Coarse aggregate	$\frac{23\ 04}{2\ 68 \times 1000}$	= 0 0086 m^3
Fine aggregate	$\frac{15\ 58}{2\ 64 \times 1000}$	= 0 0059 m^3
Total		0 0202 m^3

Since the yield was also 0 0202 m^3, there was no air in the concrete detectable within the precision of the unit weight test and significant figures of the calculations With the proportions of all components except fine aggregate established, the determination of adjusted cubic yard batch quantities can be completed as follows

Volume of water	$= \frac{200}{1000}$	= 0 200 m^3
Volume of cement	$= \frac{323}{3\ 15 \times 1000}$	= 0 103 m^3
Allowance for volume of air		= 0 000 m^3
Volume of coarse aggregate	$= \frac{1140}{2\ 68 \times 1000}$	= 0 425 m^3
Total volume exclusive of fine aggregate		= 0 728 m^3
Volume of fine aggregate required	= 1 000 − 0 728	= 0 272 m^3
Weight of fine aggregate (dry basis)	= 0 272 × 2 64 × 1000	= 718 kg

The adjusted basic batch weights per cubic meter of concrete, then, are.

Water (net mixing)	200 kg
Cement	323 kg
Coarse aggregate (dry)	1140 kg
Fine aggregate (dry)	718 kg

These differ only slightly from those given in Paragraph A2 2 9 4 for the method of assumed concrete weight Further trials or experience might indicate small additional adjustments for either method

*Saturated-surface-dry

APPENDIX 3—LABORATORY TESTS

A3.1—Selection of concrete mix proportions can be accomplished effectively from results of laboratory tests which determine basic physical properties of materials to be used, establish relationships between water-cement ratio, air content, cement content, and strength, and which furnish information on the workability characteristics of various combinations of ingredient materials The extent of investigation desirable for any given job will depend on its size and importance and on the service conditions involved Details of the laboratory program will also vary, depending on facilities available and on individual preferences

A3.2—Properties of cement

A3.2.1 Physical and chemical characteristics of cement influence the properties of hardened concrete However, the only property of cement used directly in computation of concrete mix proportions is specific gravity The specific gravity of portland cements of the types covered by ASTM C 150 and C 175 may usually be assumed to be 3 15 without introducing appreciable error in mix computations For other types such as the blended hydraulic cements of ASTM C 595, the specific gravity for use in volume calculations should be determined by test

A3.2.2 A sample of cement should be obtained from the mill which will supply the job, or preferably from the concrete supplier The sample should be ample for tests contemplated with a liberal margin for additional tests that might later be considered desirable Cement samples should be shipped in airtight containers, or at least in moisture-proof packages

A3.3—Properties of aggregate

A3.3.1 Sieve analysis, specific gravity, absorption, and moisture content of both fine and coarse aggregate and dry-rodded unit weight of coarse aggregate are physical properties useful for mix computations Other tests which may be desirable for large or special types of work include petrographic examination and tests for chemical reactivity, soundness, durability, resistance to abrasion, and various deleterious substances Such tests yield information of value in judging the long-range serviceability of concrete.

A3.3.2 Aggregate gradation as measured by the sieve analysis is a major factor in determining unit water requirement, proportions of coarse aggregate and sand, and cement content for satisfactory workability. Numerous "ideal" aggregate grading curves have been proposed, and these, tempered by practical considerations, have formed the basis for typical sieve analysis requirements in concrete standards ASTM C 33, "Specification for Concrete Aggregates," provides a selection of sizes and gradings suitable for most concrete Additional workability realized by use of air-entrainment permits, to some extent, the use of less restrictive aggregate gradations.

A3 3 3 Samples for concrete mix tests should be representative of aggregate available for use in the work For laboratory tests, the coarse aggregates should be separated into required size fractions and reconstituted at the time of mixing to assure representative grading for the small test batches Under some conditions, for work of important magnitude, laboratory investigation may involve efforts to overcome grading deficiencies of the available aggregates. Undesirable sand grading may be corrected by (1) separation of the sand into two or more size fractions and recombining in suitable proportions, (2) increasing or decreasing the quantity of certain sizes to balance the grading, or (3) reducing excess coarse material by grinding or crushing Undesirable coarse-aggregate gradings may be corrected by (1) crushing excess coarser fractions, (2) wasting sizes that occur in excess, (3) supplementing deficient sizes from other sources, or (4) a combination of these methods Whatever grading adjustments are made in the laboratory should be practical and economically justified from the standpoint of job operation Usually, required aggregate grading should be consistent with that of economically available materials.

A3 4—Trial batch series

A3 4.1 The tabulated relationships in the body of this report may be used to make rough estimates of batch quantities for a trial mix. However, they are too generalized to apply with a high degree of accuracy to a specific set of materials If facilities are available, therefore, it is advisable to make a series of concrete tests to establish quantitative relationships for the materials to be used An illustration of such a test program is shown in Table A3 4 1

A3.4.2 First, a batch of medium cement content and usable consistency is proportioned by the described methods In preparing Mix No 1, an amount of water is used which will produce the desired slump even if this differs from the estimated requirement The fresh concrete is tested for slump and unit weight and observed closely for workability and finishing characteristics In the example, the yield is too high and the concrete is judged to contain an excess of sand

A3.4.3 Mix No 2 is prepared, adjusted to correct the errors in Mix No 1, and the testing and evaluation repeated In this case, the desired properties are achieved within close tolerances and cylinders are molded to check the compressive strength The information derived so far can now be used to select proportions for a series of additional mixes, No 3 to 6, with cement contents above and below that of Mix No 2, encompassing the range likely to be needed Reasonable refinement in these batch weights can be achieved with the help of corrections given in the notes to Table 5.3 7.1.

A3.4.4 Mix No 2 to 6 provide the background, including the relationship of strength to water-cement

TABLE A3.4.1—TYPICAL TEST PROGRAM TO ESTABLISH CONCRETE-MAKING PROPERTIES OF LOCAL MATERIALS

	Cubic yard batch quantities, lb						Concrete characteristics				
Mix No	Cement	Sand	Coarse Aggregate	Water Estimated	Water Used	Total used	Slump, in	Unit wt, lb per cu ft	Yield, cu ft	28-day Compressive strength, psi	Workability
1	500	1375	1810	325	350	4035	4	147 0	27 45	—	Oversanded
2	500	1250	1875	345	340	3965	3	147 0	26 97	3350	o k
3	400	1335	1875	345	345	3955	4.5	145 5	27 18	2130	o k
4	450	1290	1875	345	345	3960	4	146.2	27 09	2610	o k
5	550	1210	1875	345	345	3980	3	147.5	26 98	3800	o.k
6	600	1165	1875	345	345	3985	3.5	148.3	26 87	4360	o.k

ACI STANDARD

ratio for the particular combination of ingredients, needed to select proportions for a range of specified requirements.

A3.4.5 In laboratory tests, it seldom will be found, even by experienced operators, that desired adjustments will develop as smoothly as indicated in Table A3.4.1 Furthermore, it should not be expected that field results will check exactly with laboratory results. An adjustment of the selected trial mix on the job is usually necessary. Closer agreement between laboratory and field will be assured if machine mixing is employed in the laboratory. This is especially desirable if air-entraining agents are used since the type of mixer influences the amount of air entrained Before mixing the first batch, the laboratory mixer should be "buttered" or the mix "overmortared" as described in ASTM C 192 Similarly, any processing of materials in the laboratory should simulate as closely as practicable corresponding treatment in the field.

A3.4.6 The series of tests illustrated in Table A3 4 1 may be expanded as the size and special requirements of the work warrant Variables that may require investigation include alternative aggregate sources, maximum sizes and gradings, different types and brands of cement, admixtures, and considerations of concrete durability, volume change, temperature rise, and thermal properties

A3.5—Test methods

A3.5.1 In conducting laboratory tests to provide information for selecting concrete proportions, the latest revisions of the following methods should be used:

A3.5.1.1 *For tests of ingredients.*

Sampling hydraulic cement—ASTM C 183

Specific gravity of hydraulic cement—ASTM C 188

Sampling stone, slag, gravel, sand, and stone block for use as highway materials—ASTM D 75

Sieve or screen analysis of fine and coarse aggregates—ASTM C 136

Specific gravity and absorption of coarse aggregates—ASTM C 127

Specific gravity and absorption of fine aggregates—ASTM C 128

Surface moisture in fine aggregate—ASTM C 70

Total moisture content of aggregate by drying—ASTM C 566

Unit weight of aggregate—ASTM C 29

Voids in aggregate for concrete—ASTM C 30

Fineness modulus—Terms relating to concrete and concrete aggregates, ASTM C 125

A3.5.1.2 *For tests of concrete*

Sampling fresh concrete—ASTM C 172

Air content of freshly mixed concrete by the volumetric method—ASTM C 173

Air content of freshly mixed concrete by the pressure method—ASTM C 231

Slump of portland cement concrete—ASTM C 143

Weight per cubic foot, yield, and air content (gravimetric) of concrete—ASTM C 138

Concrete compression and flexure test specimens, making and curing in the laboratory—ASTM C 192

Compressive strength of molded concrete cylinders—ASTM C 39

Flexural strength of concrete (using simple beam with third-point loading)—ASTM C 78

Flexural strength of concrete (using simple beam with center point loading)—ASTM C 293

Splitting tensile strength of molded concrete cylinders—ASTM C 496

A3 6—Mixes for small jobs

A3.6.1 For small jobs where time and personnel are not available to determine proportions in accordance with the recommended procedure, mixes in Table A3 6 1 will usually provide concrete that is amply strong and durable if the amount of water added at the mixer is never large enough to make the concrete overwet These mixes have been predetermined in conformity with the recommended procedure by assuming conditions applicable to the average small job, and for aggregate of medium specific gravity Three mixes are given for each maximum size of coarse aggregate. For the selected size of coarse aggregate, Mix B is intended for initial use If this mix proves to be oversanded, change to Mix C, if it is undersanded, change to Mix A It should be noted that the mixes listed in the table are based on dry or surface-dry sand If the sand is moist or wet, make the corrections in batch weight prescribed in the footnote.

A3.6 2 The approximate cement content per cubic foot of concrete listed in the table will be helpful in estimating cement requirements for the job. These requirements are based on concrete that has just enough water in it to permit ready working into forms without objectionable segregation Concrete should slide, not run, off a shovel.

TABLE A3.6.1—CONCRETE MIXES FOR SMALL JOBS

Procedure Select the proper maximum size of aggregate (see Section 5 3 2) Use Mix B adding just enough water to produce a workable consistency If the concrete appears to be undersanded, change to Mix A and, if it appears oversanded, change to Mix C

Maximum size of aggregate, in	Mix designation	Approximate weights of solid ingredients per cu ft of concrete, lb				
			Sand*		Coarse aggregate	
		Cement	Air-entrained concrete†	Concrete without air	Gravel or crushed stone	Iron blast furnace slag
½	A	25	48	51	54	47
	B	25	46	49	56	49
	C	25	44	47	58	51
¾	A	23	45	49	62	54
	B	23	43	47	64	56
	C	23	41	45	66	58
1	A	22	41	45	70	61
	B	22	39	43	72	63
	C	22	37	41	74	65
1½	A	20	41	45	75	65
	B	20	39	43	77	67
	C	20	37	41	79	69
2	A	19	40	45	79	69
	B	19	38	43	81	71
	C	19	36	41	83	72

*Weights are for dry sand If damp sand is used, increase tabulated weight of sand 2 lb and, if very wet sand is used, 4 lb

†Air-entrained concrete should be used in all structures which will be exposed to alternate cycles of freezing and thawing Air-entrainment can be obtained by the use of an air-entraining cement or by adding an air-entraining admixture If an admixture is used, the amount recommended by the manufacturer will, in most cases, produce the desired air content

APPENDIX 4— HEAVYWEIGHT CONCRETE MIX PROPORTIONING

A4.1—Concrete of normal placeability can be proportioned for densities as high as 350 lb per cu ft by using heavy aggregates such as iron ore, barite, or iron shot and iron punchings. Although each of the materials has its own special characteristics, it can be processed to meet the standard requirements for grading, soundness, cleanliness, etc. The acceptability of the aggregate should be made depending upon its intended use. In the case of radiation shielding, determination should be made of trace elements within the material which may become reactive when subjected to radiation. In the selection of materials and proportioning of heavyweight concrete, the data needed and procedures used are similar to that required for normal weight concrete except that the following items should be considered.

A4.1.1—In selecting an aggregate for a specified density, the specific gravity of the fine aggregate should be comparable to that of the coarse aggregate in order to lessen settlement of the coarse aggregate through the mortar matrix. Typical materials used as heavy aggregates include the following:

Material	Description	Specific gravity	Approx. concrete unit wt (lb/cu ft)
Limonite Goethite	Hydrous iron ores	3.4 - 3.8	180 - 195
Barite	Barium sulfate	4.0 - 4.4	205 - 225
Ilmenite Hematite Magnetite	Iron ores	4.2 - 4.8	215 - 240
Iron	Shot, pellets, punchings, etc.	6.5 - 7.5	310 - 350

A4.1.2—When the concrete in service is to be exposed to a hot, dry environment, it should be proportioned so that the fresh unit weight is at least 10 lb per cu ft higher than the required dry unit weight.

A4.1.3—When entrained air is required to resist conditions of exposure, allowance must be made for the loss in weight due to the space occupied by the air. To achieve adequate consolidation using high frequency vibrators and close insertion intervals, without the excessive loss of entrained air, plastic concrete should be designed for a high air content to offset this loss during placement.

A4.1.4—Heavyweight concrete is often used for radiation shielding. In this case, the aggregate type and concrete weight should be selected consistent with the type of radiation involved. Generally speaking the greater the mass the better are the shielding properties against gamma and beta rays. However, neutron attenuation depends more on the specific elements present in the concrete, i.e., hydrogen, carbon, boron, etc.

A4.1.5—Ferrophosphorous and ferrosilicon (heavyweight slags) materials should be used only after thorough scrutiny. Hydrogen evolution in heavyweight concrete containing these aggregates has been known to result in a reaction of a self-limiting nature, producing over 25 times its volume of hydrogen before the reaction ceases.

A4.1.6—In Section 5.3.7 (Step 7) caution must be exercised if the weight method (Section 5.3.7.1) is used to estimate the fine aggregate batch weight. The values in Table 5.3.7.1 must be corrected for overall aggregate specific gravity since the table is based on an average aggregate specific gravity of 2.7. Therefore, it is recommended that the required amount of fine aggregate be determined by the absolute volume procedure (Section 5.3.7.2).

A4.2—*Production and quality control.* The technique and equipment for producing heavyweight concrete are the same as used with normal weight concrete. In the selection of heavyweight materials and combinations thereof for the purposes of proportioning specification concretes, attention must be directed to aggregate effects on placeability, strength, and durability of the concrete. Testing and quality control measures assume greater importance than with normal weight concrete. Control of aggregate grading is essential because of the effect on the placing and consolidating properties of concrete, and on the unit weight of the concrete. In enforcing strict quality control special attention should be paid to the following:

A4.2.1—Prevention of contamination with normal weight aggregate in stockpiles and conveying equipment.

A4.2.2—Purging of all aggregate handling and batching equipment, premixers and truck mixers, before batching and mixing heavyweight concrete.

A4.2.3—Accuracy and condition of conveying and scale equipment, aggregate storage and concrete batching bins. Due to the greater weight of heavyweight aggregate, the permissible volume batched in a bin is considerably less than the design capacity. For example: a 100 ton aggregate bin designed for 75 cu yd of normal weight aggregate should not be loaded with more than 25 to 55 cu yd for the range of specific gravities shown in Section A4.1.1.

A4.2.4—Condition and loading of mixing equipment. For concrete of a weight range of approximately 4800 to 9500 lb per cu yd, the capacity of a truck mixer, without overloading, is reduced from 20 to 60 percent.

A4.2.5—Accurate aggregate proportioning to maintain w/c ratio. Degradation of some coarse heavyweight aggregates, iron ores in particular, is another production problem which should be carefully controlled. Either the coarse aggregate should be rescreened immediately prior to incorporation into the concrete or adjustments made in the mixture proportions that compensate for the increased percentage of fines in the coarse aggregate, caused by aggregate breakdown during handling. Therefore, caution should be exercised and frequent gradation checks made on stockpiled aggregates.

A4.2.6—Frequent checks of fresh unit weight.

A4.2.7—Design and construction of forms to handle additional weight of concrete.

A4.2.8—Vibrators for consolidation.

A4.3—*Example problem.* Concrete is required for counterweights on a lift bridge not subjected to freezing and thawing conditions. An average 28-day compressive strength of 4500 psi will be required. Placement conditions permit a slump of 2 to 3 in. and a maximum size aggregate of 1 in. The design of the counterweight requires a dry unit weight of 230 lb per cu ft. An investigation of economically available materials has indicated the following:

Cement	—Type I (non-air-entraining)
Fine aggregate	—Specular Hematite
Coarse aggregate	—Ilmenite

The table in Section 4.1.1 indicates that this combination of materials may result in a dry unit weight of 215 to 240 lb per cu ft. The following properties of the aggregates have been obtained from laboratory tests:

	Fine aggregate	Coarse aggregate
Fineness modulus	2 30	—
Specific gravity (Bulk SSD)	4 95	4 61
Absorption (percent)	0 05	0 08
Dry rodded weight	—	165 lb per cu ft
Maximum size	—	1 in

Employing the sequence outlined in Section 5 of this recommended practice, the quantities of ingredients per cubic yard of concrete are calculated as follows:

A4.3.1 *Step 1.* As indicated above, the desired slump is 2 to 3 in.

A4.3.2 *Step 2.* The available aggregate sources have been indicated as suitable, and the coarse aggregate will be a well-graded and well-shaped crushed ilmenite with a maximum size of 1 in.

A4.3.3 *Step 3* By interpolation in Table 5 3 3, non-air-entrained concrete with a 2 to 3 in slump and a 1 in maximum size aggregate requires a water content of approximately 310 lb per cu yd The estimated entrapped-air is 1 5 percent (Non-air-entrained concrete will be used because (1) the concrete is not exposed to severe weather, and (2) a high air content could reduce the dry unit weight of the concrete)

Note. Table 5 3 3 values for water requirement are based on the use of well-shaped crushed coarse aggregates Void content of compacted dry fine or coarse aggregate can be used as an indicator of angularity. Void contents of compacted 1 in coarse aggregate of significantly more than 40 percent indicate angular material which will probably require more water than that listed in Table 5 3 3 Conversely rounded aggregates with voids below 35 percent will probably need less water.

A4.3 4 *Step 4* From Table 5 3 4(a) the water-cement ratio needed to produce a strength of 4500 psi in non-air-entrained concrete is found to be approximately 0 52

A4.3.5 *Step 5* From the information derived in Steps 3 and 4, the required cement content is calculated to be 310/0 52 = 596 lb per cu yd.

A4.3 6 *Step 6* The quantity of coarse aggregate is estimated by extrapolation from Table 5 3 6. For a fine aggregate having a fineness modulus of 2 30 and a 1 in maximum size aggregate, the table indicates that 0 72 cu ft of coarse aggregate, on a dry-rodded basis, may be used in each cubic foot of concrete. For a cubic yard, therefore, the coarse aggregate will be 27 × 0 72 = 19 44 cu ft Since the dry-rodded unit weight of the coarse aggregate is 165 lb per cu ft, the dry weight of coarse aggregate to be used in a cubic yard of concrete would be 19 44 × 165 = 3208 lb The angularity of the coarse aggregate is compensated for in the ACI proportioning method through the use of the dry-rodded unit weight, however, the use of an extremely angular fine aggregate may require a higher proportion of fine aggregate, an increased cement content, or the use of air-entrainment to produce the required workability.

A4.3.7 *Step 7* For heavyweight concrete, it is recommended that the required fine aggregate be determined on the absolute volume basis With the quantities of cement, water, air and coarse aggregate established, the sand content can be calculated as follows:

Volume of water	$= \frac{310}{62\ 4}$	= 4 97 cu ft
Volume of air	= 0 015 × 27	= 0 40 cu ft
Solid volume of cement	$= \frac{596}{3\ 15 \times 62\ 4}$	= 3 03 cu ft
Solid volume of coarse aggregate	$= \frac{3208}{4\ 61 \times 62\ 4}$	= 11 15 cu ft
Total volume of all ingredients except sand		= 19 55 cu ft
Solid volume of sand	= 27 − 19 55	= 7 45 cu ft
Required weight of sand	= 7 45 × 4 95 × 62 4 = 2301 lb	

A4.3.8 *Step 8* Tests indicate total moisture of 0.15 percent in the fine aggregate and 0 10 percent in the coarse aggregate, therefore, the adjusted aggregate weights become:

Fine aggregate (wet) = 1 0015 × 2301 = 2304 lb
Coarse aggregate (wet) = 1 0010 × 3208 = 3211 lb

Absorbed water does not become part of the mixing water and must be excluded from the adjustment in added water Thus surface water contributed by the fine aggregate amounts to 0 15 − 0 05 = 0 10 percent, by the coarse aggregate 0 10 − 0 08 = 0 02 percent. The estimated requirement for added water, therefore becomes:

$$310 - 2301(0\,001) - 3208(0.0002) = 307 \text{ lb}$$

A4 3 9 *Step 9* The resulting estimated proportions by weight of the heavyweight concrete becomes:

Cement	= 596 lb
Fine aggregate (wet)	= 2304 lb
Coarse aggregate (wet)	= 3211 lb
Water	= 307 lb
Estimated unit wt (fresh)	= 6418/27 = 237 7 lb per cu ft

A4 4—The above heavyweight concrete proportioned mixture was actually used for approximately 5060 cu yd Field adjustments resulted in the following actual batch weights.

Cement	590 lb
Fine aggregate	2310 lb
Coarse aggregate	3220 lb
Water	285 lb (plus a water-reducing agent)

The actual field test results indicated the concrete possessed the following properties:

Unit weight (fresh)	235 7 lb per cu ft
Air content	2 8 percent
Slump	2½ in
Strength	5000 psi at 28 days

APPENDIX D

GUIDANCE FOR SELECTING TYPES OF CONCRETE MIXES FOR SPECIAL ENVIRONMENTS

6-14 Corrosion Protection for Concrete

.1 *Preliminary Report*

The minimum cover for reinforcing steel in concrete structures and the type of cement to be used is dependent upon the environment. The Districts will normally investigate the bridge sites for chloride and sulfate content of the soil. The preliminary report writer will obtain the site corrosion data from the District materials report and include it in the preliminary report.

.2 *Chloride Protection*

The minimum covering measured from the surface of the concrete to the face of any reinforcing bar shall be as shown in Table A under normal environment. For those structures subject to marine or high chloride environment the covering to the face of the bar shall be increased to the values listed under the appropriate heading.

Table A

Minimum Covering for Reinforcing Steel(1)

	Environment		
Structure Member	Normal (inch)	Marine(2) (inch)	High Chlorides (above 1000 ppm)
Piles	2	3(3)	
Footings	3	4	
Walls	2	4	
Outside face of Box Girder Webs	1	3	
I and T Beam Webs	1	3	See
Bottom Flange Bulb of I or T Beam	1½	3	
Curbs and Railings	1	2	Table B
Exposed slab Riding surface Bottom surface	 1½(4) 1 – 1½	 2 1½	
Columns Flat face Corner	 2 2	 4 5	
Caps Flat face Corner	 2 2	 4 5	

(1) A protective coating shall be applied to steam cured members in a marine environment.

(2) A marine environment is defined as in or within about 1,000 feet of the ocean or tidal water. Concrete shall contain a minimum of seven sacks of cement per cubic yard.

(3) A protective coating shall be applied to that portion of pile which will remain above the ground or water line. Limits of protective coating shall be shown on the plans.

(4) 2″ in freeze-thaw environments.

Table B

	Chlorides (PPM)[5]		
Structural Member	1000–5000 (inch)	5000–10000 (inch)	10000–10000[6] (inch)
Piles	3	3	3
Footings	3	4	5
Walls	2	3	5
Columns			
Flat face	3	4	5
Comer	3	5	6

(5) The coverings shown are for members in a chloride environment and apply only to those particular members that are in direct contact with the soil or water. Concrete shall contain a minimum of seven sacks of cement per cubic yard.

(6) Including ocean water–submerged or splash.

.3 *Sulfate Resistance*

Resistance to concrete attack by high sulfate content in the soil is provided by increasing the Type II cement factor or using Type V cement. The cement content and the cement types to be used for concrete which will be in contact with soils containing sulfates will be as shown in Table C. The limits shall be shown on the plans.

Table C

Sulfates (SO) PPM	Sacks per Cubic Yard (min)	Cement Type
0–2000	6	II
2000–15000	7	II
Over 15000	7	V

.4 *Freeze-thaw Resistance*

All concrete placed in a freeze-thaw environment shall contain a minimum of seven sacks of cement per cubic yard and 6 percent

APPENDIX E

MARINE PRODUCT MANUFACTURERS

1. Floating Piers, Gangways, Docks and Docking Systems.

American Marina Engineering Co.
3233 SW 2nd Avenue
Fort Lauderdale, FL 33315

Armco Steel Corp.
1001 Grove Street
Middleton, OH 45042

C.M. Beuthe Co.
120-B Cloverdale Avenue
Concord, CA 94518

Dock Masters Inc.
PO Box 1687
Huntington Beach, CA 92647

Hardwick Engineering and Associates
729 East Willow Street
Long Beach, CA 90806

Hallsten Supply Co.
PO Box 41036
Sacramento, CA 95841

Harbor Host Corp.
1027 East Algonquin Road
Arlington Heights, IL 60005

International Marina Systems, Inc.
PO Box 7531
Tulsa, OK 74105

Koppers Company, Inc.
Forest Products Division
750 Koppers Building
Pittsburgh, PA 15219

Marina Products Manufacturing, Inc
221 SW 14th Court
Fort Lauderdale, FL 33315

MEECO Marinas, Inc.
PO Box 66
Carrollton, TX 75006

Pacific Gangways
Pacific Pipe Co.
PO Box 4011, Bay Shore Station
Oakland, CA 94623

Poly Sintering, Inc.
Commercial Flotation Division
1624 15th Avenue West
Seattle, WA 98119

Steel-N-Foam Docks, Inc.
PO Box 737, 501 South Valley
Kansas City, KA 66119

Tomlinson Industries, Inc.
13700 Broadway
Cleveland, OH 44125

Trautwein Bros.
2410 Newport Boulevard
Newport Beach, CA 92660

United Flotation Systems
2400 Fairwood Avenue
Columbus, OH 43207

2. Perimeter Protection, Erosion Control.

Cathage Mills Inc.
Erosion Control Division
124 West 66th Street
Cincinnati, OH 45216

Construction Techniques, Inc.
1111 Superior Building
Cleveland, OH 44114

Griffolyn Company, Inc.
PO Box 33248
Houston, TX 77033

Kaiser Aluminum & Chemical Sales, Inc
Kaiser Center, 300 Lakeside Drive
Oakland, CA 94604

Macafferri-Gabions of America, Inc.
One Lefrak City Plaza
Flushing, NY 11368

Macafferri-Gabions of America, Inc.
2470 Westlake North
Seattle, WA 98109

3. Electrical Components, Corrosion Control.

The Arrow-Hart & Hegeman Electric Co.
Hartford, CT 06106

Harvey Hubbell Inc.
Bridgeport, CT 06602

Kiekhaefer Mercury
Division of Brunswick Corp.
Fond du Lac, WI 54935

Occidental Coating Co.
7755 Deering Avenue
Canoga Park, CA 91304

Pennwalk Automatic Power
213 Hutcheson Street
Houston, TX 77003

The Pyle-National Co.
1334 North Kostner Avenue
Chicago, IL 60651

Wide Lite
PO Box 191
Houston, TX 77001

4. Slings, Hoists, Lifts and Winches.

A. C. Hoyle Co.
Box 589
Iron Mountain, MI 49801

Acme Marine Hoist, Inc.
658 Rockaway Turnpike
Lawrence, Long Island, NY 11559

Clark Equipment
Industrial Truck Division
Battle Creek, MI 49016

C. M. Hoist
Division, Columbus McKinnon Corp.
Freemont Avenue
Tonawanda, NY 14150

Electrolift, Inc.
204 Sargeant Avenue
Clifton, NJ 07013

Midland-Ross Corp.
RPC Division
PO Box 400
Roxboro, NC 27573

Midwest Industries, Inc.
Marine Division
Ida Grove, IA 51445

Minuteman Sales & Service
PO Box 1
Plymouth, MA 02360

5. Wood Products.

J.H. Baxter
1700 South El Camino Real
San Mateo, CA 94402

The Dow Chemical Co.
Designed Products Department
Midland, MI 48640

Koppers Company, Inc.
Forest Products Division
750 Koppers Building
Pittsburgh, PA 15219

Permapost Products Co.
25600 SW Tualatin Valley Highway
Hillsboro, OR 97123

6. Sanitary Pumpout Systems.

American Marina Engineering Co.
3233 SW 2nd Avenue
Fort Lauderdale, FL 33315

Kenton Equipment Co.
Marine Division
3280 Kurtz Street
San Diego, CA 92110

Sani-Station
Sta-Rite Industries, Inc.
Delavan, WI 53115

7. De-Icing Systems.

Sta-Rite Industries, Inc.
Delavan, WI 53115

Schramm, Inc.
Aeration Division
612 North Garfield Avenue
West Chester, PA 19830

8. Dredging Equipment.

Ellicott Machine Corp.
1615 Bush Street
Baltimore, MD 21230

National Car Rental System, Inc.
Mud Cat, Hydro-Soil Division
5501 Green Valley Drive
Minneapolis, MN 55431

Pneuma Dredging Equipment
Civil & Marine Engineering Co., International
90 Broad Street
New York, NY 10004

Sauerman Bros., Inc.
620 South 28th Avenue
Bellwood, IL 60104

Vortex-Hydra
Ewing Sales Agency
18106 Redbud Circle
Fountain Valley, CA 92708

APPENDIX F

FEDERAL AGENCIES SPONSORING PROGRAMS WHICH MAY PROVIDE AID AND ASSISTANCE IN DEVELOPING SMALL-CRAFT FACILITIES

Information Staff
Farmers Home Administration
Department of Agriculture
Washington, DC 20250

Director
Office of Business Economics
Department of Commerce
Washington, DC 20230

Information Service
Department of Housing and
Urban Development
451 Seventh Street SW
Washington, DC 20410

Department of Interior
Conservation Education Office
Bureau of Sport Fisheries and Wildlife
18th and C Streets, NW
Washington, DC 20240

Division of Information
National Park Service
Interior Building
Washington, DC 20240

Commissioner of Reclamation
Department of Interior
Washington, DC 20240

Office of Public Affairs
National Oceanic and
Atmospheric Administration
6010 Executive Boulevard
Rockville, MD 20852

U.S. Army, Corps of Engineers
Department of the Army
Washington, DC 20314

Division of Information and Education
Forest Service
Department of Agriculture
Washington, DC 20250

Office of Administration and
Program Analysis
Economic Development Administration
Main Commerce Building
Washington, DC 20230

Environmental Protection Agency
1626 K Street, NW
Washington, DC 20460

Office of Information
Bureau of Land Management
Department of Interior
Washington, DC 20240

Organization Division
Bureau of Outdoor Recreation
Interior Building
Washington, DC 20240

Office of Water Resources
Department of Interior
Washington, DC 20240

Department of Transportation
400 Seventh Street, SW
Washington, DC 20590

U.S. Coast Guard
400 Seventh Street, SW
Washington, DC 20590

APPENDIX G

STATE AGENCIES HOLDING JURISDICTION OVER MATTERS CONCERNING SMALL CRAFT AND SMALL CRAFT FACILITIES

Department of Conservation
Water Safety Division
Montgomery, AL 36104

Department of Public Works
Division of Water and Harbors
Pouch Z
Juneau, AK 99801

Coordinator, Game and Fish Department
2222 West Greenway Road
Phoenix, AZ 85023

Information and Education Division
Game and Fish Commission
Little Rock, AR 72201

State of California Resources Agency
Dept. of Navigation and Ocean Development
1416 Ninth Street
Sacramento, CA 95814

Chief Warden
Game, Fish and Parks Department
6060 Broadway
Denver, CO 80216

Director, Boating Commission
Department of Environmental Protection
State Office Building
Hartford, CT 06115

Department of Natural Resources and Environmental Control
Dover, DE 19901

Department of Natural Resources
Division of Marine Resources
Larson Building
Tallahassee, FL 32304

Coordinator, Special Services
State Game and Fish Commission
Trinity-Washington Building
Room 710
Atlanta, GA 30334

Department of Transportation
Harbors Division
PO Box 397
Honolulu, HI 96809

Idaho Department of Parks
Statehouse
Boise, ID 83707

Department of Conservation
605 State Office Building
400 South Spring Street
Springfield, IL 62706

Enforcement Division
Department of Natural Resources
606 State Office Building
Indianapolis, IN 46209

Superintendent of Waters Section
State Conservation Commission
300 4th Street
Des Moines, IA 50319

Field Services Division
Forestry, Fish and Game Commission
Box 1028
Pratt, KS 67124

Department of Fish and Wildlife Resources
State Office Building Annex
Frankfort, KY 40601

Department of Public Safety
State Office Building Annex
Frankfort, KY 40601

APPENDIX G—Continued

Supervisor of Revenue
Wildlife and Fisheries Commission
Wildlife and Fisheries
400 Royal Street
New Orleans, LA 70130

Bureau of Watercraft
Registration and Safety
State Office Building
Augusta, ME 04330

Chief, Boating Division
Department of Chesapeake Bay Affairs
1825 Virginia Street
Annapolis, MD 21401

Director, Marine and Recreation Division
100 Nashua Street
Boston, MA 02114

Department of Natural Resources
Stevens T. Mason Building
Lansing, MI 48926

Department of Natural Resources
Centennial Office Building
St. Paul, MN 55101

Mississippi Boat and
Water Safety Commission
Robert E. Lee Building, Room 403
Jackson, MS 39201

Missouri Boat Commission
PO Box 603
Jefferson City, MO 65101

Department of Fish and Game
Helena, MT 59601

Nebraska Game and Parks Commission
2200 North 33rd Street
PO Box 30370
Lincoln, NB 68503

Department of Fish and Game
1100 Valley Road
Reno, NV 89510

Department of Safety
Division of Safety Services
Concord, NH 03301

Supervisor Motorboat Numbering
Department of Environmental Protection
PO Box 1889
Trenton, NJ 08625

State Park and Recreation Commission
141 East DeVargas Street
Santa Fe, NM 87501

New York State Parks and Recreation
State Campus, Building 2
Albany, NY 12226

Wildlife Resources Commission
Box 2919
Raleigh, NC 27602

North Dakota State Park Service
Ft. Lincoln State Park
Route 2, Box 139
Mandan, ND 58554

Department of Natural Resources
Division of Watercraft
1350 Holly Avenue
Columbus, OH 43212

Director, Lake Patrol Division
Department of Public Safety
PO Box 11415
Oklahoma City, OK 73105

State Marine Board
Agriculture Building
Salem, OR 97310

Pennsylvania Fish Commission
PO Box 1673
Harrisburg, PA 17120

Division of Coastal Resources
Veterans Memorial Building
83 Park Street
Providence, RI 02903

Wildlife Resources Department
Division of Boating
15 Lockwood Boulevard
Charleston, SC 29401

Department of Game, Fish and Parks
Pierre, SD 57501

Tennessee Game and Fish Commission
Ellington Agricultural Center
PO Box 40747
Nashville, TN 37220

Texas Parks and Wildlife Department
John H. Regan Building
Austin, TX 78701

Boating Chief
Division of Parks and Recreation
1596 W. North Temple Street
Salt Lake City, UT 84116

Agency of Environmental Conservation
Department of Water Resources
Montpelier, VT 05602

Commission of Outdoor Recreation
8th Street Office Building
803 E. Broad Street
Richmond, VA 23219

Washington State Parks and
Recreation Commission
PO Box 1128
Olympia, WA 98504

Chief, Law Enforcement Section
Department of Natural Resources
State Office Building
Charleston, WV 25305

Department of Natural Resources
PO Box 450
Madison, WI 53701

Game and Fish Commission
Cheyenne, WY 82001

Harbor Master
Metropolitan Police Department
Harbor Precinct
550 Main Avenue, SW
Washington, DC 20024

Ports Authority
GPO Box 2829
San Juan, PR 00936

Ports Authority
Maritime Division
Charlotte Amalie
St. Thomas Island, VI

APPENDIX H

Table H. Summary of Significant Services, Programs and Controls Offered or Executed within the States and Territories

State or Territory	Control, Regulation or Assistance[1]									
	A	B	C	D	E	F	G	H	I	J
Alabama	X	X	X			X	X			X
Alaska			X		227[2]					
Arizona	X		X				X			
Arkansas	X									
California	X	X	X	X		X	X	X	226[2]	
Colorado	X	X	X							
Connecticut[3]										
Delaware	X									
Florida	X	X	X					224[2]		
Georgia	X	X	X				X			
Hawaii	X	X	X		227[2]					
Idaho	X									
Illinois	X				228[2]					
Indiana	X	X	X							
Iowa[3]										
Kansas	X									
Kentucky	X	X	X					X		
Louisiana[4]										
Maine	X									
Maryland	X	X	X		X		X			X
Massachusetts	X	X	X							X
Michigan	X	X	X		X		X	X		
Minnesota	X	X	X				228[2]			
Mississippi	X	X	X							
Missouri	X									
Montana	X	X	X							
Nebraska	X	X	X					224[2]		
Nevada	X						X	X		
New Hampshire	X	X	X							

1. Alphabet Code (See end of this Table).
2. Page in text where specific item is discussed for particular State.
3. No response from State before publication.
4. State sponsors no applicable services or programs.

Table H. Summary of Significant Services, Programs and Controls Offered or Executed within the States and Territories—Continued

State or Territory	Control, Regulation or Assistance[1]									
	A	B	C	D	E	F	G	H	I	J
New Jersey	X	X	X				X			X
New Mexico	X						228[2]			
New York	X	X	X		X		X	X	224[2]	
North Carolina	X	X	X				228[2]			
North Dakota	X	X								
Ohio	X	X	X		226[2]					
Oklahoma		X	X	X	X	X	X			
Oregon	X	X	X		226[2]					
Pennsylvania	X	X	X		227[2]					
Rhode Island			X							
South Carolina	X	X	X							
South Dakota	X		X		X		X	X		
Tennessee	X									
Texas	X							X		
Utah		X								
Vermont	X	X	X					X		
Virginia		X	X					X		
Washington	X				226[2]			X		
West Virginia[3]										
Wisconsin	X	X	X				X			
Wyoming	X	X	X							
Washington, D.C.[3]										
Puerto Rico	X	X	X							
Virgin Islands[3]										

1. Alphabet Code
   - A. State exercises control over navigation on inland waters.
   - B. State sponsors navigation classes or boating safety programs.
   - C. State provides printed data on boating activities or availability of small-craft facilities.
   - D. State provides loan program for civic marina development.
   - E. State provides grant program for civic marina development.
   - F. State provides loan program for civic launching or access installations.
   - G. State provides grant program for civic launching or access installations.
   - H. State issues design standards for marina construction.
   - I. State requires economic evaluation previous to marina development.
   - J. State provides maintenance for channels, etc., into civic or private harbors.
2. Page in text where specific item is discussed for particular State.
3. No response from State before publication.
4. State sponsors no applicable services or programs.

APPENDIX I

COASTAL SMALL-CRAFT HARBORS REGULATION ORDINANCE

Submitted by
California Marine Parks and Harbors Association

An Ordinance regulating the use of–harbors and maritime facilities; setting penalties for violations of any of the provisions hereof.

This model ordinance is designed for use along ocean shores and in tidal estuaries. If adapted to use in inland waters, it should be modified with respect to water surface fluctuations and certain regional or local area terminology. Unapplicable sections should be deleted and possible others should be added to meet the local area requirements.

ARTICLE I

General Provisions

Sec. 1. Short Title: This ordinance shall be known and may be cited as the "–––––––––".

Sec. 2. Applicability: The provisions of this Ordinance and any rules and regulations adopted pursuant thereto shall be applicable, and shall govern, the harbor(s) and all other maritime facilities under the jurisdiction of–––––––––. This Ordinance shall be subordinate to existing Federal and State regulations governing the same matters and is not intended to preempt other valid laws.

Sec. 3. Invalidity of Provisions: If any provisions of this Ordinance is held invalid or inoperative, the remainder shall continue in full force and effect as though such invalid or inoperative provisions had not been made.

Sec. 4. Authority: Whenever, by the provisions of this Ordinance, a power is granted to the–––––––––or a duty is imposed upon him, the power may be exercised or duty performed by a deputy of the–––––––––or by a person authorized pursuant to law, unless it is expressly otherwise provided.

Sec. 5. Facilities, Control of Use: The–––––––––is vested with authority over and control of all floats, wharves, docks, and other facilities owned, leased, controlled, constructed or maintained by the–––––––––, or constructed or maintained by a lessee in any–––––––––harbor or any other maritime facility for the purpose of causing to be corrected any condition.

Sec. 6. Rules, Regulations and Orders: The---------shall have the power and duty to enforce the laws, ordinances, traffic and safety regulations covering usage of------ harbors and other maritime facilities, under his jurisdiction.

Sec. 7. Chief of Harbor (or Other Designation): The---------officer of the Harbor---------or authorized agent acting under the orders and jurisdiction of the---------shall have full authority in enforcement of all laws, ordinances and regulations affecting the---------harbor or other maritime facility and waterways and beaches within such harbors and facilities, and he may cite alleged offenders to appear before the---------Court.

Sec. 8. Violations: Violation of this Ordinance is a misdemeanor punishable by a fine of not more than Five Hundred Dollars ($500) or by imprisonment in the---------jail for not more than six (6) months or by both such fine and imprisonment. A repetition of continuation of any violation of any provisions of this ordinance or of any order or direction of the---------on successive days constitutes a separate offense for each day during any portion of which such violation is committed, continued, or permitted.

ARTICLE II

Definitions

Access Service Route: Shall mean any access roads and/or easements designated or identified by---------for use by authorized emergency or utility vehicles.

Auxiliary: Shall mean any vessel having both sails and either an inboard or outboard motor and which may be propelled by its sails or by its motor, or both.

Basin: Shall mean a naturally or artificially enclosed or nearly enclosed body of water where small craft may lie.

Beach: Shall mean a public or private beach area bordering the waters of a--------- harbor or maritime facility.

Camp Cars: Shall mean a vehicle with or without motor power which is designated for permanent or temporary human habitation and which contains sleeping facilities, plumbing, heating, cooking (whether attached or portable) or electrical equipment. Any such camp car shall be subject to the provisions of Article VIII, Section 66.

Carrying Passengers for Hire: Shall mean the carriage of a person by vessel for valuable consideration, whether directly or indirectly flowing to the owner, charterer, operator, agent or any other person interested in the vessel.

Commercial Vessel: Shall mean any vessel used or engaged for any type of commercial venture, including but not limited to the display of advertising or the carrying of cargo and/or passengers for hire.

Distress: Shall mean a state of disability or a present or obviously imminent danger which if unduly prolonged could endanger life or property.

Emergency: Shall mean a state of imminent or proximate danger to life or property in which time is of the essence.

Entrance Channel: Shall mean all that portion described as follows: (Insert own description.)

Facilities: Shall mean any and all facilities of a harbor or maritime facility either publicly or privately owned that are intended primarily to be used by or for the service of small craft (including ramps, hoists, parking areas, leased water areas, concessions and service facilities) located on land or in the waters of the---------under jurisdiction of the---------in either---------or---------territory.

Fairway: Shall mean the parts of a waterway kept open and unobstructed for navigation.

Fire Department: Shall mean the---------of the---------.

Float: Shall mean any floating structure normally used as a point of transfer for passengers and goods and/or for mooring purposes.

Harbormaster: Shall mean the Chief Officer of the Harbor Patrol or competent member of the Harbor Patrol that he may designate to act in his stead in his absence.

Harbor Patrol: Shall mean the organization comprising all members regularly employed by the---------as Harbor Patrolmen or Harbor Patrol Officers.

Harbor Patrolman: Shall mean a harbor policeman as referred in Section---------of the State Harbors and Navigation Code, who, when qualified, shall have the authority of "Peace Officer."

Head of Operating Agency: Shall mean the---------.

Live Bait Receiver: Shall mean a water-ventilated container immersed in water, the purpose of which is to confine live bait fish.

Maritime Facility: Shall mean any facility affecting the use and operations of pleasure or commercial vessels bordering on, concerned with, related to a protected water area of the---------ocean that is owned, managed or controlled by the---------or under the jurisdiction of the---------in either incorporated or unincorporated territory.

Moor: Shall mean to secure a vessel other than by anchoring.

Mooring: Shall mean (1) a place where buoyant vessels are secured other than a pier; (2) the equipment used to secure a vessel; and (3) the process of securing a vessel other than by anchoring.

Mooring Buoy: Shall mean an appliance used to secure to the bottom by anchors and provided with attachments to which a vessel may be secured by use of its anchor chain or mooring lines.

Operating Agency: Shall mean---------.

Public Agency: Shall mean---------.

Public Area: Shall mean all areas of any harbor except those areas under specific lease to private persons of firms or owned privately.

Regulatory Marker or Waterway Marker: Shall mean any of the waterway markers defined as "regulatory markers" in the State Administrative Code, Title---------; Article---------.

Slip: Shall mean berthing space for a single vessel alongside a pier, finger float, or walkway.

Shore: Shall mean that part of the land in immediate contact with a body of water, including the area between high and low water lines.

Shall and May: "Shall" is mandatory "May" is permissive.

State: Shall mean the State of---------.

Stray Vessel: Shall mean (1) an abandoned vessel; (2) a vessel the owner of which is unknown; or (3) a vessel underway without a competent person in command.

To Anchor: Shall mean to secure a vessel to the bottom within a body of water by dropping an anchor or anchors or other ground tackle.

Underway: Shall mean the condition of a vessel not at anchor; without moorings; and not made fast to the shore nor aground.

Waterway: Shall mean any water area providing access from one place to another, principally a water area providing a regular route for water traffic, that is owned, managed, or controlled by the---------or under the jurisdiction of the---------either in incorporated or unincorporated territory.

Waters of a Harbor: Means all waters of any harbor that is owned, managed, or controlled by the---------or under the jurisdiction of the---------in which the tide ebbs and flows, whether or not the ordinary or mean high tide line of the--------- Ocean has been fixed by ordinance, statute, court action or otherwise and whether or not the lands lying under said tidal water are privately or publicly owned.

ARTICLE III

General Boating and Traffic Control Regulations

Sec. 9. Traffic Control Authority: The---------shall have authority to control water-borne traffic in any portion of the waters of a harbor or maritime facility under his jurisdiction by use of authorized State regulatory markers, signal, orders or directions any time preceding, during and after any race, regatta, parade or other special event held in any portion of the waters of a harbor or maritime facility or at any time when the--------- deems it necessary in the interest of safety of persons and vessels or other property, and it shall be unlawful for any person to willfully fail or refuse to comply with any authorized State regulatory marker utilized by---------, or with any signal, orders or directions of the---------.

Sec. 10. Basic Speed Law: The operation of any vessel within the harbor area in excess of posted speed limits or, in the absence of such limits, in a manner to create a wash which endangers persons or property, shall constitute a violation of this Ordinance; provided that special written permission may be granted to conduct and engage in water sports and regattas in specific designated areas.

Sec. 11. Permits for Races and Special Events: It shall be a violation of this Ordinance for any person to engage or participate in a boatrace, watersport, exhibition, or other special event unless especially authorized by permit from the---------who shall have authority to issue such permits and to attach such conditions thereto as, in his opinion, are necessary and reasonable for the protection of life and property.

Sec. 12. Reverse Gears: It shall be unlawful for any person to operate on the waters of any harbor or maritime facility any power or motor driven vessel that does not have a means to reverse or stop the vessel, except when participating as a contestant per Section 10. Motor vessels shall have aboard and ready for instant use a suitable anchor with chain or line affixed.

Sec. 13. Rafting or Nesting: A permit shall be obtained from---------for the purpose of rafting or nesting vessels during a regatta or boatrace. This permit must be obtained from---------at least---------hours before the regatta schedule to be held.

Sec. 14. Insurance: As a condition for granting such permit, the---------shall determine as a condition of a permit for racing whether insurance shall be carried by participants and if required shall be for the total period of time and shall be in an amount not less than $---------for bodily injury to one person; and $---------for bodily injury in any one accident; and $---------for property damage in any one accident. The policy providing such insurance shall be in the name of---------as additional insured (insureds). A certificate evidencing that such insurance is in force and will remain in force for the period of time such permit shall be in effect.

ARTICLE IV

General Regulations

Sec. 15. Liability:

(a) Boat Owner: Any person using the facilities within the limits of a harbor or maritime facility shall assume all risk of damage or loss to his property and he-------- assumes no risk on account of fire, theft, Act of God, or damages of any kind to vessels within the harbor or maritime facility.

(b) Marina Owner and/or Operator: It shall be the responsibility of the owner, licensee, lessee, or operator of any marina, anchorage, repair yard, or other marine facility, located within any harbor, waterway or other maritime facility, to maintain the physical improvements under his jurisdiction in a safe, clean, and visually attractive condition at all times, to provide adequate security and fire prevention measures and appropriate fire fighting equipment as may be directed by————————, and to rent or lease available accommodations on a first-come first-served basis without regard to color, race or creed upon payment of established fees. Failure to initiate within 30 days of receipt of written notice from————————to correct unsafe or otherwise unsatisfactory conditions and to pursue same to completion to the satisfaction of————————shall be a violation of this section.

Sec. 16. Launching and Recovery of Vessels: None other than the driver may occupy a motor vehicle while it is present upon the area known as the launching ramp located within the————————. All motor vehicles using said ramp area must securely block at least one wheel of the said motor vehicle while it is standing upon said ramp.

Sec. 17. Permits, Suspensions or Revocations: All permits granted under the authority of this Ordinance shall be valid only for such period as may be determined by————————and permits of unqualified duration of validity shall not be granted. A violation of the provisions of this Ordinance or of any other applicable Ordinance by any permittee shall be grounds for suspension or revocation of such permit or permits.

Sec. 18. Lost and Found Property: The finder of lost property within the harbor shall deliver it to the————————and report its identity and location to the————————. Unless promptly claimed by the owner, the————————may remove the property, store it, advertise and sell it or otherwise dispose of it, all at the expense of the owner in conformance with————————of ————————Code.

Sec. 19. Damage to Harbor or Other Property: It shall be unlawful to willfully or carelessly destroy, damage, disturb, deface or interfere with any public property in the Harbor area.

Sec. 20. Tampering with or Boarding Vessels without Permission: It shall be a violation of this Ordinance for any person willfully to board, break in, enter, damage, move or tamper with any vessel or part thereof, located within the harbor unless authorized by the rightful owner of such vessel. Violation of this provision shall constitute a misdemeanor, punishable by the penalties hereinabove provided for violations of this Ordinance and to additional penalties not to exceed in the aggregate $1,000 and six months imprisonment for each offense. Any person violating this provision shall, in addition, be responsible to the rightful owner of any such vessel for any damages caused by such violation and to the reasonable cost of any attorneys fees, necessarily incurred as a result thereof.

Sec. 21. Obstruction of Facilities: It shall be a violation of this Ordinance for any person willfully to prevent any other person from the use and enjoyment of the harbor facilities.

Sec. 22. Place of Abode: It shall be unlawful for any person other than one specifically authorized by permit, license or lease from the---------, to camp, lodge, sleep or tarry overnight upon any public portion of a harbor or maritime facility, or to erect, maintain, use or occupy any tent, lodge, shelter, structure, housetrailer, trailer coach, or conveyance used as a place of abode, exclusive of public waters.

Sec. 23. Signs, Erection and Maintenance: The---------may place and maintain, or cause to be placed and maintained, either on land or water, such signs, notices, signals buoys or control devices as he deems necessary to carry out the provisions of this Ordinance, or to secure public safety and the orderly and efficient use of a harbor or maritime facility. *For Sale* signs shall be limited to a size of eight and one-half inches (8 ½") by eleven inches (11") and must be posted on the vessel.

Sec. 24. Securing Permission for Debarkation: It shall be a violation of this Ordinance to disembark passengers or discharge cargo from a commercial vessel onto any public or privately owned float, pier or wharf within the harbor, without the consent of the owner thereof or of the---------, as the case may be, except at piers and wharfs expressly designed for commercial purposes.

Sec. 25. Protected Swimming Area: It shall be a violation of this Ordinance to operate or navigate any vessel within a designated swimming area. The---------may identify swimming areas by signs, buoys, or other means.

Sec. 26. Record of Vessels: The---------shall keep an accurate record of the number, size, type and description of all vessels within a---------harbor or maritime facility using public facilities and which remain more than---------hours, and it shall be unlawful for any person having knowledge thereof to fail or refuse to provide said information to---------on demand.

Sec. 27. Underwater Diving:

(a) Permit for skin diving or underwater diving by persons having certificates of competence may be authorized in designated areas by the---------. Such action shall otherwise not be engaged in except in cases of emergency or for the purpose of underwater inspection.

(b) When a person or persons are engaged in an underwater diving activity other than an emergency or inspection, there shall be present an attendant not less than sixteen (16) years of age who shall be on the surface of the water close over the person or persons engaged in the underwater activity, and such attendant shall conspicuously display the *Divers Flag* during diving activities. The—————————shall be given prior notice of the time of all such diving operations.

Sec. 28. Swimming, a Hazard to Navigation: All swimming and bathing shall be in those areas designated by the—————————for such purposes and such areas may be defined or properly marked by competent authority.

ARTICLE V

Regulations Concerning Anchoring, Mooring and Security of Vessels

Sec. 29. Placement of Private Moorings: It shall be a violation of this Ordinance to place any mooring in the harbor without a permit from the—————————.

Sec. 30. Anchoring: It shall be a violation of this Ordinance to berth or anchor a vessel in the harbor without obtaining a permit from the—————————, or from the berthing facility operator, except in designated anchoring areas. Vessels in distress are excepted from this prohibition, but as soon as practicable, the person in charge of any such vessel shall report the situation to the—————————. Except in specially designated anchorage areas, proper anchor lights must be displayed and fog signal sounded when appropriate.

Sec. 31. Use of Harbor Facilities: It shall be a violation of this Ordinance to berth or anchor any vessel within the harbor without obtaining a permit from the—————————, who will require proper ground tackle and a sighting of anchors, when used, at intervals not exceeding three (3) months.

Sec. 32. Obstructing Channels: It shall be a violation of this Ordinance knowingly or willfully to obstruct the free use of any channel or waterway within the harbor or to fail to report to the—————————any collision between vessels or other accident or incident causing damage to persons or property.

Sec. 33. Abandoned Vessels: When, in the opinion of the—————————, a vessel has been abandoned in the harbor, he may take custody and control of such vessel and remove it, store it or otherwise dispose of it, all at the expense and sole risk of the vessel owner. Reasonable notice of such disposal shall be publicly given.

Sec. 34. Vessels Making Fast: No person shall make fast or secure a vessel to any mooring already occupied by another vessel, or to a vessel already moored, except that a rowboat, dinghy or yacht tender regularly used by a larger vessel for transportation of persons or property to or from shore may be secured to such larger vessel or to the mooring regularly used by such larger vessel. If tied within a slip, such rowboat, dinghy, or tender shall not extend into the fairway beyond the larger vessel if such larger vessel is also occupying the slip, or otherwise beyond the slip itself.

Sec. 35. Docking or Berthing at---------Facilities: A person having charge of any vessel shall not make it fast or secure it to any---------jetty, breakwater, bulkhead, wharf, pier or mooring buoy without the consent of---------except in an emergency, in which case such person shall forthwith report the emergency to the---------and thereafter act in accordance with the---------instructions.

Sec. 36. Secure Berthing and Anchoring of Vessels: The owner of any vessel moored or anchored within a---------harbor or maritime facility shall be responsible for causing such vessel to be tied and secured or anchored with proper care and equipment and in such manner as may be required to prevent breakaway and resulting damage, and shall thereafter provide for periodic inspection by---------, maintenance, replacement and adjustment of anchor, mooring or tie lines at reasonable intervals.

Sec. 37. Unseaworthy Vessels Prohibited in Harbor: Exception: A person shall not moor or permit to be moored in any harbor a vessel of any kind whatsoever which is unseaworthy or in a badly deteriorated condition or which is likely to sink or to damage docks, wharves, floats or other vessels or which may become a menace to navigation, except in cases of emergency.

Sec. 38. Correcting an Unsafe Berthing: If any vessel shall be found in the judgment of---------to be anchored or moored within any harbor or maritime facility in an unsafe or dangerous manner, or in such a way as to create a hazard to other vessels or to persons or property,---------shall order and direct necessary measures to eliminate such unsafe or dangerous condition. Primary responsibility for compliance with such orders and directions or---------shall rest with the owner of the improperly anchored or moored vessel or his authorized agent; in the absence of such owner or agent, said responsibility shall rest with the authorized operator of the facility at which the vessel is anchored or moored. In an emergency situation and in the absence of any such responsible person,---------shall forthwith board such vessel and cause the improper situation to be corrected, and the owner of the vessel shall be liable for any costs incurred by---------in effecting such correction.

Sec. 39. Removal and Custody of Illegally Berthed or Abandoned Vessels: If any unattended vessel shall be found to be anchored or moored illegally within a harbor or maritime facility, or if---------has reasonable grounds to believe that a vessel has been adandoned within a---------harbor or maritime facility, the---------may assume custody of such vessel and cause it to be removed and held or placed in storage. ---------or his---------shall not be held liable for any damage to such vessel nor liable to its owners before or after assuming custody. Vessels so taken into custody shall be released to the owner by the---------only after satisfactory proof of ownership has been presented and full reimbursement made to---------for all costs incident to recovery, movement and storage as set forth in Article V, Sec. 40. If proof of ownership cannot be established within a reasonable amount of time, said vessel shall be dealt with in accordance with Article IV, Sec. 18.

Sec. 40. Fees Incidental to Recovery, Movement and Storage: Charges imposed by---------for recovery and/or movement of vessels shall be in accordance with the "Schedule of Charges for services rendered and supplies Furnished by the Harbor Patrol" as approved by the---------or as subsequently amended, and whenever a vessel is impounded or held for safekeeping there shall be in addition a charge for storage at the rate of---------.

Sec. 41. Obstructions of Fairways, Channels or Berthing Spaces and Removal of Sunken Vessels:

(a) It shall be unlawful to tie up or anchor a vessel in a---------harbor or maritime facility in such a manner as to obstruct the fairways or channels or to prevent or obstruct the passage of other vessels; or to voluntarily or carelessly sink or allow to be sunk any vessel in any channel, fairway, berthing space; or to float loose timbers, debris, logs or piles in any channel, fairway, or berthing space in such a manner as to impede navigation or cause damage to vessels therein. It is understood that wrecked or sunken vessels within a harbor are subject to the published rules and regulations of the United States Coast Guard and any applicable State law, rules or regulations.

(b) Whenever the navigation of any waters within---------harbor or maritime facility, including anchorages and berths therein, shall be obstructed or endangered by any sunken vessel or other obstruction and the obstruction or danger has existed for a period of more than ten (10) days, the vessel or obstruction shall be subject to removal, sale or other disposition in accordance with Article 4, Section 18. The owner or owners of such vessel or other property causing said obstruction or danger shall be liable to the---------for all costs incident to said removal and disposition, and the---------, its employees, agents, and officers, shall not be liable for damages of any nature whatsoever arising out of or in any way connected with removal, sale or disposition of such vessel or other property.

Sec. 42. Dangerous or disabled Vessels: Any vessel that may enter a---------harbor or maritime facility in a disabled condition, or any vessel within a harbor or maritime facility which may for any reason be rendered disabled, shall immediately become subject to the orders and directions of the---------and it shall be unlawful for any person to fail or refuse to comply with his orders or directions with regard to the disposition of such vessel.

Sec. 43. Unseaworthy Vessels: No person shall secure or permit to be anchored or moored in a harbor, waterway, or maritime facility a vessel of any kind whatsoever which is unseaworthy or in a badly deteriorated condition, or which is likely to sink or to damage docks, wharves, floats, and/or other vessels, or which may become a menace to navigation. Such vessels shall be removed from the water and/or be otherwise disposed of as directed by---------.

ARTICLE VI

Regulations Concerning Commercial Activity

Sec. 45. Vessels for Hire–Passenger Information: The owners, master or person in charge of or operating any vessel using and-------harbor or maritime facility-------be required to furnish to---------information regarding the number of passengers carried and the charges or other considerations paid by such passengers. Failure to provide such information to the---------on demand shall be a violation of this article.

Sec. 46. Soliciting: Soliciting is prohibited within the harbor, except as may be specially authorized by permit issued by the---------, and subject to terms and conditions prescribed in such permit.

Sec 47. Bait Receivers: Non-Conforming: Removal of:

(a) All unattended live bait receivers in the waters of a---------harbor or maritime facility shall have a screen, solid cover, or lids which shall fit closely over the well of the receiver, unless the receiver is within and completely enclosed by a larger structure.

(b) Storage of bait in any receiver not conforming to the requirements of this Article is prohibited; non-conforming bait receivers may be sealed, removed, stored, sold or otherwise disposed of by---------at his discretion without liability for any damage to receivers or death or loss of bait, and the owner of such non-conforming receivers shall be liable for any cost incurred by---------in effecting removal, storage, sale, or other disposition.

Sec. 48. Commercial Bait Tanks: Bait tanks on commercial vessels containing bait shall, when said vessels are in the waters of a————————harbor or maritime facility, be covered by a screen cover or other cover which shall fit closely over the top of all said bait tanks except while bait is actually being transferred to or from said tank, and the operators of such commercial vessels shall at all times have aboard a covered can, box or other additional receptacle for dead bait.

Sec. 49. Bait, Transfer: No person shall transfer live bait from one vessel to another within the limits of a————————harbor or maritime facility except when all vessels involved are anchored or berthed, or such vessels are outside navigational channels.

Sec. 50. Sale of Live Bait: No person shall sell live bait from a vessel within the limits of a————————harbor or maritime facility. This section shall not apply to the delivery of live bait by vessel to a commercial live bait receiver which has been authorized by————————lease or written permit of————————to dispense live bait.

Sec. 51. Water Taxi and Rental Vessels: No person shall operate a water taxi within a harbor or maritime facility without first obtaining a permit from the————————and complying with any rules and regulations of Ordinances of the————————including any licensing requirement.

Sec. 52. Disposal of Bait in Harbor Prohibited: No person shall put, place or allow live bait to be put or placed in the waters of the harbor, except when the same is actually used for the purpose of fishing, and at no time, for any purpose, shall any person put dead bait or any portion thereof in the waters of the harbor unless the same be attached to a hook or hooks in the act of fishing.

Sec. 53. Commercial Vessels Providing Sleeping Accommodations—Watchman Required: Whenever any person other than the owner or members of the regular crew is allowed or permitted to sleep in or otherwise occupy accommodations aboard a commercial vessel or vessel regularly carrying passengers for hire, it shall be the duty of the owner or other person in charge of such vessel to maintain on duty a competent watchman, guard, or crew member, and failure to so maintain such a watchman, guard, or crew member on duty while the accommodations of such vessel are so occupied shall be a violation of this Ordinance.

ARTICLE VII

Sanitation Regulations

Sec. 54. Discharge of Refuse: It shall be a violation of this Ordinance to discharge or permit the discharge into the waters of the harbor of any refuse or waste matter, petroleum or petroleum matter, paint, varnish or any other foreign matter, including dead animals, fish and bait.

Sec. 55. Toilet Fixtures:

(a) Vessel's Toilet Fixtures Not to be Used: No person shall operate the toilet fixtures of a vessel within a harbor or maritime facility at any time so as to cause or permit to pass or to be discharged into the waters of such harbor or maritime facility any sewage or other waste matter or contaminant of any kind.

(b) Acceptable Devices: Upon application to the---------, persons operating, maintaining or possessing vessels using a---------harbor or maritime facility may be authorized by---------to use and operate toilet fixtures equipped with approved and acceptable devices that will prevent contaminants from entering the waters of such harbor of maritime facility.

(c) Toilet fixtures of a vessel which are equipped with a device or devices for the purpose of preventing contaminants from entering into the waters of a harbor or maritime facility shall not be used for the disposal of sewage or other contaminants unless a permit in writing has first been issued by---------.

Sec. 56. Use of Vessel as Abode: Living aboard vessels in the harbor is prohibited except as may be specially authorized by permit issued by the---------. For the purpose of this Section, the term "living aboard" means the continuous use of a vessel for a period in excess of three days, including use of the vessel for overnight lodging.

Sec. 57. Responsibility for Sanitation of Facilities: The lessee, agent, manager or person in charge of a facility or water area under lease from the---------harbor or maritime facility shall at all times maintain the premises under his charge in a clean, sanitary condition, free from malodorous materials and accumulations of garbage, refuse, debris and other waste materials. Should---------find that any facility or water area under lease is not so maintained, he shall in writing notify said lessee, agent, manager or other person in charge of said facility or area to immediately commence and diligently prosecute to completion the necessary correction of the unsanitary condition to the satisfaction of---------. Failure to do so with reasonable dispatch shall be a violation of this Article, and the---------may then cause condition to be corrected and the cost of such correction shall be charged to said lessee, agent, manager or person in charge.

ARTICLE VIII

Safety and Maintenance

Sec. 58. Welding and Burning: Except at specially designated areas, open fires are prohibited within the harbor, except for stoves or fireplaces permanently installed onboard and below decks on vessels or hibaches or barbeques used for cooking and/or heating purposes. Repairs to vessels requiring welding or other open flame devices may be performed only upon special authorization by the---------and within the time period stipulated in such authorization.

Sec. 59. Flammable and Combustible Liquids and/or Materials: Within a--------- harbor or maritime facility no person shall sell, offer for sale, or deliver in bulk any class of flammable liquid or combustible material, nor dispense any flammable or combustible liquids into the fuel tanks of a vessel except when in compliance with all requirements of the N.F.P.A. *Fire Code* and any other laws or regulations applicable thereto.

Sec. 60. Obstruction to Walkways: Obstructing walkways within the harbor by mooring lines, waterhoses, electrical cables, boarding ladders, permanently fixed stairs or any other materials is strictly prohibited. Dinghys may not be left on the floats and piers, but may be stored only in areas designated for that purpose.

Sec. 61. Defective or Dangerous Conditions: Whenever any buildings, structures or floating facilities within a harbor or maritime facility either on land or water are found to be defective or damaged so as to be unsafe or dangerous to persons or property, it shall be the duty of the owner, agent, lessee, operator or person in charge thereof to immediately post a proper notice and/or fence or barricade and at night to adequately light such unsafe area or areas, and such unsafe area or areas shall be kept posted and lighted and/or fenced or barricaded until the necessary repairs are made. In the event an owner, agent, lessee, operator or person in charge fails or neglects to repair or to put up fences or other barriers to prevent persons from using or going upon the unsafe area or areas,---------may then take such measures as he may deem necessary for the protection of the public and charge the cost of same to such owner, lessee, agent, person or persons having charge of the buildings, structures, or floating facilities that are defective or dangerous.

Sec. 62. Time, Fees and Permit Requirements for Use of Mooring or Slip:

(a) Permission may be granted by the---------for a private vessel to use a---------mooring or slip for---------without charge and for up to--------- without charge if it is determined the vessel may be secured or moored for such longer time without using space otherwise needed. Private vessels moored at---------facilities for a period in excess of---------shall pay mooring fees as hereinafter provided.

(b) Visiting Vessels–Transients: A vessel will be considered transient if the vessel remains in the slip or at the dock designated for such transient for not more than---------calendar months. A person shall not berth or dock a vessel, except on official business, for more than---------hours in any---------slip or mooring unless the---------says that vessel may be berthed or moored for a longer time without using space otherwise needed, in which case he may be granted permission for a longer stay without paying the transient fees. Total time shall be at the discretion of the---------but not to exceed--------hours. The transient rental fees shall be as follows:--------- upon entering a harbor, owner or operator of vessel shall proceed directly to harbor headquarters in order that he may be assigned a berth. Berthing fees are usually payable in advance.

(c) Subletting of Berthing Space: The owner of any vessel having space shall not sublet said space to another user or boat owner; however by agreement between the---------and the transient, another boat owner may use the slip, provided the original transient pays the fee or fees and has requested permission from the---------to berth said vessel.

Sec. 63. Vessel Extending Beyond Slip:

(a) No part of any vessel shall extend more than---------feet beyond the end of any slip without permission of---------including but not limited to boats with davits, booms, boomkin, or bowsprit.

(b) No part of any vessel shall extend over the main walkway so as to be a hazard.

Sec. 64. Mooring at Termini of Main Walks: Vessels may be moored or secured at the terminus of any main walk within a basin, except that any such vessel shall not extend into the fairway more than---------measured at right angles from the pierhead line of a basin. Any such vessel shall be secured parallel to such pierhead line.

Sec. 65. Commercial Sportfishing Activities Allowed Only at Specific Areas: No owner or operator of any commercial sportfishing boat or any other boat, licensed or unlicensed, shall conduct, maintain, or engage in any sportfishing activity for hire from any premises within---------except from those leaseholds specifically permitted to conduct such activities, nor shall any lessee or any boat mooring operator in---------permit, authorize, or allow the operation of a commercial sportfishing activity from within the area of their control or tenancy unless specifically authorized by written permit of--------- or by terms of their lease.

Sec. 66. Parking of Camp Cars and Boat Trailers: Camp cars herein defined shall park in an area set aside for such purpose and no other except they may be parked for a period not to exceed---------hours in the regular parking areas provided they not be used for eating, sleeping, cooking, preparation of foods or personal toilet. Boat trailers used for carrying a boat shall be parked only in the areas set aside for such parking. They may be parked for a reasonable period of time on the public streets or other parking areas not to exceed---------hours for the purpose of removing a boat from such trailer or placing a boat upon such trailer.

APPENDIX J

PRIVATE ORGANIZATIONS OFFERING INFORMATION

American Boat & Yacht Council, Inc.
15 East 26th Street
New York, NY 10010

American Wood Preservers Institute
1651 Old Meadow Road
McLean, VA 22101

Boating Industry Association
401 North Michigan Avenue
Chicago, IL 60611

Marine Accessories & Services Association
401 North Michigan Avenue
Chicago, IL 60611

National Association of Engine & Boat Manufacturers
537 Steamboat Road
Greenwich, CT 06830

National Recreation & Park Association
1601 North Kent Street
Arlington, VA 22209

Outboard Boating Club of America
401 North Michigan Avenue
Chicago, IL 60611

Portland Cement Association
Old Orchard Road
Skokie, IL 60076

APPENDIX K

CERC SMALL-CRAFT HARBOR QUESTIONNAIRE

Respondent: Please submit as much of the following information as possible. Where exact figures are not readily available, give approximate answers.

Owner:_______________ **Designer:**_______________

Site Location: River_____ Lake_____ Bay_____ Ocean_____
Other_______________

Boat Capacity (Wet Storage): Private Powerboats_____ Private Sailboats_____
Commercial Boats_____ Rental Boats_____

Type of Docks: Fixed_____ Floating_____ Manufacturer_____
If Floating, anchored by: Piles_____ Cable_____

Boats per Slip: 1_____ 2_____ No. of Covered Slips_____ Depth of Basin_____

Dry Storage Capacity: Trailers or Dollys_____ Open_____ Racks_____
Trailers or Dollys_____ Covered_____ Racks_____

Average Range of Water Surface Fluctuation (feet): Daily_____ Monthly_____ Yearly_____

Most Extreme Range of Water Surface Fluctuation on Record for Body of Water Served:
Feet_____

Major Activities of Users (percent): Fishing_____ Cruising_____ Skiing_____
Other_______________

Power Supply to Docks: None_____ Free_____ Metered_____ Included in Slip Rental_____
Number of Outlets_____ Amps. per Outlet_____ 120 V._____ 240 V._____
Other_______________

Water Supply to Docks: None_____ Free_____ Included in Slip Rental_____

Lighting on Docks: None_____ Located: Above Waist Level_____ Below Wasist Level_____

Public Address System: Yes_____ No_____

Phones: None___ Provided in Office___ Provided at Docks by: Marina____________
Phone Company_______
Other____

Launching Facilities: None____ Ramp Width____ Ramp Slope____
Boat Size Handled____ Type of Hoist and Capacity__________

Number of Trash or Garbage Collections per Week: ________

Sanitary Holding Tank Pumpout Facilities: Yes_____ No_____

Car Parking Spaces Provided per Slip:________________

Ancillary Facilities Available: Fuel Station______ Snackbar______ Restaurant______
Bait and Tackle_____ Boat Rentals______ Boat Sales______
Boat Repair: Minor_____ Major_______
Other____

Are Live-Aboards Permitted?______________________________

Any Wave or Surge Problems?______________________________

Any Troublesome Maintenance Problems?________________________

APPENDIX K–Continued

GENERAL INFORMATION, DATA, AND VISUAL AIDS

We would appreciate the enclosure (with your submittal) of as many items listed below as possible; items will be returned if so designated.

1. An aerial photo of your marina and any other photos of significant or unique features; all photos to be black and white.

2. Prints of the general plan or layout of the marina (if no aerial photo is available).

3. Construction drawings that show general features or typical sections of piers, revetments, bulkheads, anchoring system, etc. (Note: If not readily available, it may be possible for our draftsmen to draw them up from rough, dimensioned sketches, so please include.)

4. Standards of design issued to prospective lessees, engineering firms or contractors for construction or expansion of facility. Names and addresses of constructors or manufacturers. (This normally applies only to large, publicly operated marinas.)

5. Copies of charts or tables that show slip occupancy, entrance-traffic counts, maintenance and operation costs, revenues, water level fluctuation, etc. A breakdown of construction cost per slip and maintenance cost per slip.

6. Copies of articles that have been written narrating the history of project development, i.e., origin, designer, financing, etc.

7. Sales brochures or fact sheets giving general data on the installation, i.e., types and numbers of slips, ancillary facilities, etc.

8. A narrative concerning any design feature that has worked out better or worse than expected, or problems you feel could have been avoided by better planning. Mention special systems such as ice protection, wave protection, anchorage in extreme water drawdown, antihurricane design, etc.

www.ingramcontent.com/pod-product-compliance
Ingram Content Group UK Ltd.
Pitfield, Milton Keynes, MK11 3LW, UK
UKHW051456290926
15471UKWH00008B/1241

9 781589 639652